Y0-CKG-332

HEREDITY AND ENVIRONMENT

LONDON : HUMPHREY MILFORD
OXFORD UNIVERSITY PRESS

HEREDITY AND ENVIRONMENT
IN THE DEVELOPMENT OF MEN

BY
EDWIN GRANT CONKLIN
PROFESSOR OF BIOLOGY PRINCETON UNIVERSITY

Fifth Edition Revised

PRINCETON
PRINCETON UNIVERSITY PRESS
1922

FIRST EDITION
Copyright, 1915, by
Princeton University Press
Published February, 1915
Second Printing, June, 1915

REVISED SECOND EDITION
Copyright, 1916
Published May, 1916
Second Printing, August, 1917
Third Printing, March, 1918

REVISED THIRD EDITION
Copyright, 1919
Published September, 1919
Second Printing, March, 1920

REVISED FOURTH EDITION
Copyright, 1922
Published January, 1922

REVISED FIFTH EDITION
Copyright, 1922
Published September, 1922
Second Printing, September, 1923
Third Printing, September, 1924
Fourth Printing, September, 1925
Fifth Printing, July, 1927

TRANSLATIONS
Japanese, from 2d English Ed.
Published September 15, 1916, Tokyo.
French, from 3d English Ed.
Published, December, 1920, Paris

Printed at the Princeton University Press Princeton U. S. A.

PREFACE TO THE FIRST EDITION

THE origin of species was probably the greatest biological problem of the past century; the origin of individuals is the greatest biological subject of the present one. The many inconclusive attempts to determine just how species arose led naturally to a renewed study of the processes by which individuals came into existence, for it seems probable that the principles and causes of the development of individuals will be found to apply also to the evolution of races. As the doctrine of evolution wrought great change in prevalent beliefs regarding the origin and past history of man, so present studies of development are changing opinions as to the personality of man and the possibilities of improving the race. The doctrine of evolution was largely of theoretical significance, the phenomena of development are of the greatest practical importance; indeed there is probably no other subject of such vast importance to mankind as the knowledge of and the control over heredity and development. Within recent years the experimental study of heredity and development has led to a new epoch in our knowledge of these subjects, and it does not seem unreasonable to suppose that in time it will produce a better breed of men.

The lectures which comprise this volume were given at Northwestern University in February, 1914, on the Norman W. Harris Foundation and were afterward repeated at Princeton University. I gladly take this opportunity of expressing to the faculties, students and friends of both institutions my deep appreciation of their interest and courtesy. In attempting to present to a general audience the results of recent studies on heredity and development, with special reference to their application to man, the author has had to choose between simplicity and sufficiency of statement, between apparent dogmatism and scientific caution, between a popular and a scientific presentation. These are hard alternatives, but the first duty of a lecturer is to address his audience and to make his subject plain and interesting, if he can, rather than to talk to the scientific gallery over the heads of the audience. In preparing the lectures for publication it has not been possible to avoid the technical treatment of certain objects, but in the main the lectures are still addressed to the audience rather than to the scientific gallery. Unfortunately biology is still a strange subject to many intelligent people and its terminology is rather terrifying to the uninitiated; but it is hoped that the glossary at the end of the volume may rob these unfamiliar terms of many of their terrors.

The first three chapters of this book appeared in *Popular Science Monthly,* June to November 1914, by previous arrangement with the Princeton University Press, and a portion of the last chapter was first published in *Science* in January 1913. The illustrations are from many sources: [1]Figures 9, 10, 19, 20, 22, 23, 26-29, 35, 40-43, 51-54, 58, 72-75, 77, 83 are original; Figures 1-8, 11-18, 21, 24, 25, 30-34, 36-39, 44-47, 49, 55-57, 59-62, 70 were redrawn with more or less modification from original sources which are indicated in every case; the remaining figures, namely Figures 48, 50, 63-69, 71, 76, 78-80, 84-96 were copied with little or no modification from various papers, photographs and books, the original source being given in each case. The writer wishes to express his obligation to all authors and publishers upon whose works he has drawn, and he thanks especially the Columbia University Press for permission to use Figures 48 and 50, and the Open Court Publishing Company for Figures 84-87.

I take this opportunity of thanking Dr. W. E. Castle and Dr. J. H. McGregor for the use of photographs which are reproduced in Figures 81, 82 and 96; and I wish especially to thank my assistant, Marguerite Ruddiman, for her aid in preparing figures and manuscript for publication.

Princeton, December, 1914

[1] These figure numbers apply to the first and second editions but not to later ones.

PREFACE TO THE SECOND EDITION

THE interest of the public and the kindness of the publishers have made possible a revised edition of this book within a year after its first publication. This has permitted the rewriting of certain passages which were not well expressed and the introduction of considerable new material which will serve, it is hoped, to further elucidate some of the questions discussed, as well as to give results of more recent work. In particular Figs. 4, 5, 9, 10, 18, 19 have been added and besides minor corrections and additions to all the chapters considerable changes have been made in Chapter III, section on Maternal Inheritance, Chapter IV, section on Inheritance or Non-Inheritance of Acquired Characters, and Chapter V, sections on Artificial Selection, Origin of Mutations, Past Evolution of Man, and Eugenics.

January, 1916.

PUBLISHER'S NOTE.—In the second edition, second printing, August, 1917, the text was corrected by the author so as to embody some of the results of the latest investigations in this subject.

PREFACE TO THE THIRD EDITION

THE exhaustion of previous editions of this book and the necessity of re-setting the whole of it presents the opportunity for a rather thorough revision of the entire work. In addition to minor changes which have been made in all the chapters there has been a rearrangement and enlargement of the material of the chapter on the "Cellular Basis of Heredity and Development" and since this is the most technical chapter in the book it is placed after the chapter on "Phenomena of Inheritance" which deals with subjects which are more familiar to the average reader. Although the cellular phenomena of heredity are less familiar and more technical than other aspects of this subject they cannot be omitted or slighted if one wishes to understand the most recent and significant discoveries in this field. Figures and descriptions have been added of typical cells, of cell division, of the origin and maturation of the germ cells, of sex determination and especially of the mechanism of heredity in that most famous of all objects for the study of inheritance, the fruit fly *Drosophila melanogaster*. The author is indebted to Professor Morgan and his associates for permission to use certain figures from their books and papers on this subject.

In this edition a much stronger position has been taken for the "chromosomal theory" of heredity than in former editions, for this theory is now so well established that it deserves a prominent place in even an elementary book.

In spite of these additions the object of keeping the presentation as simple as possible has been adhered to and those who desire a more complete account should consult the "Mechanism of Mendelian Inheritance" by Morgan and his associates, or "Genetics in Relation to Agriculture" by Babcock and Clausen.

September, 1919.

PUBLISHER'S NOTE. The second printing of this edition, (March 1920), has been corrected and revised in a few places.

PREFACE TO THE FOURTH EDITION

I HAD intended that the previous revision of this book should have been the last, but I cannot resist the opportunity, which is again offered me by the necessity of reprinting it, to make certain corrections and alterations, and in particular to refer to some recent discoveries of great importance. Knowledge is advancing so rapidly in the field of Genetics that it is not possible to keep such a book as this strictly up to date, but at least it should record the golden milestones that are passed. Most of these additions will be found in the chapter on the Cellular Basis of Heredity and Development; others deal with the Phenomena of Heredity, Internal Secretions, the Inheritance of Acquired Characters, and Mutations; while minor changes occur in all the chapters.

I am especially indebted to Dr. Theophilus S. Painter for the drawing of human chromosomes shown in Figure 57, and to him and to Professor Michael F. Guyer for information regarding the number of chromosomes in man. To Dr. Calvin B. Bridges and to Professor T. H. Morgan I am under obligations for information embodied in the new map of *Drosophila* chromosomes shown in Figure 69, and I am particularly indebted to Professor Morgan for his kindness in loaning me the figures of *Drosophila* mutants which are reproduced in Figures 103-105. Finally I am glad to thank publicly my colleague Professor George H. Shull for many valuable criticisms and suggestions.

Princeton, December, 1921.

PREFACE TO THE FIFTH EDITION

THE present edition of this book follows so soon after the previous one that few changes have been made except in the figures illustrating the text. Figures 26, 28, 35, 36, 37, 38, 39, 40, 41a, 47, 61, 62, 66, 74, 82, 86, 87, 95, 96, 97, 98, 99, 104 of the previous edition have been remade or replaced by new figures. Again I am especially indebted to Professor Morgan for his kindness in furnishing figures 63, 64, and 68, and for many valuable suggestions.

Woods Hole, August, 1922.

Revised for Second Printing, July 1923
Revised for Fourth Printing, August 1925

CONTENTS

CHAPTER I. FACTS AND FACTORS OF DEVELOPMENT

INTRODUCTION — 3

A. PHENOMENA OF DEVELOPMENT — 5

I. DEVELOPMENT OF THE BODY — 6
1. The Germ Cells — 6
2. Fertilization — 11
3. Cell Division — 16
4. Embryogeny — 23
5. Organogeny — 23
6. Oviparity and Viviparity — 26
7. Development of Functions — 30

II. DEVELOPMENT OF THE MIND — 32
1. Sensitivity — 37
2. Reflexes, Tropisms, Instincts — 39
3. Memory — 43
4. Intellect, Reason — 45
5. Will — 50
6. Consciousness — 53
7. Parallel Development of Body and Mind — 55

B. FACTORS OF DEVELOPMENT — 56
1. Preformation — 57
2. Epigenesis — 58
3. Endogenesis and Epigenesis — 58
4. Heredity and Environment — 60

CHAPTER II. PHENOMENA OF INHERITANCE

A. OBSERVATIONS ON INHERITANCE — 63
Individuals and their Characters — 63
Hereditary Resemblances and Differences — 64

I. HEREDITARY RESEMBLANCES — 65
1. Racial Characters — 65

    2. Individual Characters   65
       a. Morphological Features
       b. Physiological Peculiarities
       c. Teratological and Pathological Peculiarities
       d. Psychological Characters

II. HEREDITARY DIFFERENCES   72

    1. New Combinations of Characters   72
    2. New Characters or Mutations   74
    3. Mutations and Fluctuations   75
    4. Every Individual Unique   75

B. STATISTICAL STUDY OF INHERITANCE   76

    1. Galton's "Law of Ancestral Inheritance"   77
    2. Galton's "Law of Filial Regression"   80

C. EXPERIMENTAL STUDY OF INHERITANCE   82

I. MENDELISM   82

    1. Results of Crossing Individuals with One Pair of Contrasting Characters   84
       Other Mendelian Ratios
    2. Results of Crossing Individuals with More than One Pair of Contrasting Characters   91
       Dihybrids and Trihybrids
    3. Inheritance Formulae   95
    4. Presence and Absence Hypothesis   97
    5. Summary of Mendelian Principles   99
       a. The Principle of Unit Characters
       b. The Principle of Dominance
       c. The Principle of Segregation

II. MODIFICATIONS AND EXTENSIONS OF MENDELIAN PRINCIPLES   100

    1. The Principle of Unit Characters and Inheritance Factors   100
    2. Modifications of the Principle of Dominance   106
    3. The Principle of Segregation   108
       "Blending" Inheritance
       Maternal Inheritance

III. MENDELIAN INHERITANCE IN MAN   116

CHAPTER III. CELLULAR BASIS OF HEREDITY AND DEVELOPMENT

A. INTRODUCTORY — 123

1. Confused Ideas of Heredity — 123
2. The Transmission Hypothesis — 124
3. Germinal Continuity and Somatic Discontinuity — 124
4. Germplasm and Somatoplasm — 126
5. The Units of Living Matter — 129
6. Heredity and Development — 132

B. THE GERM CELLS — 133

1. Fertilization — 134
2. Cleavage and Differentiation — 138
3. The Origin of the Sex Cells — 148
   a. The Division Period
   b. The Growth Period
   c. The Maturation Period

C. SEX DETERMINATION — 157

1. Chromosomal Determination — 158
2. Environmental Influence — 165

D. THE MECHANISM OF HEREDITY — 171

I. THE SPECIFICITY OF GERM CELLS — 171

II. CORRELATIONS BETWEEN GERMINAL AND SOMATIC ORGANIZATION — 178

1. Chromosomal Inheritance Theory — 179
2. Linkage of Characters and Chromosomal Localization — 185
   a. Sex Linked Inheritance
   b. Other Cases of Linkage
3. Cytoplasmic Inheritance — 196
   a. Polarity
   b. Symmetry
   c. Inverse Symmetry
   d. Localization Pattern

E. THE MECHANISM OF DEVELOPMENT — 204

1. The Formation of Different Substances — 205
2. Segregation and Isolation of Substances in Cells — 206
   a. By Protoplasmic Movements
   b. By Differential Cell Divisions
3. The Chromosome Theory and Embryonic Differentiation — 208

CHAPTER IV. INFLUENCE OF ENVIRONMENT

A. RELATIVE IMPORTANCE OF HEREDITY AND ENVIRONMENT — 213
1. Former Emphasis on Environment — 213
2. Present Emphasis on Heredity — 215
3. Both Indispensable to Development — 215

B. EXPERIMENTAL MODIFICATIONS OF DEVELOPMENT — 216

I. DEVELOPMENT STIMULI — 216
1. Physical Stimuli — 217
2. Chemical Stimuli — 217
3. General vs. Specific Stimuli — 217

II. DEVELOPMENTAL RESPONSES — 217
Dependent upon (a) Nature of Organism, (b) Nature of Stimulus, (c) Stage of Development
1. Modifications of Germ Cells before Fertilization — 218
2. During Fertilization — 221
3. After Fertilization — 221

C. FUNCTIONAL ACTIVITY AS A FACTOR OF DEVELOPMENT — 231

D. INHERITANCE OR NON-INHERITANCE OF ACQUIRED CHARACTERS — 237

E. APPLICATIONS TO HUMAN DEVELOPMENT: EUTHENICS — 249

CHAPTER V. CONTROL OF HEREDITY: EUGENICS

A. DOMESTICATED ANIMALS AND CULTIVATED PLANTS — 259
I. INFLUENCE OF ENVIRONMENT IN PRODUCING NEW RACES — 265
II. ARTIFICIAL SELECTION — 266
III. METHODS OF MODERN GENETICS — 272
1. Mendelian Association and Dissociation of Characters — 272
2. Mutations — 276
3. Causes of Mutation — 280

B. CONTROL OF HUMAN HEREDITY — 287
I. PAST EVOLUTION OF MAN — 287

II. CAN HUMAN EVOLUTION BE CONTROLLED? 292

1. Selective Breeding the only Method of Improving the Race 292
2. No Improvement in Human Heredity within Historic Times 293
3. Why the Race Has not Improved 294

III. EUGENICS 295

1. Possible and Impossible Ideals 298
2. Negative Eugenical Measures 302
3. Positive Eugenical Measures 305
4. Contributory Eugenical Measures 308
5. The Declining Birthrate 310

CHAPTER VI. GENETICS AND ETHICS

I. THE VOLUNTARISTIC CONCEPTION OF NATURE AND OF HUMAN RESPONSIBILITY 319

II. THE MECHANISTIC CONCEPTION OF NATURE AND OF PERSONALITY 320

1. The Determinism of Heredity 321
2. The Determinism of Environment 324

III. DETERMINISM AND RESPONSIBILITY 327

1. Determinism not Fatalism 328
2. Control of Phenomena and of Self 328
3. Birth and Growth of Freedom 330
4. Responsibility and Will 331
5. Our Unused Talents 333
6. Self Knowledge and Self Control 335

IV. THE INDIVIDUAL AND THE RACE 339

1. The Conflict between the Freedom of the Individual and the Good of Society 339
2. Perpetuation and Improvement of the Race the Highest Ethical Obligation 340

REFERENCES 345

GLOSSARY 353

INDEX 363

CHAPTER I

FACTS AND FACTORS OF DEVELOPMENT

CHAPTER I

FACTS AND FACTORS OF DEVELOPMENT

INTRODUCTION

Man's Place in Nature.—One of the greatest results of the doctrine of organic evolution has been the determination of man's place in nature. For many centuries it has been known that in bodily structure man is an animal; that he is born, nourished and developed, that he matures, reproduces and dies just as does the humblest animal or plant. For centuries it has been known that man belongs to that group of animals which have backbones, the vertebrates; to that class which have hair and suckle their young, the mammals, and to that order which have grasping hands, flat nails, and thoracic mammæ, the primates, a group which includes also the monkeys and apes. But as long as it was supposed that every species was distinct in its origin from every other one, and that each arose by a special divine fiat, it was possible to maintain that man was absolutely distinct from the rest of the animal world and that he had no kinship to the beasts, though undoubtedly he was made in their bodily image. But with the establishment of the doctrine of organic evolution this resemblance between man and the lower animals has come to have a new significance. The almost universal acceptance of this doctrine by scientific men, the many undoubted resemblances between man and the lower animals, and the discovery of the remains of lower types of man, real "missing links," have inevitably led to the conclusion that man also is a product of evolution, that he is a part of the great world of living things and not a being who stands apart in solitary grandeur in some isolated sphere.

Oneness of All Life.—But wholly aside from the doctrine of evolution, the fact that essential and fundamental resemblances exist among all kinds of organisms can not fail to impress thoughtful men. Life processes are everywhere the same in principle, though varying greatly in detail. All the general laws of life which apply to animals and plants apply also to man. This is no mere logical inference from the doctrine of evolution, but a fact which has been established by countless observations and experiments. The essential oneness of all life gives a direct human interest to all living things. If "the proper study of mankind is man," the basic study of man is the lower organisms in which life processes are reduced to their simplest terms, and where alone they may be subjected to conditions of rigid experimentation. Upon this fundamental likeness between the life processes of man and those of other animals are based the wonderful advances in experimental medicine, which may be counted among the greatest of all the achievements of science.

Control of Development and Evolution.—The experimental study of heredity, development and evolution in forms of life below man must certainly increase our knowledge of and our control over these processes in the human race. If human heredity, development and evolution may be controlled to even a slight extent we may expect that sooner or later the human race will be changed for the better. At least no other scheme of social betterment and race improvement can compare for thoroughness, permanency of effect, and certainty of results, with that which attempts to change the natures of men by establishing in the blood the qualities which are desired. We hear much nowadays about man's control over nature, though in no single instance has he ever changed any law or principle of nature. What he can do is to put himself into such relations to natural phenomena that he may profit by them, and all that can be done toward the improvement of the human race is consciously and purposively to apply to man those great principles of development and evolution which have been at work, unknown to man, through all the ages.

A. PHENOMENA OF DEVELOPMENT

Ontogeny and Phylogeny.—One of the greatest and most far-reaching themes which has ever occupied the minds of men is the problem of development. Whether it be the development of an animal from an egg, of a race or species from a pre-existing one, or of the body, mind and institutions of man, this problem is everywhere much the same in fundamental principles, and knowledge gained in one of these fields must be of value in each of the others. Ontogeny and phylogeny are not wholly distinct phenomena, but are only two aspects of the one general process of organic development. The evolution of races and of species is sufficiently rare and unfamiliar to attract much attention and serious thought; while the development of an individual is a phenomenon of such universal occurrence that it is taken as a matter of course by most people, something so evident that it seems to require no explanation; but familiarity with the fact of development does not remove the mystery which lies back of it, though it may make plain many of the processes concerned. The development of a human being, of a personality, from a germ cell is the climax of all wonders, greater even than that involved in the evolution of a species or in the making of a world.

The fact of development is everywhere apparent; its principal steps or stages are known for thousands of animals and plants; even the precise manner of development and its factors or causes are being successfully explored. Let us briefly review some of the principal events in the development of animals, and particularly of man, and then consider some of the chief factors and processes of development. Most of our knowledge in this field is based upon a study of the development of animals below man, but enough is now known of human development to show that in all essential respects it resembles that of other animals, and that the problems of heredity and differentiation are fundamentally the same in man as in other animals.

1. Development of the Body

The entire individual—structure and functions, body and mind—develops as a single indivisible unity, but for the sake of clarity it is desirable to deal with one aspect of the individual at a time. For this reason we shall consider first the development of the body, and then the development of the mind.

1. *The Germ Cells.*—In practically all animals and plants individual development begins with the fertilization of a female sex cell, or egg, by a male sex cell, or spermatozoon. The epigram of Harvey, "*Omne vivum ex ovo,*" has found abundant confirmation in all later studies. Both egg and spermatozoon are alive and manifest all the general properties of living things. How little this fact is appreciated by the public is shown by the repeated announcements of the newspapers that someone who has made an egg develop without fertilization "has created life." An egg or a spermatozoon is as much alive as is any other cell; as characteristically alive as is the adult animal into which it develops.

What is Life?—It is difficult to define life, as it is also to define matter, energy, electricity, or any other fundamental phenomenon, but it is possible to describe in general terms what living things are and what they do. Every living thing whatever, from the smallest and simplest micro-organism to the largest and most complex animal, from the microscopic egg or spermatozoon to the adult man, manifests the following distinctive properties:

(a) *Protoplasmic and Cellular Organization.*—It contains protoplasm, "the material basis of life," which is composed of the most complex substances known to chemistry. Protoplasm is not a homogeneous substance, but it always exists in the form of cells, which are minute masses of protoplasm composed of many distinct parts, the most important of these being the nucleus and the cell-body (Fig. 1.)

The nucleus is a central rounded body usually denser than the surrounding cytoplasm from which it is separated by a thin membrane. It contains granules or threads of a substance which has

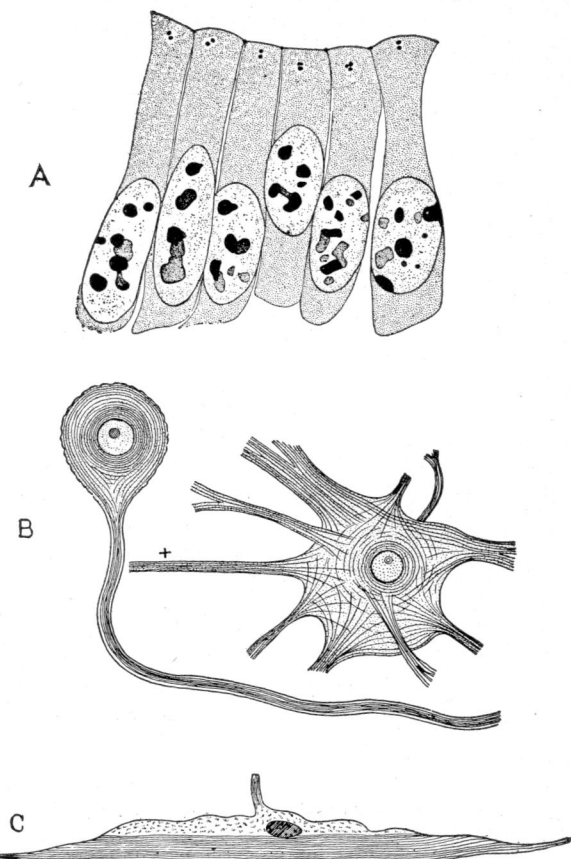

FIG. 1. TYPICAL TISSUE CELLS FROM DIFFERENT ORGANS. *A,* Epithelial cells from intestine of duck embryo, showing nuclei with dark chromatic masses and clear achromatin surrounded by nuclear membrane, centrosomes as two or three dark granules at the free border of the cells, and cytoplasm filling the cell body. *B.* Two nerve cells, *a.* of a ringed worm, *b.* of a fish, showing cell-body containing nucleus and nucleolus, many neurofibrils and in *a* one nerve fiber, in *b* many processes, one of which (+) is the nerve fiber. *C.* Muscle cell from a round worm showing nucleus and cytoplasm in upper part of cell and contractile fibrils in lower part.

a strong chemical affinity for certain dyes and hence is called chromatin; it also frequently contains one or more rounded bodies which look like little nuclei and are called nucleoli. The chromatin and nucleoli are imbedded in a substance which does not stain readily with dyes and which is therefore called achromatin. Surrounding the nucleus is the substance of the cell-body or cytoplasm and in this the various products of differentiation such as muscle or nerve fibrils, secretion products and food substances are found. The cytoplasm often contains also a centrosome which is a deeply-staining granule surrounded by radiating lines and which is an organ for causing intra-cellular movements, especially in connection with the division of the nucleus and cell body. The nucleus and cytoplasm also contain more or less water and inorganic salts, and all of these things taken together constitute what is known as protoplasm (Fig. 1).

Protoplasm is therefore organized, that is composed of many parts all of which are integrated into a single system, the cell. Higher animals and plants are composed of multitudes of cells, differing more or less from one another, which are bound together and integrated into a single organism. Living cells and organisms are not static structures that are fixed and stable in character, but they are systems that are undergoing continual change. They are like the river, or the whirlpool, or the flame, which are never at two consecutive moments composed of the same particles but which nevertheless maintain a constant general appearance; in short they are complex systems in dynamic equilibrium.

The principal physiological processes by which all living things maintain this equilibrium are:

(b) *Metabolism,* or the transformation of matter and energy within the living thing in the course of which some substances are oxidized into waste products, with the liberation of energy, while other substances are built up into protoplasm, each part of every cell converting food substances into its own particular substance by the process of assimilation.

FACTS AND FACTORS OF DEVELOPMENT

(c) *Reproduction,* or the capacity of organisms to give rise to new organisms, of cells to give rise to other cells, and of parts of cells to give rise to similar parts by the process of division.

(d) *Irritability,* or the capacity of receiving and responding to impinging energies, or stimuli, in a manner which is usually, but not invariably, adaptive or useful.

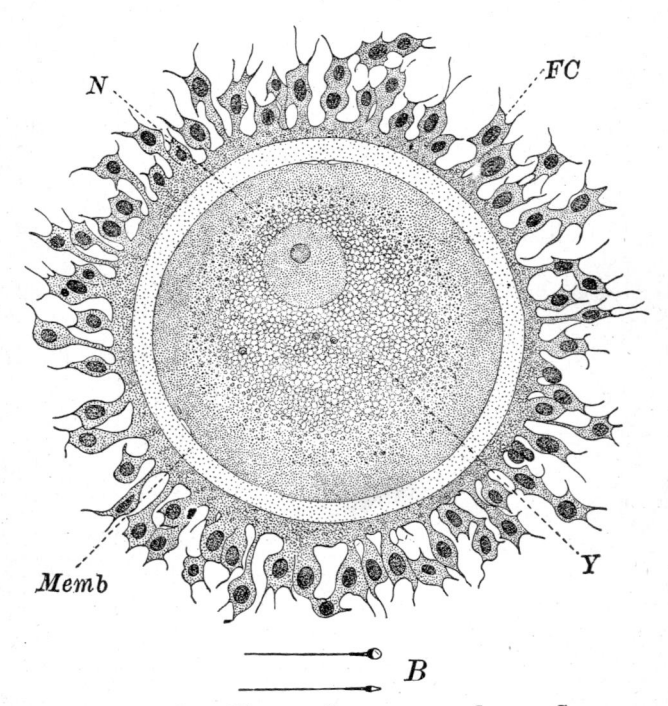

FIG. 2. A NEARLY RIPE HUMAN OVUM IN THE LIVING CONDITION. The ovum is surrounded by a series of follicle cells (*FC*) inside of which is the clear membrane (*Memb.*) and within this is the ovum proper containing yolk granules (*Y*) and a nucleus (*N*) embedded in a clear mass of protoplasm. Magnified 500 diameters (x 500). (From O. Hertwig.) *B*, two human spermatozoa drawn to about the same scale of magnification. (After G. Retzius.)

Germ Cells Alive.—Both the egg and the sperm are living cells with typical cell structures and functions, but with none of the parts of the mature organism into which they later develop. But although they do not contain any of the differentiated structures and functions of the developed organism, they differ from other cells in that they are capable under suitable conditions of producing these structures and functions by the process of development or differentiation, in the course of which the general structures and functions of the germ cells are converted into the specific structures and functions of the mature animal or plant.

Gametes and Zygotes.—In both plants and animals the sex cells are alike in many respects, though they differ greatly in appearances. The female sex cells of animals are called ova, the male cells spermatozoa. Corresponding male and female sex cells are found in plants also. Collectively all kinds of sex cells are called gametes, and the individual formed by the union of a male and a female gamete is known as a zygote, while the cell formed by the union of egg and sperm is frequently called the oosperm.

The egg cell of animals is usually spherical in shape and contains more or less food substance in the form of yolk; it varies greatly in size, depending chiefly upon the quantity of yolk, from the great egg of a bird, in which the yolk or egg proper may be hundreds of millimeters in diameter, to the microscopic eggs of oysters, worms, etc., which may be no more than a few thousandths of a millimeter in diameter. The human ovum (Fig. 2) is microscopic in size (about 0.2 mm. in diameter) but it is not as small as is found in many other animals. It has all the characteristic parts of any egg cell, and can not be distinguished microscopically from the eggs of several other mammals, yet there is no doubt that the ova of each species differ from those of every other species, and later we shall see reasons for concluding that the ova produced by each individual are different from those produced by any other individual.

The sperm, or male gamete, is among the smallest of all cells

FACTS AND FACTORS OF DEVELOPMENT

and is usually many thousands of times smaller than the egg. In most animals, and in all vertebrates, it is an elongated, thread-like cell with an enlarged head which contains the nucleus, a smaller middle-piece, and a very long and slender tail or flagellum, by the lashing of which the spermatozoon swims forwards in the jerking fashion characteristic of many monads or flagellated protozoa. In different species of animals the spermatozoa differ more or less in size and appearance, and there is every reason to believe that the spermatozoa of each species are peculiar in certain respects even though we may not be able to distinguish any structural difference under the microscope. The human spermatozoa (Fig. 3) closely resemble those of other primates but are still slightly different, and the conclusion is logically inevitable, as we shall see later, that the spermatozoa as well as the ova of each individual differ slightly from those of every other individual.

2. *Fertilization.*—If a spermatozoon in its swimming comes into contact with a ripe but unfertilized egg, the head and middle-piece of the sperm sink into the egg while the tail is usually broken off and left outside (Fig. 4). About the time of the entrance of the spermatozoon into the egg the latter divides twice, giving off two minute cells known as polar bodies which lie at the upper or animal pole of the egg. The nucleus in the head of the sperm, after it has entered the egg, begins to absorb material from the egg and to grow in size and at the same time a minute granule, the centrosome, appears, either from the middle-piece or from the

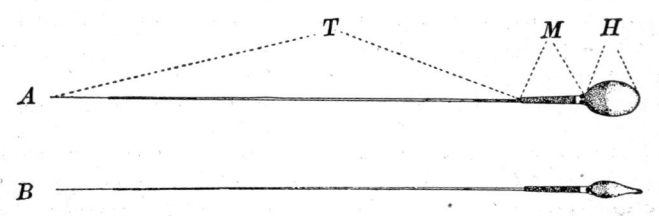

FIG. 3. TWO HUMAN SPERMATOZOA. *A*, showing the side of the flattened head; *B*, its edge; *H*, head; *M*, middle-piece; *T*, tail. (After G. Retzius.)

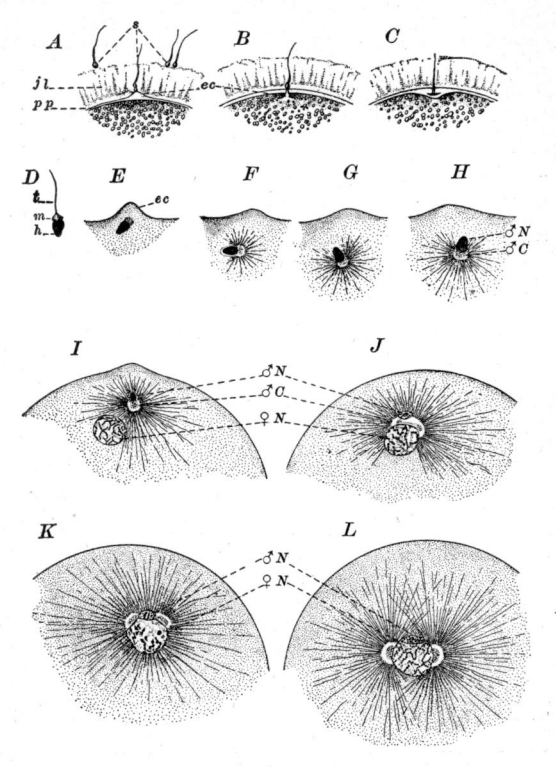

FIG. 4. FERTILIZATION OF THE EGGS OF STAR-FISH AND SEA-URCHIN. *A-C*, Successive stages in the entrance of a spermatozoon into the egg of the star-fish, *Asterias glacialis*. Only one sperm has penetrated the jelly layer (*jl*) which surrounds the egg and the peripheral protoplasm (*pp*) of the egg protrudes as an entrance cone (*ec*) to meet it. (After Fol.) *D*, Mature spermatozoon of the sea-urchin, *Toxopneustes*, showing head (*h*); middle-piece (*m*); and tail (*t*). *E-H*, Successive stages in the penetration of the sperm nucleus ($♂N$) and centrosome ($♂C$) into the egg of *Toxopneustes*. *I-L*, Stages in the approach of the sperm nucleus ($♂N$) to the egg nucleus ($♀N$), and in the division of the sperm centrosome ($♂C$) and the formation of the first cleavage spindle. (*D-L* after Wilson.)

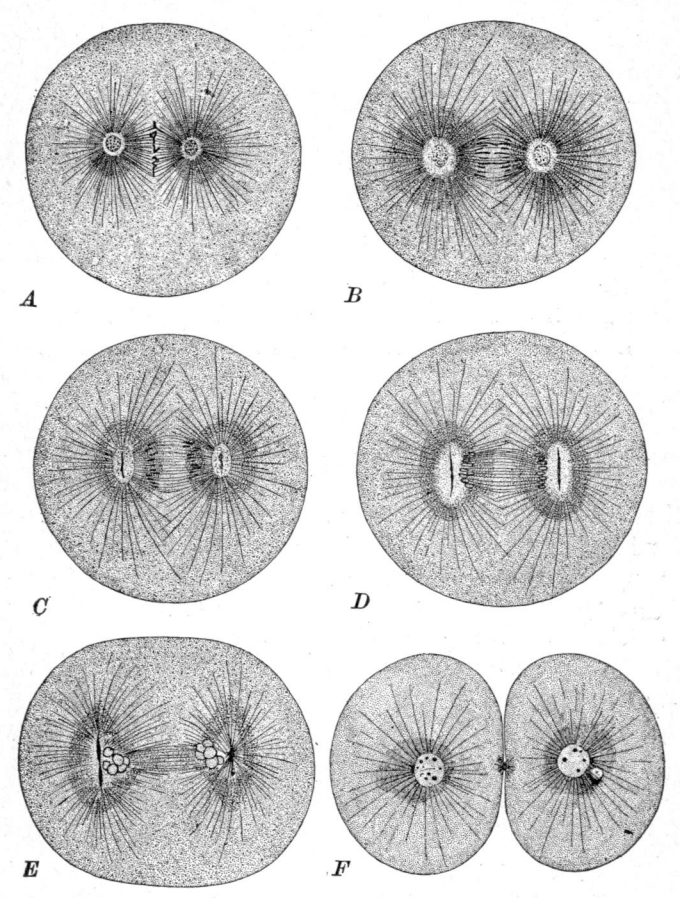

FIG. 5. FIRST CLEAVAGE OF THE EGG OF THE SEA-URCHIN, *Echinus microtuberculatus*. *A*, Nuclear division figure, or mitotic spindle, with a centrosome at each pole and with the chromosomes from the egg and sperm nuclei at the equator of the spindle. *B* and *C*, Later stages showing the separation of the daughter chromosomes from one another and their movement toward the two poles of the spindle. *D* and *E*, Still later stages showing the swelling of the chromosomes and their fusion to form nuclear vesicles. *F*, Complete division of egg into two cells, each containing one daughter nucleus and centrosome. (After Boveri.)

head of the sperm, and radiating lines run out from the centrosome into the substance of the egg. The sperm nucleus and centrosome then approach the egg nucleus and ultimately the two nuclei come to lie side by side (Fig. 4 *I-L*). Usually when one spermatozoon has entered an egg all others are barred from entering, probably by some change in the surface layer of the egg or in the chemical substances given out by the egg.

Oosperm or Zygote a Double Being.—This union of a single spermatozoon with an egg is known as fertilization. Whereas egg cells are usually, but not invariably, incapable of development unless fertilized, there begins, immediately after fertilization, a long series of transformations and differentiations of the fertilized egg which leads to the development of a complex animal, even of a person. In the fusion of egg and sperm a new individual, the oosperm, comes into being. The oosperm, formed by the union of the two sex cells, is really a double cell, since parts of the egg and sperm never lose their identity, and the individual which develops from this oosperm is a double being; even in the adult man this double nature of every cell, caused by the union of egg and sperm, is never lost.

A New and Distinct Individual.—In by far the larger number of animal species the oosperm, either just before or shortly after fertilization, is set free to begin its own individual existence, and in such cases it is perfectly clear that the fertilization of the egg marks the beginning of the new individual. But in practically every class of animals there are some species in which the fertilized egg is retained within the body of the mother for a varying period during which development is proceeding. In such cases it is not quite so evident that the new individual comes into being with the fertilization of the egg; rather the moment of birth, or separation from the mother, is generally looked upon as the beginning of the individual's existence. And yet in all cases the new individual is always distinguishable from the body of the mother since there is no protoplasmic connection between the two.

FACTS AND FACTORS OF DEVELOPMENT

In mammals generally, including also the human species, not a strand of protoplasm, not a nerve fiber, not a blood vessel passes over from the mother to the embryo; the latter is from the moment of fertilization onward a distinct individual with particular individual characteristics, and this is just as true of viviparous animals in which the egg undergoes a part of its development within the body of the mother as it is of oviparous ones in which the eggs are laid before development begins.

The fertilized egg of a star-fish or frog or man is not a different individual from the adult form into which it develops, rather it is a star-fish, a frog, or a human being in the one-celled stage. This fertilized egg fuses with no other cells, it takes into itself no living substance, but manufactures its own protoplasm from food substances; it receives food and oxygen from without and it gives out carbonic acid and other waste products; it is sensitive to certain alterations in the environment such as thermal, chemical and electrical changes—it is, in short, a distinct living thing, an individual or person. Under proper environmental conditions this fertilized egg cell develops, step by step, without the addition of anything from the outside except food, water, oxygen, and such other raw materials as are necessary to the life of any adult animal, into the immensely complex body of a star-fish, a frog, or a man. At the same time, from the relatively simple reactions and activities of the fertilized egg there develop, step by step, without the addition of anything from without except raw materials and environmental stimuli, the multifarious activities, reactions, instincts, habits, and intelligence of the mature animal.

Is not this miracle of development more wonderful than any possible miracle of creation? And yet as one watches this marvellous process by which the fertilized egg grows into the embryo, and this into the adult, each step appears relatively simple, each perceptible change is minute; but the changes are innumerable and unceasing and in the end they accomplish this miracle of transforming the fertilized egg cell into the fish or frog or man—a thing which would be incredible were it not for the fact that it has been seen by hundreds of observers and can be verified at any time by those who will take the trouble to study the process for themselves.

3. *Cell Division.*—After fertilization the first step in development is the cleavage or division of the egg (Fig. 5). This is in the main like any typical cell division and since the details of this process are of extraordinary interest in the study of the mechanism of heredity and development it is desirable to give at once a rather detailed account of the way in which the nucleus and cell-body divide.

a. *Mitosis or Indirect Division of the Nucleus.*—It was once supposed that both the nucleus and the cell-body divide by a simple process of constriction or direct division, but it is now known that the nucleus rarely divides in this manner, and that the nuclei of germ cells never do so. On the contrary the nucleus almost always divides by a complex process known as *mitosis* or indirect division (Figs. 6 and 7). During this process the chromatin granules of the "resting" nucleus become arranged in lines like beads on a thread (Fig. 8); these threads, which are called *chromosomes,* are at first long and slender and much coiled, but afterwards they grow shorter, thicker and straighter and it can then be seen that in each species of animal or plant there is a definite and constant number of these threads or chromosomes (Fig. 6, *A-D*); this number varies from 2 to 200 in different species of animals, the most usual number being somewhere between 10 and 30, but so far as is known each species has a constant number of chromosomes in every cell of the body.

The nucleolus and the nuclear membrane then disappear, the chromosomes move into the equator of the cell forming the *equatorial plate* (Fig. 6, *F*) and each one splits lengthwise into two daughter chromosomes which move apart toward the two poles of the cell (Fig. 7, *G, H*) where all the daughter chromosomes come together to form the two daughter nuclei. The cell body then divides by a process of constriction into two daughter cells (Fig. 7, *I, J*).

The formation, splitting and separation of the chromosomes is the most constant and characteristic feature of indirect nuclear division, but there are other important features which must now be mentioned. In all animals and in many of the lower plants there is present in the cell-body just outside the nuclear membrane a small deeply-staining granule, the centrosome, which is usually surrounded by radiating lines. In the early stages of mitosis this granule divides into two which move apart until they come to lie on opposite sides of the nucleus (Fig. 6, *A-C*). When the nuclear membrane dissolves the radiating lines which surround these two centrosomes increase greatly in length forming two *asters* and those rays which run through the nuclear area constitute a *spindle* with the chromosomes in its equator and the centrosomes at its two poles (Fig. 6, *D-F*). Later the chromosomes move along the spindle toward its poles where the daughter nuclei are formed. The centrosomes, asters and spindle, known collectively as the *amphiaster,* constitute an apparatus for the accurate separation of the daughter chromosomes and for the division of the cell-body.

The chromosomes are most compact and deeply-staining at the metaphase or equatorial plate stage of division; after they have moved to the poles of the spindle they begin to absorb achromatin from the surrounding plasma thus swelling up and becoming *chromosomal vesicles* with clear contents and chromatic walls (Fig. 8, *E, F*). These vesicles then continue to enlarge and their chromatin takes the form of threads or granules. After the formation of the daughter nuclei the different vesicles are so closely pressed together that it is usually impossible to see the partition walls between them; however in several different animals and plants the chromosomal vesicles are recognizable even in the resting nucleus (Fig. 8, *G*), and in every organism the same

18   HEREDITY AND ENVIRONMENT

number of chromosomes, having the same relative shapes and sizes, comes out of a nucleus at the time of division as went into it at the preceding mitosis, each new chromosome coming out of a

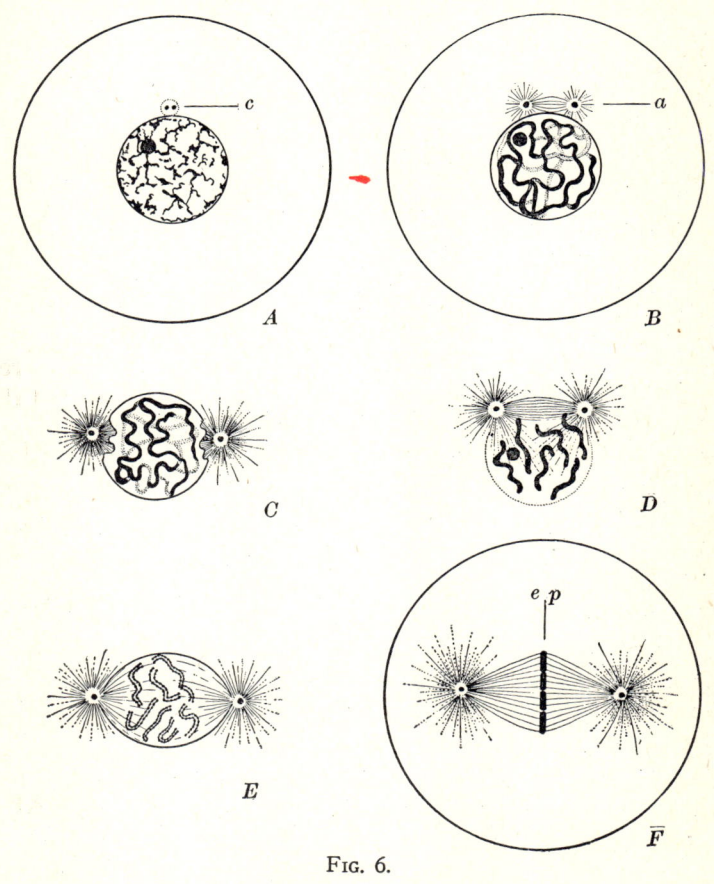

Fig. 6.

Figs. 6, 7. Diagrams of Successive Stages in the Division of a Cell by Mitosis. *A*, Cell with "resting" nucleus and centrosome (*c*); *B-E*, Early stages of division during which the chromatin takes the form of short thick threads, the chromosomes (*C, D, E*) while the centrosome gives

chromosomal vesicle (Fig. 8, *A, B, C*). Whenever one can trace individual chromosomes or chromosomal vesicles through the resting stage it is certain that every chromosome preserves its

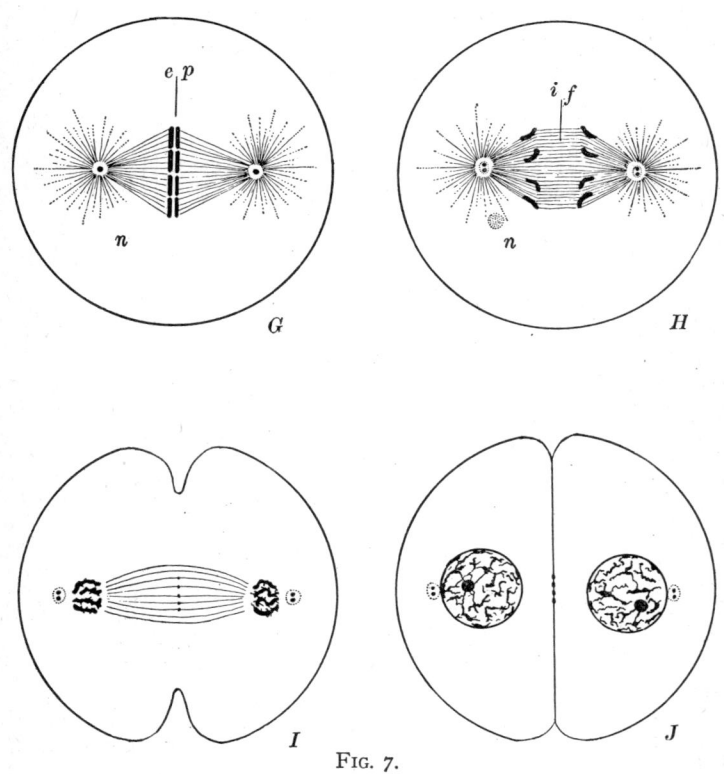

FIG. 7.

rise to a spindle (*a*) with astral radiations at its poles; *F*, Middle stage of division in which the chromosomes lie in the equator of the spindle forming the equatorial plate (*ep*); *G, H*, Stages showing the splitting of each chromosome and the movement of the halves toward the poles of the spindle; *ep*, Equatorial plate; *if*, Interzonal filaments; *n*, Nucleolus: *I*, Complete separation of daughter chromosomes and formation of daughter nuclei; beginning of division of cell body; *J*, Complete separation of daughter cells and return of nucleus and centrosome to resting condition. (After Wilson.)

20   HEREDITY AND ENVIRONMENT

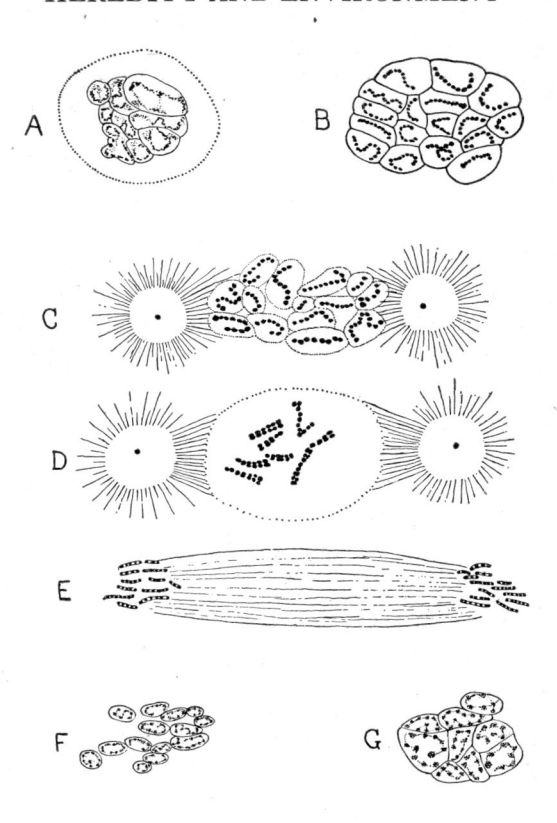

FIG. 8. SUCCESSIVE STAGES OF MITOSIS IN THE CLEAVAGE OF THE EGG OF A FISH (*Fundulus*) showing that new chromosomes are formed inside of old ones (chromosomal vesicles), that chromosomes or chromosomal vesicles persist from one division to the next and that even the "resting" nucleus is composed of chromosomal vesicles. *A*, nucleus at beginning of mitosis, having shrunk from dotted outline and showing chromosomal vesicles containing chromatin granules; *B*, each chromosomal vesicle contains granules or chromomeres which are condensing to form a chromosome; *C*, amphiaster showing faint outlines of the chromosomal vesicles with their contained chromosomes: *D*, amphiaster showing each chromosome beginning to split and the chromosomes dividing; *E*, late phase of division showing daughter chromosomes at the poles of the spindle and each chromosome becoming vesicular; *F*, still later phase, each chromosome a vesicle containing chromatin granules; *G*, daughter nucleus showing chromosomal vesicles containing scattered chromatin granules. (After Richards.)

individuality or identity, and even where this cannot be done the fact that the same number of chromosomes, having the same peculiarities of shape and size comes out of a nucleus as went into it, is evidence that here also each chromosome has preserved its identity.

b. *Cleavage of the Egg.*—After the entrance of the spermatozoon into the egg the sperm nucleus moves toward the egg nucleus until the two meet when they divide by mitosis (Figs. 4 *I-L*, 5 *A-F*). The centrosome, which usually accompanies the sperm nucleus in its passage through the egg, divides and forms a spindle-shaped figure with astral radiations at its two poles (Fig. 4). The chromatin, or stainable substance of the egg and sperm nuclei, takes the form of threads or chromosomes (Fig. 5). Each chromosome then splits lengthwise, its two halves moving to opposite ends of the spindle, where the daughter chromosomes form the chromosomal vesicles of the daughter nuclei. In this way the chromatin of the egg and sperm nuclei is exactly halved.

After the germ nuclei have divided in this manner the entire egg divides by a process of constriction into two cells (Fig. 5 *F*). This is the beginning of a long series of cell divisions, each of them essentially like the first, by which the egg is subdivided successively into a constantly increasing number of cells. During the earlier divisions there is little or no increase in the volume of the egg, consequently successive generations of cells continually grow smaller (Figs. 9-11). This process is known as the cleavage of the egg, and by it the egg is not only split up into a considerable number of small cells, but a much more important result is that the different kinds of protoplasm in the egg become isolated in different cleavage cells, so that these substances can no longer freely commingle. The cleavage cells, in short, come to contain different kinds of substance, and thus to differ from one another. The differentiations of the cleavage cells appear much earlier in some forms than in others, but in all

22   HEREDITY AND ENVIRONMENT

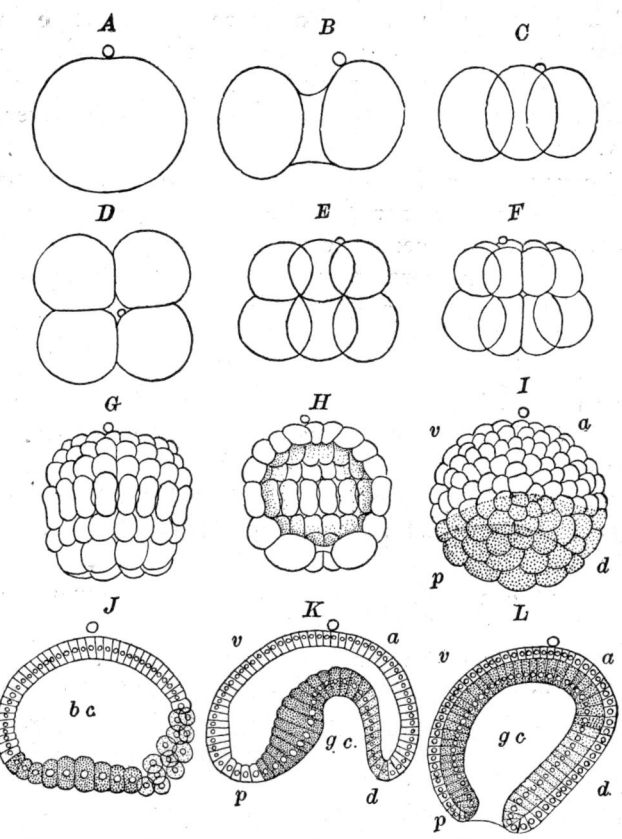

FIG. 9. SUCCESSIVE STAGES IN THE CLEAVAGE AND GASTRULATION OF *Amphioxus*. *A*, one cell; *B*, two cells; *C* and *D*, four cells; *E*, eight cells; *F*, sixteen cells; *G*, blastula stage of about ninety-six cells; *H*, section through the same showing the cleavage cavity; *I*, blastula seen from the left side showing three zones of cells, viz., an upper clear zone of ectoderm, a middle (faintly shaded) zone of mesoderm and a lower (deeply shaded) zone of endoderm cells; *J*, section through the same showing these three types of cells; *K* and *L*, successive stages in the infolding of the endoderm; cells indicated as in the preceding figure. The polar body is shown at the upper pole. *a*, anterior; *p*, posterior; *v*, ventral; *d*, dorsal; *bc*, blastocœl; *gc*, gastrocœl.

cases such differentiations appear during early or late cleavage (Figs. 9-11).

4. *Embryogeny.*—From this stage onward the course of development differs in different classes of animals to such an extent that it is difficult to formulate any general description that will apply to all of them. Usually the many cleavage cells form a hollow sphere, the blastula (Figs. 9, 11, H), and this in turn becomes a gastrula (Figs. 9 K, 11 J), in which at first two, and later three, groups or layers of cells may be recognized; the outer layer, which is formed from cells nearest the upper pole of the egg, is the ectoderm; the inner layer, or endoderm, is formed from cells nearest the lower pole; a middle layer, or group of cells, the mesoderm, is formed from cleavage cells which in vertebrates lie between the upper and lower poles (Fig. 11, m).

5. *Organogeny.*—By further differentiation of the cells of these layers and by dissimilar growth and folding of the layers themselves the various organs of the embryo begin to appear. From the ectoderm are formed the outer layer of the skin and the whole nervous system; from the endoderm arise the lining of the alimentary canal and its outgrowths; from the mesoderm come, in whole or in part, the skeletal, muscular, vascular, excretory, and reproductive systems. In vertebrates the nervous system appears as a plate of rather large ectoderm cells (Fig. 11, n); this plate rolls up at its sides to form a groove and then a tube; and by enlargement of certain portions of this tube and by foldings and thickenings of its walls the brain and spinal cord are formed (Figs. 11, K, L; 13, C, D). The retina or sensory portion of the eye is formed as an outgrowth from the fore part of the brain (Fig. 13, D); the sensory portion of the ear comes from a cup-shaped depression of the superficial ectoderm which covers the hinder portion of the head (Fig. 13, E and F). The back-bone begins to appear as a delicate cellular rod (Fig. 11, c), which then in higher vertebrates becomes surrounded successively by a fibrous, a cartilaginous, and a bony sheath. And so one might

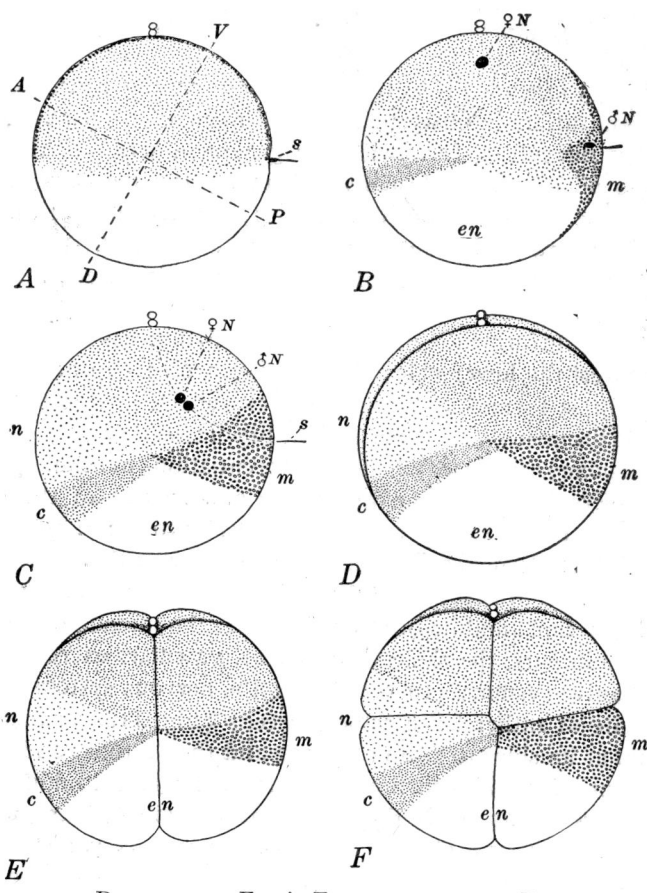

FIGS. 10, 11. DIAGRAMS OF FROG'S EGGS SHOWING THE RELATIONS OF THE AXES AND SUBSTANCES OF THE EGG TO THE AXES AND PRINCIPAL ORGANS OF THE EMBRYO. All eggs viewed from right side, polar bodies above; *A*, anterior; *P*, posterior; *D*, dorsal; *V*, ventral; *s*, spermatozoon; ♂*N*, sperm nucleus, ♀*N*, egg nucleus; *m*, mesodermal crescent where mesoderm will form; *c* and *n*, gray crescent, where chorda (*c*) and nervous system (*n*) will form; *en*, area of endoderm; area around polar bodies will form ectoderm of skin; *sc*, segmentation cavity; *bp*, blastopore; *E*, enteron; *M*, region of mouth.

FACTS AND FACTORS OF DEVELOPMENT 25

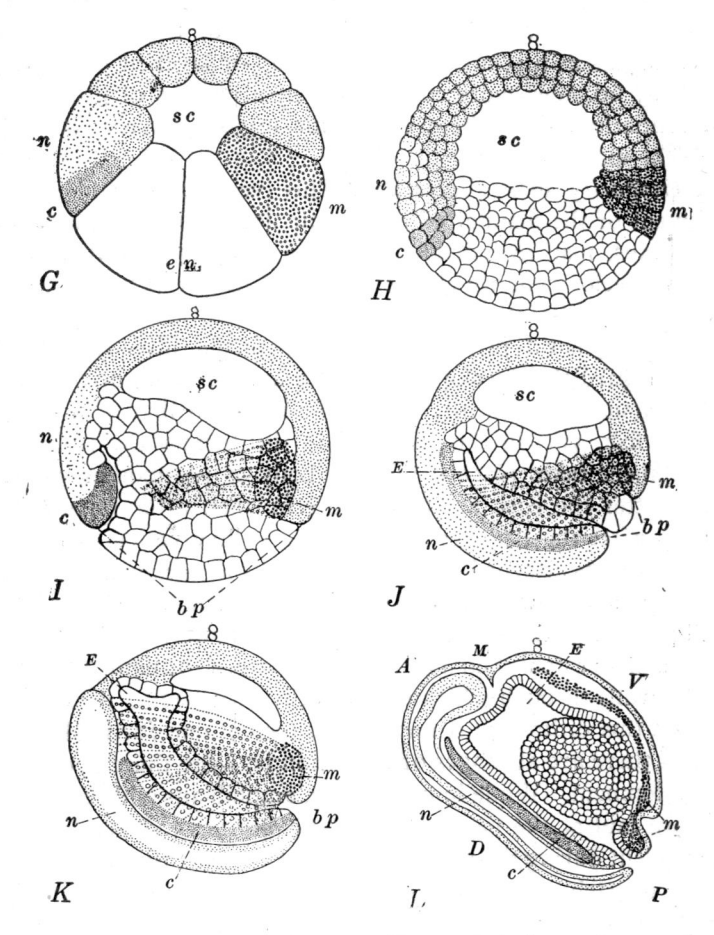

Fig. 10. Surface Views of Entire Eggs. *A*, before entrance of spermatozoon; *B*, just after entrance of sperm; *C*, union of egg and sperm nuclei; *D*, 2-cell stage; *E*, 4-cell stage; *F*, 8-cell stage.

Fig. 11. Sections in Median Plane of Embryos. *G*, 16-32-cell stage; *H*, blastula; *I*, early gastrula; *J*, late gastrula; *K*, early embryo, *L*, late embryo.

go on with a description of all the organs of the body, each of which begins as a relatively simple group or layer of cells, which gradually becomes more complicated by a process of growth and differentiation, until these embryonic organs assume more and more the mature form.

6. *Oviparity and Viviparity.*—This very brief and general statement of the manner of embryonic development applies to all vertebrates, man included. There are many special features of human development which are treated at length in works on embryology, but which need not detain us here since they do not affect the general principles of development already outlined. In one regard the development of the human being or of almost any mammal is apparently very different from that of a bird or frog or fish, viz., in the fact that in the former the embryonic development takes place within the body of the mother whereas in the latter the eggs are laid before or soon after fertilization. In man, after the cleavage of the egg, a hollow vesicle is formed, which becomes attached to the uterine walls by means of processes or villi which grow out from it (Fig. 12, *D*, *E*, *F*) while only a small portion of the vesicle becomes transformed into the embryo. There is thus established a connection between the embryo and the uterine walls through which nutriment is absorbed by the embryo. And yet this difference is not a fundamental one for in different animals there are all stages of transition between these two modes of development. While in most fishes, amphibians and reptiles the eggs are laid at the beginning of development and are free and independent during the whole course of ontogeny, there are certain species in each of these classes in which the development takes place within the body of the mother. Even in birds a portion of the development takes place within the body of the female before the eggs are laid, and there are mammals (monotremes) which lay eggs, while in others (marsupials) the young are born in a very imperfect condition.

Mother always Distinct from Child.—These facts indicate that

FACTS AND FACTORS OF DEVELOPMENT

there is no fundamental difference between oviparity and viviparity. In the latter the union between the embryo and the mother is a nutritive but not a protoplasmic one. Blood plasma passes from one to the other by a process of soakage, and the only maternal influences which can affect the developing embryo are such as may be conveyed through the blood plasma and are chiefly nutritive in character. Careful studies have shown that sup-

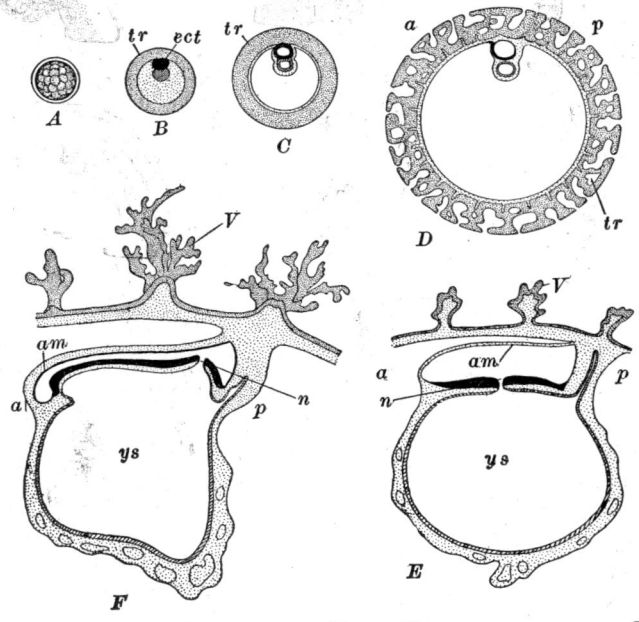

FIG. 12. DIAGRAMS SHOWING THE EARLY DEVELOPMENT OF THE HUMAN OOSPERM. *A*, cleavage stage which has just come into the uterus; *B* and *C*, blastodermic vesicles embedded in the mucous membrane of the uterus; *D*, *E* and *F*, longitudinal sections of later stages, the anterior and posterior poles being marked by the axis *a p*. In *C* cavities have appeared in the ectoderm, endoderm and mesoderm. *D*, villi forming from the trophoblast (nutritive layer, *tr*); black indicates ectoderm (*ect*); oblique lines, endoderm; few stipples, mesoderm; *V*, villi; *am*, amnion; *ys*, yolk sac; *n*, neurenteric canal; x 25. (After Keibel.)

28   HEREDITY AND ENVIRONMENT

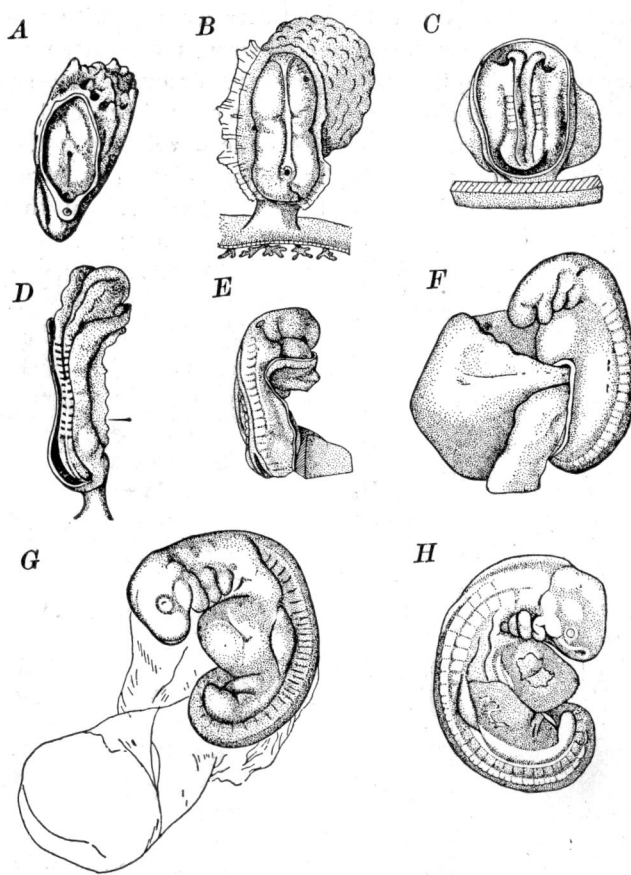

Fig. 13. *A-H*, successive stages in the early development of the human embryo; *A*, blastodermic vesicle showing primitive axis in embryonic area, age unknown; *B*, blastodermic vesicle attached to uterine wall at the posterior pole, showing neural groove, age unknown; *C*, later stage in which the neural folds are closing and five pairs of somites have appeared, age ten to fourteen days; *D*, stage of fourteen somites showing enlargements of the neural folds at the anterior end which will form the brain, age fourteen to sixteen days; *E* and *F*, later stages, the latter with twenty-three somites and three visceral clefts; the ear shows as a depression at the dorsal angle of the second cleft; *G*, embryo, of thirty-five somites, showing eye, branchial arches and limb buds; *H*, embryo of thirty-six somites showing nasal pit, eye, branchial arches and clefts, limb buds and heart. (After Keibel.)

FACTS AND FACTORS OF DEVELOPMENT

posed "maternal impressions" of the physical, mental, or emotional conditions of the mother upon the unborn child have no existence in fact, except in so far as the quality of the mother's

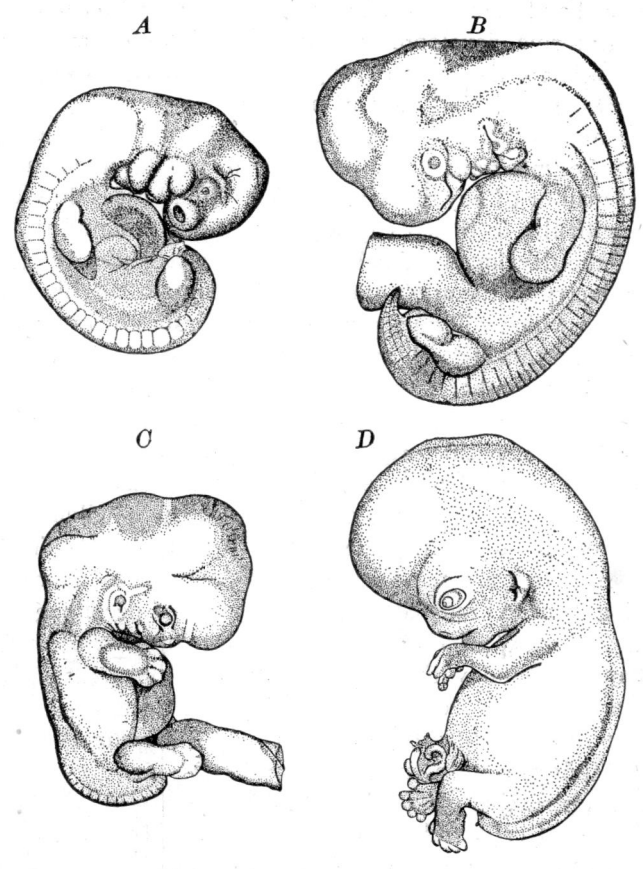

FIG. 14. *A*, human embryo of forty-two somites, age twenty-one days; *B*, embryo of about four weeks; *C*, still older embryo showing the beginnings of the formation of digits; *D*, embryo of about two months; *C* and *D* are drawn on a smaller scale than *A* and *B*. (After Keibel.)

blood may be changed and may affect the child. At no time, whether before or after birth, is the mother more than nurse to the child. Hereditary influences are transmitted only through the egg cell and the sperm cell and these influences are not affected by intra-uterine development. The principles of heredity and development are the same in oviparous and in viviparous animals —in fishes, frogs, birds and men.

Summary.—This is a very brief and incomplete statement of some of the important stages or phases of the development of the body of man or of any other vertebrate. In all cases development begins with the fertilized egg which contains none of the structures of the developed animal, though it may exhibit the polarity and symmetry of the adult and may also contain specific kinds of protoplasm which will give rise to specific tissues or organs of the adult. From this egg cell arise by division many cells which differ from one another more and more as development proceeds, until finally the adult animal results. A specific type of development is due to a specific organization of the germ cells with which development begins, but the earlier differentiations of the egg are relatively few and simple as compared with the bewildering complexities of the adult, and the best way of understanding adult structures is to trace them back in development to their simpler beginnings and to study them in the process of becoming.

7. *Development of Functions.*—The development of functions goes hand in hand with the development of structures; indeed function and structure are merely different aspects of one and the same thing, namely organization. All the general functions of living things are present in the germ cells, viz., (1) Constructive and destructive metabolism, (2) Reproduction, as shown in the division of cells and cell constituents, (3) Irritability, or the capacity of receiving and responding to stimuli. All these general functions of living things are manifested by germ cells, but as development advances each of these functions becomes more specialized, more complicated and more perfect. A cell which at

an early stage was protective, locomotor and sensory in function may give rise to daughter cells in which these functions are distributed to different cells; cells which at an early stage were sensitive to many kinds of stimuli give rise to daughter cells which are especially sensitive to one particular kind of stimulus, such as vibration, light, or chemicals.

Differentiation of Functions and Structures.—Functions develop from a generalized to a specialized condition by the process of "physiological division of labor" which accompanies morphological division of substance. But just as in the union of hydrogen and oxygen a new substance, water, appears which was not present before, by a process of "creative synthesis," and as in the development of structures new parts appear, which were not present in the germ, so new functions appear in the course of development, which are not merely sorted out of the general functions present at the beginning, but which are created by the interaction and synthesis of parts and functions previously present. For example, Lane has shown that young rats are quite insensitive to light until several days after birth although the eye begins to form at a very early stage of development. Doubtless every part of the eye is functioning in one way or another during the entire development but not until all parts are formed and connected and all their functions are synthesized does the new function, vision, spring into existence. Undoubtedly the same is true of many other complex functions which have no existence until all their constituents are present and integrated, when they suddenly appear.

Living Functions and Structures Inseparable.—Much less attention has been paid to the development of functions than to the development of structures, and consequently it is not possible to describe the former with the same degree of detail as the latter. But in spite of the lack of detailed knowledge regarding the development of particular functions the general fact of such development is well established. To what extent structures may modify functions or functions structures, in the course of development, is a problem which has been much discussed, and upon the answer to it depends the fate of certain important theories, for example Lamarckism; but this problem can be solved only by thoroughgoing experimental and analytical work. In the meantime it seems safe to conclude that living structures and functions are inseparable and that anything which modifies one of these must of necessity modify the other also; they are merely different aspects of organization, and are dealt with separately by the morphologists and physiologists only as a matter of convenience. At the same time there can be no doubt that minute changes of function can frequently be detected where no corresponding change of structure can be seen, but this shows only that physiological tests may be more delicate than morphological ones. In certain lines of modern biological work such as bacteriology, cytology, and genetics, many functional distinctions are recognizable between organisms that are morphologically indistinguishable. But this does not signify that functional changes precede structural ones, but only that the latter are more difficult to see than the former. For every change of function it is probable that an "unlimited microscopist" could discover a corresponding change of structure.

11. Development of the Mind

The development of the mind parallels that of the body: whatever the ultimate relations of the mind and body may be, there can be no reasonable doubt that the two develop together from the germ. It is a curious fact that many people who are seriously disturbed by scientific teachings as to the evolution or gradual development of the human race accept with equanimity the universal observation as to the development of the human individual,— mind as well as body. The animal ancestry of the race is surely no more disturbing to philosophical and religious beliefs than the germinal origin of the individual, and yet the latter is a fact of universal observation which can not be relegated to the domain of hypothesis or theory, and which can not be successfully denied. If we admit the fact of the development of the entire individual, surely it matters little to our philosophical or religious beliefs to admit the development or evolution of the race.

Ancient Speculations.—The origin of the mind, or rather of the soul, is a topic upon which there has been much speculation by philosophers and theologians. One of the earliest hypotheses was that which is known as transmigration or metempsychosis. This doctrine probably reached its greatest development in ancient India, where it formed an important part of Buddhistic belief; it was also a part of the religion of ancient Egypt; it was embodied in the philosophies of Pythagoras and Plato. According to these teachings, the number of souls is a constant one; souls are neither made nor destroyed, but at birth a soul which had once tenanted another body enters into the new body. This doctrine was generally repudiated by the Fathers of the Christian Church. Jerome and others adopted the view that God creates a new soul for each body that is generated, and that every soul is thus a special divine creation. This has become the prevailing view of the Christian Church and is known as creationism. On the other hand Tertullian taught that souls of children are generated from the souls of parents as bodies are from bodies. This doctrine, which is known as traducianism, has been defended by certain modern theologians, but has been formally condemned by the Roman Catholic Church.

Traducianism undoubtedly comes nearer the scientific teachings as to the development of the mind than does either of the other doctrines named, but it is based upon the prevalent but erroneous belief that the bodies of the parents generate the body of the child, and that correspondingly the souls of the parents generate the soul of the child. Now we know that the child comes from germ cells and not from the highly differentiated bodies of the parents, and furthermore that these cells are not made by the parents' bodies but have arisen by the division of antecedent germ cells (see

p. 125). Consequently it is not possible to hold that bodies generate bodies or even germ cells, nor that souls generate souls. The only possible scientific position is that the mind (or soul) as well as the body develops from the germ.

Certainty of Mental Development.—No fact in human experience is more certain than that the mind develops by gradual and natural processes from a simple condition which can scarcely be called mind at all; no fact in human experience is fraught with greater practical and philosophical significance than this, and yet no fact is more generally disregarded. We know that the greatest men of the race were once babies, embryos, germ cells, and that the greatest minds in human history were once the minds of babies, embryos and germ cells, and yet this stupendous fact has had but little influence on our beliefs as to the nature of man and of mind. We rarely think of Plato and Aristotle, of Shakespeare and Newton, of Pasteur and Darwin, except in their full epiphany, and yet we know that when each of these was a child he "thought as a child and spake as a child," and when he was a germ cell he behaved as a germ cell.

Wonders of this Development.—The development of the mind from the activities of the germ cells is certainly most wonderful and mysterious, but probably no more so than the development of the complicated body of the adult animal from the structures of the germ. Both belong to the same order of phenomena and there is no more reason for supposing that the mind is supernaturally created than that the body is. Indeed, we know that the mind is formed by a process of development, and the stages of this development are fairly well known. There is nowhere in the entire course of mental development a sudden appearance of psychical processes, but rather a gradual development of these from simpler and simpler beginnings. No detailed study has been made of the reactions of human germ cells and embryos, but there is every reason to believe that these reactions are simpler in the embryo and germ cell than in the infant, and that they are generally similar to the reactions of the germ cells and embryos of other animals, and to the behavior of many lower organisms.

Matter and Mind.—A few years ago such a statement would have been branded as "materialism" and promptly rejected without examination by those who are frightened by names. But the general spread of the scientific spirit is shown not only by the growing regard for evidence but also by the decreasing power of epithets. "Materialism," like many another ghost, fades away into thin air or at least loses many of its terrors, when closely scrutinized. But the statement that mind develops from the germ cells is not an affirmation of materialism, for while it identifies the origin of the entire individual, mind and body, with the development of the germ, it does not assert that "matter" is the cause of "mind" either in the germ or in the adult. It must not be forgotten that germ cells are living things and that we go no further in associating the beginnings of mind with the beginnings of body in the germ than we do in associating mind and body in the adult. It is just as materialistic to hold that the mind of the mature man is associated with his body as it is to hold that the beginnings of mind in the germ are associated with the beginnings of the body, and both of these tenets are incontrovertible.

Body and Mind.—It seems to me that the mind is related to the body as function is to structure; there are those who maintain that structure is the cause of function, that the real problem in evolution or development is the transformation of one structure into another, and that the functions which go with certain structures are merely incidental results; on the other hand are those who maintain that function is the cause of structure and that the problem of evolution or development is the change which takes place in functions and habits, these changes causing corresponding transformations of structure. Among adherents of the former view may be classed many morphologists and Neo-Darwinians; among proponents of the latter, many physiologists and Neo-Lamarckians. It seems to me that the defenders of each of these views fail to recognize the essential unity of the entire organism, structure as well as function; that neither of these precedes the other as cause precedes effect, though each may modify or condition the other, but that they are two aspects of one common thing, viz., organization. In the same way I think that the body or brain is not the cause of mind, nor mind the cause of body or brain, but that both are inherent in one common organization or individuality.

In asserting that the mind develops from the germ as the body does, no attempt is made to explain the fundamental properties of body or mind. As the structures of the body may be traced back to certain fundamental structures of the germ cell, so the characteristics of the mind may be traced back to certain fundamental properties and activities of the germ. Many of the psychical processes may be traced back in their development to properties of sensitivity, reflex motions, and persistence of the effects of stimuli. All organisms manifest these properties and for aught we know to the contrary they may be original and necessary characteristics of living things. In the simplest protoplasm we find organization, that is, structure and function, and in germinal protoplasm we find the elements of the mind as well as of the body, and the problem of the ultimate relation of the two is the same whether we consider the organism in its germinal or in its adult stage.

Germinal Bases of Mind

In some way the mind as well as the body develops out of the germ. What are the germinal bases of mind? What are the psychical *Anlagen* in embryos and how do they develop? In this case, even more than in the development of the body, we are compelled to rely upon comparisons between human development and that of other animals, but the great principle of the oneness of life, as respects its fundamental processes, has never yet failed to hold true and will not fail us here. In the study of the psychical processes of organisms other than ourselves we are compelled to rely upon a study of their activities, their reactions to stimuli, since we can not approach the subject in any other way. The reactions and behavior of organisms under normal and experimental conditions give the only insight which we can get into their psychical processes; and this applies to men no less than to protozoa.

1. *Sensitivity*.—The most fundamental phenomenon in the behavior of organisms is irritability or sensitivity, which is the ability of receiving and responding to stimuli: this is one of the fundamental properties of all protoplasm. But living matter is not equally sensitive to all stimuli, nor to all strengths of the same stimulus. Many of the simplest unicellular plants and animals show that they are differentially sensitive; they often move toward weak light and away from strong light, away from extremes of heat and cold, into certain chemical substances and away from others; in short, all organisms, even the simplest, may respond differently to different kinds of stimuli or to different degrees of the same stimulus. This is what is known as *differential sensitivity* (Figs. 15-19). On the other hand, many organisms respond in the same way to different stimuli, and this may be taken to indicate generally that they are not differentially sensitive to such stimuli; it is not to be concluded because organisms respond differently to certain stimuli that they are therefore capable of distinguishing between all kinds of stimuli, for this is certainly not true. Even in adult men the capacity of distinguishing between different kinds of stimuli is far from perfect.

Sensitivity of Germ Cells.—Egg cells and spermatozoa show this property of sensitivity. The egg is generally incapable of locomotion, and since the results of stimulation must usually be detected by movements it is not easy to determine to what extent the egg is sensitive; but though the egg lacks the power of locomotion, it possesses in a marked degree the power of intra-cellular movement of the cell contents. When a spermatozoon comes into contact with the surface of the egg the cortical protoplasm of the egg flows toward that point and may form a cone or protoplasmic prominence into which the sperm is received (Fig. 4, *ec*). It is an interesting fact that the same sort of response follows when a frog's egg is pricked by a needle, thus showing that in this case the egg does not distinguish between the prick of the needle and that of the spermatozoon. The spermatozoon is usually a locomotor cell and it responds differently to certain stimuli, just as many bacteria and protozoa do; spermatozoa are strongly stimulated by weak alkali and alcohol, they gather in certain chemical substances and not in others, they collect in great numbers around fertilizable egg cells, etc.

Sensitivity of Oosperm and of Embryo.—The movements of fertilized egg cells, cleavage cells, and early embryonic cells are usually limited to flowing movements within the individual cells. These movements, which are of a complicated nature, are of the greatest significance in the differentiation of the egg into the embryo; they are caused chiefly by internal stimuli and by non-localized external ones. Modifications of the external stimuli often lead to modifications of these intracellular movements and to abnormal types of cleavage and development—in short, these movements show that the fertilized egg is differentially sensitive.

In the further course of development particular portions of the embryo become especially sensitive to some kinds of stimuli, while other portions become sensitive to others. In this way the differ-

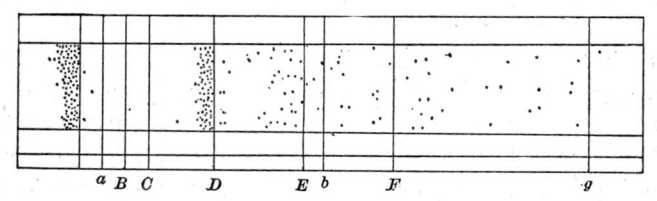

FIG. 15. DISTRIBUTION OF BACTERIA IN THE SPECTRUM. The largest group is in the ultra-red at the left; the next largest group is in the yellow-orange close to the line *D*. (From Jennings, after Engelmann.)

FACTS AND FACTORS OF DEVELOPMENT

ent sense organs, each especially sensitive to one particular kind of stimulus, arise from the generalized sensitivity of the oosperm, and thus general sensitivity, which is a property of all protoplasm, becomes differential sensitivity and special senses in the process

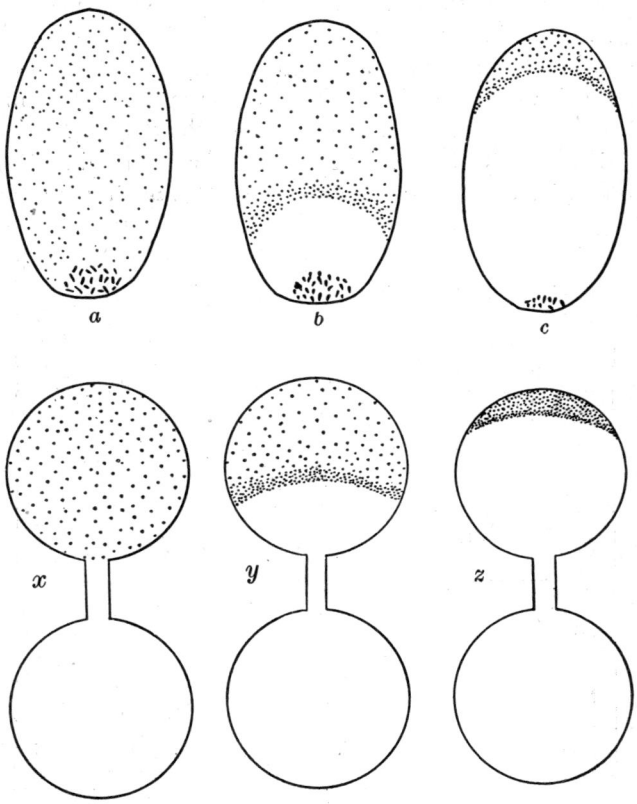

FIG. 16, *a, b, c*. REPULSION OF *Spirilla* BY COMMON SALT. *a,* condition immediately after adding crystals; *b* and *c,* later stages in the reaction.

x, y, z, repulsion of *Spirilla* by distilled water. The upper drop consists of sea-water containing *Spirilla,* the lower drop, of distilled water. At *x* these have just been united by a narrow neck; at *y* and *z,* the bacteria have retreated before the distilled water. (From Jennings, after Massart.)

of embryonic differentiation. Such sensitivity is the basis of all psychic processes; sensations are the elements of the mind.

2. *Reflexes, Tropisms, Instincts.*—All the responses of germ cells, and of the simplest organisms, to stimuli are in the nature of reflexes or tropisms, that is, relatively simple, machine-like responses. "Reflex motions" originally referred to those responses of higher animals in which the peripheral stimuli were reflected, as it were, from the spinal cord to the appropriate muscles without the participation of the brain. But at present the word "reflex" has come to have a much broader application and is used for all simple, automatic responses, even though there are no

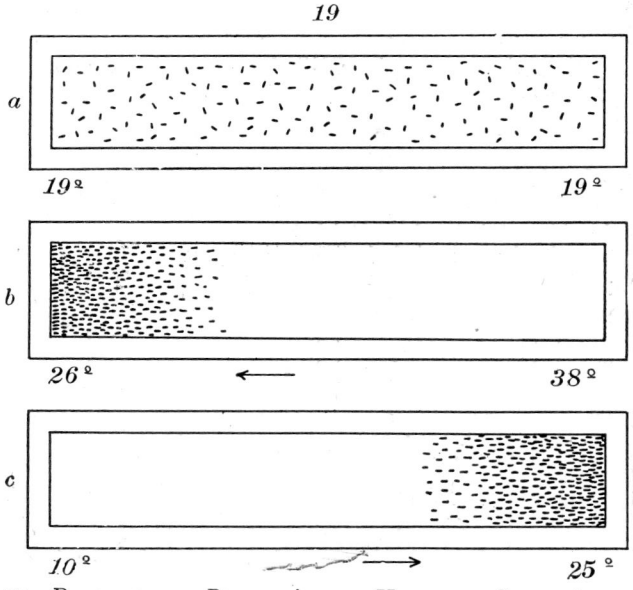

Fig. 17. Reactions of *Paramecium* to Heat and Cold. At *a* the infusoria are uniformly distributed in a trough, both ends of which have a temperature of 19°; at *b* the infusoria are shown collected at the cooler end of the trough; at *c* they have collected at the warmer end of the trough. (From Jennings, after Mendelssohn.)

nerves and even when the response is not movement but secretion, metabolism or any other activity. "Tropisms," on the other hand, is a more specific term and refers to the movements of organisms toward or away from a source of stimulus, the former being

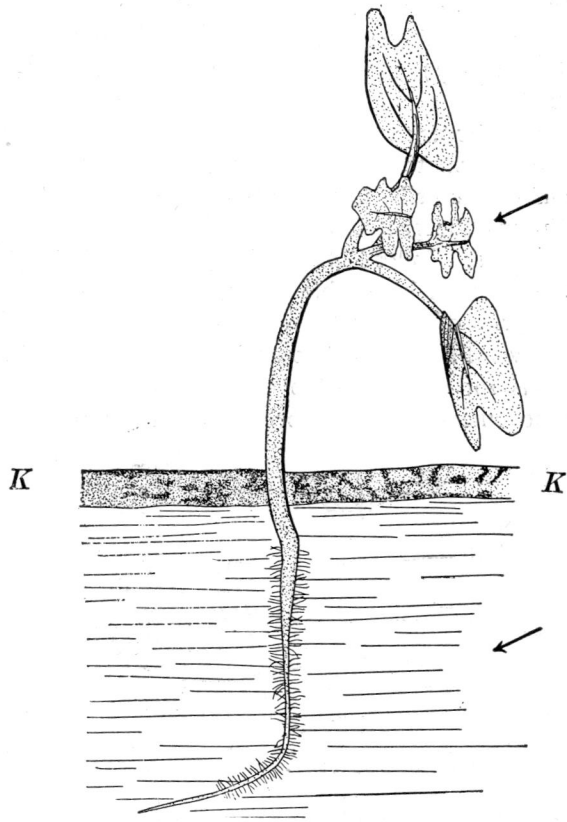

FIG. 18. PHOTOTROPISM OF SEEDLING OF WHITE MUSTARD supported by a sheet of cork (*K, K*) floating on the water. The direction from which light comes is shown by the arrows; the stem and leaves are turned toward the light (positive phototropism), the root away from the light (negative phototropism). (After Strasburger.)

known as positive and the latter as negative tropism. When responses are very complex, one response calling forth another and involving many complex reflexes or chains of reflexes, as is frequently the case in animals, they are known as "instincts." Reflexes and tropisms occur in the simplest organisms, such as bacteria, protozoa, and single cells, as well as in higher plants and animals (Figs. 15-19), but instincts are limited to animals with a nervous system.

Reflexes and Tropisms of Germ Cells and Embryos are seen in movements of spermatozoa, movements of the protoplasm in egg cells and embryonic cells, movements of cells and cell masses in the formation of the gastrula, alimentary canal, nervous system and other organs. Indeed the entire process of development, whether accompanied by visible movements or not, may be regarded as a series of automatic responses to stimuli. When the embryo becomes differentiated to such an extent as to have specialized organs for producing movement its capacity for making responsive movements to stimuli becomes much increased. In the

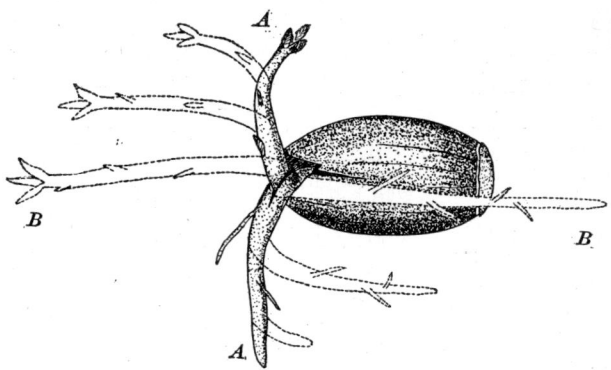

FIG. 19. GEOTROPISM OF SEEDLING OAK. After starting to grow with the axis *A-A* in vertical position the seedling was gradually turned through 90° until the axis *B-B* was vertical; during this change of position the stem continues to grow upward (negative geotropism), the root downward (positive geotropism) as indicated by dotted outlines.

FACTS AND FACTORS OF DEVELOPMENT

embryo the rhythmic contractions of heart, amnion and intestine are early manifestations of reflex motions. These appear chiefly in the involuntary muscles before nervous connections are formed, the protoplasm of the muscle cells probably responding directly to the chemical stimulus of certain salts in the body fluids, as Loeb has shown in other cases. Reflexes which appear later are the "random movements" of the voluntary muscles of limbs and body, which are called forth by nerve impulses. Tropisms are manifested only by organisms capable of considerable free movement and hence are absent in the foetus though present in many free-living larvae.

Development of Instincts.—Some instincts are present immediately after birth, such as the instinct of sucking or crying in the human infant, though these are so simple when compared with some instincts which develop later that they might be classed as reflexes; it is doubtful whether any of the activities before birth could properly be designated as instincts. Reflexes, tropisms and instincts have had a phylogenetic as well as an ontogenetic origin, and consequently we might expect that they would in general make for the preservation of the species; as a matter of fact we usually find that they are remarkably adapted to this end. For instance the instincts of the human infant to grasp objects, to suck things which it can get into its mouth, to cry when in pain, are complicated reflexes which have survived in the course of evolution, probably because they serve a useful purpose.

Very much has been written on the nature and origin of instincts, but the best available evidence strongly favors the view that instincts are complex reflexes, which, like the structures of an organism, have been built up, both ontogenetically and phylogenetically, under the stress of the elimination of the unfit, so that they are usually adaptive.

3. *Summation of Stimuli, Memory.*—Another general characteristic of protoplasm is the capacity of storing up or registering the effects of previous stimuli. A single stimulus may produce

changes in an organism which persist for a longer or shorter time, and if a second stimulus occurs while the effect of a previous stimulus still persists, the response to the second stimulus may be very different from that to the first. Macfarlane found that if the sensitive hairs on the leaf of *Dionaea,* the Venus fly-trap (Fig. 20, *SH*), be stroked once no visible response is called forth, but if they be stroked a second time within three minutes the leaf instantly closes. If a longer period than three minutes elapses after the first stimulus and before the second no visible response follows, *i.e.,* two successive stimuli are necessary to cause the leaves to close, and the two must not be more than three minutes apart; the effects of the first stimulus are in some way stored or registered in the leaf for this brief time. This kind of phenome-

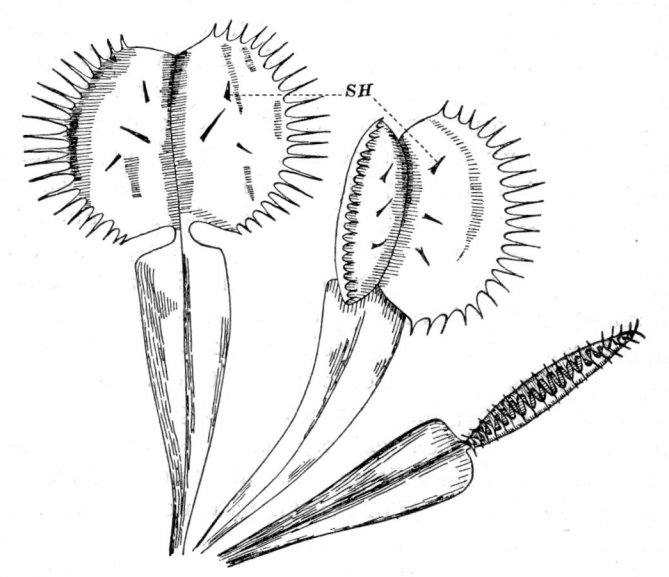

FIG. 20. *Dionaea muscipula* (VENUS' FLY-TRAP). Three leaves showing marginal teeth and sensitive hairs (*SH*). The leaf at the left is fully expanded, the one at the right is closed.

non is widespread among living things and is known as "summation of stimuli." In all such cases the effects of a former stimulus are in some way stored up for a longer or shorter time in the protoplasm. It is possible that this is the result of the formation of some chemical substance which remains in the protoplasm for a certain time, during which time the effects of the stimulus are said to persist, or it may be due to some physical change in the protoplasm analogous to the "set" in metals which have been subjected to mechanical strain.

Organic Memory.—Probably of a similar character is the persistence of the effects of repeated stimuli and responses on any organ of a higher animal. A muscle which has contracted many times in a definite way ultimately becomes "trained" so that it responds more rapidly and more accurately than an untrained muscle; and the nervous mechanism through which the stimulus is transmitted also becomes trained in the same way. Indeed such training is probably chiefly a training of the nervous mechanism. The skill of the pianist, of the tennis player, of the person who has learned the difficult art of standing and walking, or the still more difficult art of talking, is probably due to the persistence in muscles and nerves of the effects of many previous activities. All such phenomena were called by Hering "organic memory," to indicate that this persistence of the effects of previous activities in muscles and other organs is akin to that persistence of the effects of previous experiences in the nervous mechanism which we commonly call memory.

Associative Memory.—It seems probable that this ability of protoplasm in general to preserve for a time the effects of former stimuli is fundamentally of the same nature as the much greater power of nerve cells to preserve such effects for much longer periods and in complex associations, a faculty which is known as associative memory. The embryos, and indeed even the germ cells of higher animals, may safely be assumed to be endowed with protoplasmic and organic memory, out of which, in all probability, develops associative and conscious memory in the mature organism.

4. *Intellect, Reason.*—Even the intellect and reason which so strongly characterize man have had a development from relatively simple beginnings. All children come gradually to an age of intelligence and reason. In its simpler forms at least intelligence is the capacity of consciously profiting by experience, while reasoning consists in the comparison of past experiences with new and more or less different phenomena. In the absence of individual experience young children have none of this power, but it comes gradually as a result of remembering past experiences and of fitting such experiences into new conditions.

Useful Responses.—Young infants and many lower animals lack intelligence or reason, though their behavior is frequently of such a sort as to suggest that they are reasoning. Even the lowest animals avoid injurious substances and conditions and find beneficial ones; more complex animals learn to move objects, solve problems, and find their way through labyrinths in the shortest and most economical way; but this apparently intelligent and purposive behavior has been shown to be due to the gradual elimination of all sorts of useless activities, and to the persistence of the useful ones.

The ciliated infusorian, *Paramecium,* moves by the beating of cilia, which are arranged in such a way that they drive the animal forward in a spiral course. However, when it is strongly irritated, the normal forward movement is reversed; the cilia beat forward instead of backward and the animal is driven backward for some distance (Fig. 21, 1, 2, 3); it then stands nearly still, merely rolling over and swerving toward the aboral side, and finally it goes ahead again, usually on a new course (Fig. 21, 3, 4, 5, 6). These movements seem to be conditioned rather rigidly by the organization of the animal: they are more or less fixed and mechanical in character, though to a certain extent they may be modified by experience or physiological states. *Para-*

mecium behaves as it does by virtue of its constitution, just as an egg develops in a particular way because of its particular organization.

"Trial and Error."—But although limited in its behavior to these relatively simple motor reactions, *Paramecium* does many things which seem to show intelligence and purpose. It avoids many injurious substances, such as strong salts or acids, and it collects in non-injurious or beneficial substances, such as weak acids, masses of bacteria upon which it feeds, etc. It avoids extremes of heat and cold and if one end of a dish containing *Paramecia* is heated and the other end is cooled by ice, the *Paramecia* collect in the region somewhere between these two extremes (Fig. 17). Jennings, by studying carefully the behavior of single individuals, established the fact that this apparently intelligent action is due to differential sensitivity and to the single motor reaction of the animal. If in the course of its swimming a *Paramecium* comes into contact with an irritating substance or condition, it backs a short distance, swerves toward its aboral side, and goes ahead in a new path; if it again comes in contact with the irritating conditions this reaction is repeated, and so on indefinitely until finally a path is found in which the cause of irritation is avoided altogether. In short, *Paramecium* continually tries its environment, and backs away from irritating substances or conditions. Its apparently intelligent reactions are thus explained as due to a process of "trial and error."[1]

[1] In *Paramecium,* there is certainly no consciousness of trial and error, and probably no unconscious attempt on the part of the animal to attain certain ends. Its responses are reflexes or tropisms, which are determined by the nature of the animal and the character of the stimulus. The fact that these responses are in the main self-preservative is due to the teleological organization of *Paramecium* which has been evolved, according to current opinion, as the result of long ages of the elimination of the unfit. If, in the opinion of any one, the expression "trial and error" necessarily involves a striving after ends, it would be advisable to replace it in this case by some such term as "useful or adaptive reactions."

The behavior of worms, star-fishes, crustaceans, mollusks, as well as of fishes, frogs, reptiles, birds and mammals, has been studied and in all cases it is found that their method of responding to stimuli is not at first really purposive and intelligent but by the gradual elimination of useless responses and the preservation (or remembering) of useful ones the behavior may come to be purposive and intelligent.

Intelligence Develops from Trial and Error.—Thorndike found that when dogs, cats or monkeys were confined in cages which could be opened from the inside by turning a button, or pressing upon a lever, or pulling a cord, they at first clawed around all sides of the cage until by chance they happened to operate the mechanism which opened the door. Thereafter they gradually learned by experience, that is, by trial and error, and finally by trial and success, just where and how to claw in order to get out at once. When a dog has learned to turn a button at once and open a door we say he is intelligent, and if he can learn to apply his knowledge of any particular cage to other and different cages, a thing which Thorndike denies, we should be justified in saying

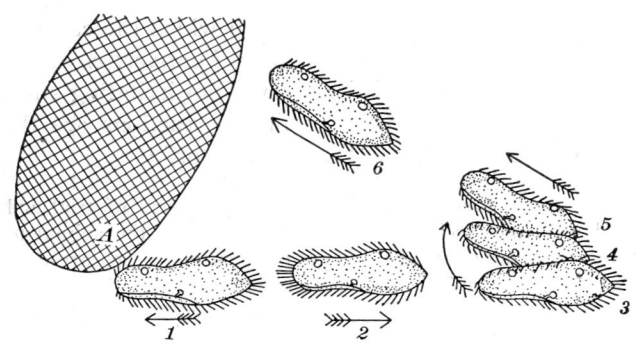

FIG. 21. DIAGRAM OF THE AVOIDING REACTION OF *Paramecium*. *A* is a solid object or other source of stimulation. *1-6*, successive positions occupied by the animal. The rotation on the long axis is not shown. (After Jennings.)

that he reasons, though in this case intelligence and reason are founded upon memory of many past experiences, of many trials and errors and of a few trials and successes.

There is every evidence that human beings arrive at intelligence and reason by the same process, a process of many trials and errors and a few trials and successes, a remembering of these past experiences and an application of them to new conditions. A baby grasps for things which are out of its reach, until it has learned by experience to appreciate distances; it tests all sorts of pleasant and unpleasant things until it has learned to avoid the latter and seek the former; it experiments with its own body until it has learned what it can do and what it can not do. Is not this learning by experience akin to the same process in the dog and more remotely to the trial and error of the earthworm or the adaptive reflexes of *Paramecium*? Is not intelligence and reason in all of us, and upon all subjects, based upon the same processes of trial and error, memory of past experiences and application of these to new conditions? Surely this is true in all experimental and scientific work. Indeed the scientific method is the method of trial and error, and finally trial and success—the method recommended by St. Paul to 'prove all things and hold fast that which is good.'

Learning by Experience.—In *Paramecium* the reflex type of behavior is relatively complete; there is no associative memory and no ability to learn by experience. In the earthworm associative memory is but slightly developed and the animal learns but little by experience and can make no application of past experiences to new conditions. In the dog associative memory is well developed; the animal learns by experience and can, to a limited extent, apply such memory of past experiences to new conditions. In adult man all of these processes are fully developed and particularly the last, viz., the ability to reason. But in his development the human individual passes through the more primitive stages of intelligence, represented by the lower animals named; the germ cells and embryo represent only the stages of reflex behavior, to these trial and error and associative memory are added in the infant and young child, and to these the application of past experience to new conditions, or reason, is added in later years.

5. *Will.*—Another characteristic, which many persons regard as the supreme psychical faculty, is the will. This faculty also undergoes development and from relatively simple beginnings. The will of the child has developed out of something which is far less perfect in the infant and embryo than in the child. Observations and experiments on lower animals and on human beings as well as introspective study of our own activities, appear to justify the following conclusions:

(1.) *No Activity Without Stimuli.*—Every activity of an organism is a response to one or more stimuli, external or internal in origin. These stimuli are in the main, if not entirely, energy changes outside or inside the organism. In lower organisms as well as in the germ cells and embryos of higher animals the possible number of responses are few and prescribed owing to the relative simplicity of these organisms, and the response follows the stimulus directly. In more complex organisms the number of possible responses to a stimulus is greatly increased, and the visible response may be the end of a long series of internal changes which are started by the original stimulus.

(2.) *Inhibitions.*—The response to a stimulus may be modified or inhibited in the following ways:

(a) *Conflicting Stimuli.*—Through conflicting stimuli and changed physiological states, due to fatigue, hunger, etc. Many stimuli may reach the organism at the same time and if they conflict they may nullify one another or the organism may respond to the strongest stimulus and disregard the weaker ones. When an organism has begun to respond to one stimulus it is not easily diverted by another. Jennings found that the attached infusorian, *Stentor,* which usually responds to strong stimuli by closing

FACTS AND FACTORS OF DEVELOPMENT

up, may, when repeatedly stimulated, loosen its attachment and swim away, thus responding in a wholly new manner when its physiological state has been changed by repeated stimuli and responses. Whitman found that leeches of the genus *Clepsine* prefer shade to bright light, and other things being equal they always seek the under sides of stones and shaded places; but if a turtle from which they normally suck blood is put into an aquarium with leeches, they at once leave the shade and attach themselves to the turtle. They prefer shade to bright light but they prefer their food to the shade. The tendency to remain concealed is inhibited by the stronger stimulus of hunger. On the other hand he found that the salamander, *Necturus,* is so timid that it will not take food, even though starving, until by gradual stages and gentle treatment its timidity can be overcome to a certain extent. Here fear is at first a stronger stimulus than hunger and unless the stimulus of fear can be reduced the animal will starve to death in the presence of the most tempting food.

(b) *Compulsory Limitations.*—Responses may also be modified through compulsory limitation of many possible responses to a particular one, and the consequent formation of a habit. This is the method of education employed in training all sorts of animals. Thus Jennings found that a star-fish could be trained to turn itself over, when placed on its back, by means of one pair of arms simply by persistently preventing the use of the other arms. Many responses of organisms are modified in a similar way, not only by artificial limitations but also by natural ones.

(3) *Fixed and Plastic Behavior.*—Responses which have become fixed and constant through natural selection, or other means of limitation, may become more varied and general when the compulsory limitation is relaxed. Behavior in the former case is fixed and instinctive, in the latter more varied and plastic. Thus Whitman found that the behavior of domesticated pigeons is more variable and their instincts are less rigidly fixed than in wild species. If the eggs of the wild passenger pigeon are removed to a little

distance from the nest the pigeon returns to the nest and sits down as if nothing had happened. She soon finds out, not by sight but by feeling, that something is missing, and she leaves the nest after a few minutes without heeding the eggs. The ring-neck pigeon also misses the eggs and sometimes rolls one of them back into the nest, but never attempts to recover more than one. The dovecote pigeon generally tries to recover both eggs. According to Whitman:

In these three grades the advance is from extreme blind uniformity of action, with little or no choice, to a stage of less rigid uniformity . . . . Under conditions of domestication the action of natural selection has been relaxed, with the result that the rigor of instinctive co-ordination, which bars alternative action, is more or less reduced. Not only is the door to choice thus unlocked, but more varied opportunities and provocations arise, and thus the internal mechanism and the external conditions and stimuli work both in the same direction to favor greater freedom of action. When choice thus enters no new factor is introduced. There is greater plasticity within and more provocation without, and hence the same bird, without the addition or loss of a single nerve cell, becomes capable of higher action and is encouraged and even constrained by circumstances to learn to use its privileges of choice. Choice, as I conceive it, is not introduced as a little deity encapsuled in the brain. . . . But increased plasticity invites greater interaction of stimuli and gives more even chances for conflicting impulses.

(4) *Conscious Choice or Will.*—Finally in all animals behavior is modified through previous experience, just as structure is also. Where several responses to a stimulus are possible and where experience has taught that one response is more satisfactory than another, action may be limited to this particular response, not by external compulsion but by the internal impulse of experience and intelligence. This is what we know as conscious choice or will. Whitman says:

Choice runs on blindly at first and ceases to be blind only in proportion as the animal learns through nature's system of compulsory education. The teleological alternatives are organically provided; one is taken and fails to give satisfaction, another is tried and gives contentment. This little freedom is the dawning grace of a new dispensation, in which education by experience comes in as an amelioration of the law of elimination. . . . Intelligence implies varying degrees of freedom of choice, but never complete emancipation from automatism.

Freedom of action does not mean action without stimuli, but rather the introduction of the results of experience and intelligence as additional stimuli. The activities which in lower animals are "cabined, cribbed, confined," reach in man their fullest and freest expression; but the enormous difference between the relatively fixed behavior of a protozoan or a germ cell and the relatively free activity of a mature man is bridged not only in the process of evolution, but also in the course of individual development.

6. *Consciousness.*—The most complex of all psychic phenomena, indeed the one which includes many if not all of the others, is consciousness. Like every other psychic process this has undergone development in each of us; we not only came out of a state of unconsciousness, but through several years we were gradually acquiring consciousness by a process of development. Whether consciousness is the sum of all the psychic faculties, or is a new product dependent upon the interaction of the other faculties, it must pass through many stages in the course of its development, stages which would commonly be counted as unconscious or subconscious states, and complete consciousness must depend upon the complete development and activity of the other faculties, particularly associative memory and intelligence.

Germ Cells Not Conscious.—The question is sometimes asked whether germ cells, and indeed all living things, may not be conscious in some vague manner. One might as well ask whether water is present in hydrogen and oxygen. Doubtless the elements out of which consciousness develops are present in the germ cells, in the same sense that the elements of the other psychic processes or of the organs of the body are there present; not as a miniature of the adult condition, but rather in the form of elements or factors, which by a long series of combinations and transformations, due to interactions with one another and with the environment, give rise to the fully developed condition.

Continuity of Consciousness.—Finally there seems good reason for believing that the continuity of consciousness, the continuing sense of identity, is associated with the continuity of organization, for in spite of frequent changes of the materials of which we are composed our sense of identity remains undisturbed. However, the continuity of protoplasmic and cellular organization generally remains undisturbed throughout life, and the continuity of consciousness is associated with this continuity of organization, especially in certain parts of the brain. It is an interesting fact that in man, and in several other animals which may be assumed to have a sense of identity, the nerve cells, especially those of the brain, cease dividing at an early age, and these identical cells persist throughout the remainder of life. If nerve cells continued to divide throughout life, as epithelial cells do, there would be no such persistence of identical cells, and one is free to speculate that in such cases there would be no persistence of the sense of identity.

Organization includes both structure and function, and continuity of organization implies not only persistence of protoplasmic and cellular structures but also persistence of the functions of sensitivity, reflexes, memory, instincts, intelligence, and will; the continuity of consciousness is associated with the continuity of these activities as well as with the structures of the body in general and of the brain in particular. It is well known that things which interrupt or destroy these functions or structures interrupt or destroy consciousness. Lack of oxygen, anesthetics, normal sleep cause in some way a temporary interruption of these functions and consequently temporary loss of consciousness; while

FACTS AND FACTORS OF DEVELOPMENT

certain injuries or diseases of the brain which bring about the destruction of certain centers or association tracts may cause permanent loss of consciousness.

7. *Parallel Development of Body and Mind.*—The development of all of these psychical faculties runs parallel with the development of bodily structures and apparently the method of development in the two cases is similar, viz., progressive differentiation of complex and specialized structures and functions from relatively simple and generalized beginnings. Indeed the entire organism, structure and function, body and mind, is a unity and the only justification for dealing with these constituents of the organism as if they were separate entities, whether they be regarded in their adult condition or in the course of their development, is to be found in the increased convenience and effectiveness of such separate treatment.

Development, like many other vital phenomena, may be considered from several different points of view, such as (1) physico-chemical events involved, (2) physiological processes, (3) morphological features, (4) ecological correlations and adaptations, (5) psychological phenomena, (6) social and moral characteristics. All of these phases of development are correlated; indeed they are parts of one general process, and a complete account of this process must include them all. General considerations may lead us to the belief that each of the succeeding aspects of development named above may be causally explained in terms of the preceding ones, and hence all be reducible to physics and chemistry. But this is not now demonstrable and may not be true. Function and structure may be related causally, or they may be two aspects of one substance. The same is true of body and mind or of matter and energy. But even if each of these different phases in the development of personality may not be causally explained by the preceding ones, at least the principle of explanation employed for any aspect of development ought to be consistent and harmonious with that employed for any other aspect.

The phenomena of mental development in man and other animals may be summarized as follows:

DEVELOPMENT OF PSYCHICAL PROCESSES IN ONTOGENY AND PHYLOGENY OUT OF GENERAL IRRITABILITY OR SENSITIVITY, THAT IS THE CAPACITY OF RESPONDING TO STIMULI

ALL LIVING THINGS, INCLUDING GERM CELLS AND EMBRYOS, SHOW:

1. *Differential Sensitivity* =
   Different Responses to Stimuli differing in Kind or Quantity.
2. *Reflexes, Tropisms* =
   Relatively Simple, Mechanical Responses.
3. *Organic Memory* =
   Results of Previous Experience registered in General Protoplasm.
4. *Adaptive Responses* =
   Results of Elimination of Useless Responses through Trial and Error.
5. *Varied Responses* =
   Dependent upon Conflicting Stimuli and Physiological States.
6. *Identity* =
   Continuity of Individual Organization.
7. *Subjective Phenomena*, if any, Accompanying preceding processes.

MATURE FORMS OF HIGHER ANIMALS SHOW:

1. *Special Senses and Sensations* =
   Differentiated out of General Senses and Sensations.
2. *Instincts (Inherited), Habits (Acquired)* =
   Complex Reflexes, involving Nerve Centers.
3. *Associative Memory* =
   Results of Experience registered in Nerve Centers and Association Tracts.
4. *Intelligence, Reason* =
   Results of Trial and Error plus Associative Memory, *i.e.*, Learning by Experience.
5. *Inhibition, Choice, Will* =
   Dependent Upon Associative Memory, Intelligence, Reason.
6. *Consciousness* =
   Continuity of Memory, Intelligence, Reason, Will.
7. *Feelings, Emotions* =
   Accompanying one or more of preceding processes.

B. FACTORS OF DEVELOPMENT

THESE are some of the facts of development,—a very incomplete résumé of some of the stages through which a human being passes in the course of his development from the germ. What are the factors of development? By what processes is it possible to derive from a relatively simple germ cell the complexities of an adult animal? How can mind and consciousness develop out of the relatively simple psychical elements of the germ? These are some of the great problems of development—one of the greatest and most far-reaching themes which has ever occupied the minds of men.

1. *Preformation.*—When the mind is once lost in the mystery of this ever-recurring miracle it is not surprising to find that there have been those who have refused to believe it possible and who have practically denied development altogether. The old doctrine of "evolution," as it was called by the scientists of the eighteenth century, or of preformation as we know it to-day, held that all the organs or parts of the adult were present in the germ in a minute and transparent condition as the leaves and stem are present in a bud, or as the shoot and root of the little plant are present in the seed.[1] In the case of animals it was generally impossible to see the parts of the future animal in the germ, but this was supposed to be due to the smaller size of the parts and to their greater transparency, and with poor microscopes and good imaginations some observers thought they could see the little animal in the egg or sperm, and even the little man, or "homunculus," was described and figured as folded up in one or the other of the sex cells.

This doctrine of preformation was not only an attempt to solve the mystery of development, but it was also an attempt to avoid the theological difficulties supposed to be involved in the view that individuals are produced by a process of natural development rather than by supernatural creation. If every individual of the race existed within the germ cells of the first parents, then in the creation of the first parents the entire race with its millions of individuals was created at once. Thus arose the theory of "emboîtement," or infinite encasement, the absurdities of which

[1] The little plant in the seed is itself the product of the development of a single cell, the ovum, in which no trace of a plant is present, but of course this fact was not known until after careful microscopical studies had been made of the earliest stages of development.

contributed to the downfall of the entire doctrine of preformation, which, in the form given it by many naturalists of the eighteenth century, is now only a curiosity of biological literature.

2. *Epigenesis.*—As opposed to this doctrine of preformation, which was founded largely on speculation, arose the theory of epigenesis, which was in its main features founded upon the direct observation of development, and which maintained that the germ contains none of the adult parts, but that it is absolutely simple and undifferentiated, and that from these simple beginnings the individual gradually becomes complex by a process of differentiation. We owe the theory of epigenesis at least so far as its main features are concerned, to William Harvey, the discoverer of the circulation of the blood, and to Caspar Friedrich Wolff, whose doctoral thesis, published in 1759 and entitled *"Theoria Generationis,"* marked the beginning of a great epoch in the study of development. Wolff demonstrated that adult parts are not present in the germ, either in animals or in plants, but that these parts gradually appear in the process of development. He held, erroneously, that the germ is absolutely simple, homogeneous and undifferentiated, and that differentiation and organization gradually appear in this undifferentiated substance. How to get differentiations out of non-differentiated material, heterogeneity out of homogeneity, was the great problem which confronted Wolff and his followers, and they were compelled to assume some extrinsic or environmental force, some *vis formativa* or *spiritus rector,* which could set in motion and direct the process of development.

The doctrine of preformation, by locating in the germ all the parts which would ever arise from it, practically denied development altogether; epigenesis recognized the fact of development, but attributed it to mysterious and purely hypothetical external forces; the one placed all emphasis upon the germ and its structures, the other upon outside forces and conditions.

3. *Endogenesis and Epigenesis.*—Modern students of develop-

ment recognize that neither of these extreme views is true—adult parts are not present in the germs, nor in the latter homogeneous—but there are in germ cells many different structures and functions which are, however, very unlike those of the adult, and by the transformation and differentiation of this germinal organization the complicated organization of the adult arises. Development is not the unfolding of an infolded organism, nor the mere sorting of materials already present in the germ cells, though this does take place, but rather it consists in the formation of new materials and qualities, of new structures and functions, by the combination and interaction of the germinal elements present in the oosperm. In similar manner the combination and interaction of chemical elements yield new substances and qualities which are not to be observed in the elements themselves. Such new substances and qualities, whether in the organic or in the inorganic world, do not arise by the gradual unfolding of what was present from the beginning, but they are produced by a process of "creative synthesis."

Modern studies of germ cells have shown that they are much more complex than was formerly believed to be the case; they may even contain different "organ-forming substances" which in the course of development give rise to particular organs; these substances may be so placed in the egg as to foreshadow the polarity, symmetry and pattern of the embryo, but even the most highly organized egg is relatively simple as compared with the animal into which it ultimately develops. Increasing complexity, which is the essence of development, is caused by the combination and interaction of germinal substances under the influence of the environment. The organization of the oosperm may be compared to the arrangement of tubes and flasks in a complicated chemical operation; they stand in a definite relation to one another and each contains specific substances. The final result of the operation depends not merely upon the substances used, nor merely upon the way in which the apparatus is set up, but upon both of these things as well as upon the environmental conditions represented by temperature, pressure, moisture or other extrinsic factors.

4. *Heredity and Environment.*—Unquestionably the factors or causes of development are to be found not merely in the germ but also in the environment, not only in intrinsic but also in extrinsic forces; but it is equally certain that the directing and guiding factors of development are in the main intrinsic, and are present in the organization of the germ cells, while the environmental factors exercise chiefly a stimulating, inhibiting or modifying influence on development. In the same dish and under similar environmental conditions, one egg will develop into a worm, another into a sea urchin, another into a fish, and it is certain that the different fate of each egg is determined by conditions intrinsic in the egg itself, rather than by environmental conditions. We should look upon the germ as a living thing, and upon development as one of its functions. Just as the character of any function is determined by the organism, though it may be modified by environment, so the character of development is determined by heredity, *i.e.*, by the organization of the germ cells, though the course and results of development may be modified by environmental conditions.

Summary

In conclusion, we have briefly reviewed in this chapter the well known fact that every living thing in the world has come into existence by a process of development; that the entire human personality, mind as well as body, has thus arisen; and that the factors of development may be classified as intrinsic in the organization of the germ cells, and extrinsic as represented in environmental forces and conditions. The intrinsic factors are those which are commonly called heredity, and they direct and guide development in the main; the extrinsic or environmental factors furnish the conditions in which development takes place and they modify, more or less, its course.

CHAPTER II

PHENOMENA OF INHERITANCE

CHAPTER II

PHENOMENA OF INHERITANCE

A. OBSERVATIONS ON INHERITANCE

THE observations of men for ages past have established the fact that in general "like produces like," and that, in spite of many exceptions, children are in their main characteristics like their parents. And yet offspring are never exactly like their parents, and this has led to the saying that "like does not produce like but only somewhat like." What is meant is that there are general resemblances but particular differences between parents and offspring.

INDIVIDUALS AND THEIR CHARACTERS

IN considering organic individuals one may think of them as wholes or as composed of parts, as indivisible units or as constituent characters; either aspect is a true one and yet neither is complete in itself. Formerly in discussions on heredity the individual was regarded in its entirety and when all hereditary resemblances and differences were averaged it was said that one child resembled the father, another child the mother. This method of lumping together and averaging resemblances and differences led to endless confusion. In heredity no less than in anatomy it is necessary to deal with the constituents of organisms; in short, the organism must be analyzed and each part studied by itself.

Method of Galton and Mendel.—Francis Galton was one of the first to bring order out of chaos by dealing with traits or characters singly instead of treating all together. He made careful studies on the inheritance of weight and size in the seeds of sweet peas, and on the inheritance of stature, eye-color, intellectual capacity, artistic ability and certain diseases in man. At the same time that Galton was thus laying the foundations for a scientific study of heredity by dealing with characters separately, another and an even greater student of heredity, Gregor Mendel, was doing the same thing in his experiments with garden peas, but inasmuch as Mendel's work remained practically unknown for many years, Galton has been rightly recognized as the founder of the scientific study of heredity.

Of course, neither Galton nor anyone else who has followed his method of dealing with the characters of organisms singly, ever supposed that such characters could exist independently of other characters and apart from the entire organism. This is such a self-evident fact that it may seem needless to mention it, and yet there have been critics who have believed, or have assumed to believe, that modern students of heredity attempt to analyze organisms into independently existing characters, whereas in most cases they have done only what the anatomist does in treating separately the various organs of the body.

Hereditary Resemblances and Differences

The various characters into which an organism may be analyzed show a greater or smaller degree of resemblance to the corresponding characters of its parents. Whenever the differential cause of a character is a germinal one the character is, by definition, inherited; on the other hand, whenever this differential cause is environmental the character is not inherited. While it is true that inheritance is most clearly recognized in those characters in which offspring resemble their parents, even characters in which they differ from their parents may be inherited, as is plainly seen when, in any character, a child resembles a grandparent or a more distant ancestor more than either parent. Sometimes actually new characters arise in descendants which were not present in ascendants, but which are thereafter inherited. Ac-

PHENOMENA OF INHERITANCE

cordingly inherited characters may be classified as resemblances and differences, though both are determined by germinal organization, or heredity. There is therefore no fundamental difference between inherited similarities and dissimilarities. Heredity and variation are not opposing nor contrasting tendencies which make offspring like their parents in one case and unlike them in another; really inherited characters may be like or unlike those of the parents.

On the other hand many resemblances and differences between parents and offspring are due not to heredity at all, but to environmental conditions. By means of experiment it is possible to distinguish between hereditary and environmental resemblances and differences, but among men where experiments are generally out of the question it is often difficult or impossible to make this distinction.

I. HEREDITARY RESEMBLANCES

1. *Racial Characters.*—All peculiarities which are characteristic of a race, species, genus, order, class and phylum are of course inherited, otherwise there would be no constant characteristics of these groups and no possibility of classifying organisms. The chief characters of every living thing are unalterably fixed by heredity. Men do not gather grapes of thorns nor figs of thistles. Every living thing produces offspring after its own kind. Men, horses, cattle; birds, reptiles, fishes; insects, mollusks, worms; polyps, sponges, micro-organisms,—all of the million known species of animals and plants differ from one another because of inherited peculiarities, because they have come from different kinds of germ cells or protoplasm.

2. *Individual Characters.*—Many characters which are peculiar to certain individuals are known to be inherited, and in general use the word "inheritance" refers to the repetition in successive generations of such individual peculiarities. Among such individual characters are the following:

(a) *Morphological Features.*—Hereditary resemblances are especially recognizable in the gross and minute anatomy of every organism, in the form, structure, location, size, color, etc., of each and every part. The number of such individual peculiarities which are inherited is innumerable and only a few of the more striking of these can be mentioned.

It is a matter of common knowledge that unusually great or small stature runs in certain families, and Galton developed a formula for determining the approximate stature of children from the known stature of the parents and from the mean stature of the race (Fig. 25). However, his statistical and mathematical formulæ give only general or average results, from which there are many individual departures and exceptions.

In the same way the color of the skin, the color and form of hair and the color of eyes are in general like those of one or more of the parents or grandparents. We all know that certain facial features such as the shape and size of eyes, nose, mouth and chin are generally characteristic of certain families.

But the inheritance of anatomical features extends to much more minute characters than those just mentioned. In certain families a few hairs in the eyebrows are longer than the others, or there may be patches of parti-colored hair over the scalp, or dimples in the cheek, chin, or other parts of the skin may occur, and these trifling peculiarities are inherited with all the tenacity shown in the transmission of more important characters. Johannsen has found races of beans in which the average weight of individual seeds differed only by .02 to .03 gram, and yet these minute differences in weight were characteristic of each race and were of course inherited. Jennings has found races of *Paramecium* which show hereditary differences of .005 mm. in average length (Fig. 22). Nettleship says that the lens of the human eye weighs only 175 milligrams, or about one three-millionth part of the body weight, and in hereditary cataract only about one twentieth part of the lens becomes opaque, and yet this minute fraction of the body weight shows the influence of heredity. Even the size, shape and number of the cells in certain organs, and in given embryonic stages, may be repeated generation after generation; and if our analysis were sufficiently complete we should doubtless find that even the minute parts of cells, such as nuclei, chromosomes and centrosomes, show individual peculiarities which are inherited.

(b) *Physiological peculiarities* are inherited as well as morphological ones; indeed function and structure are only two aspects

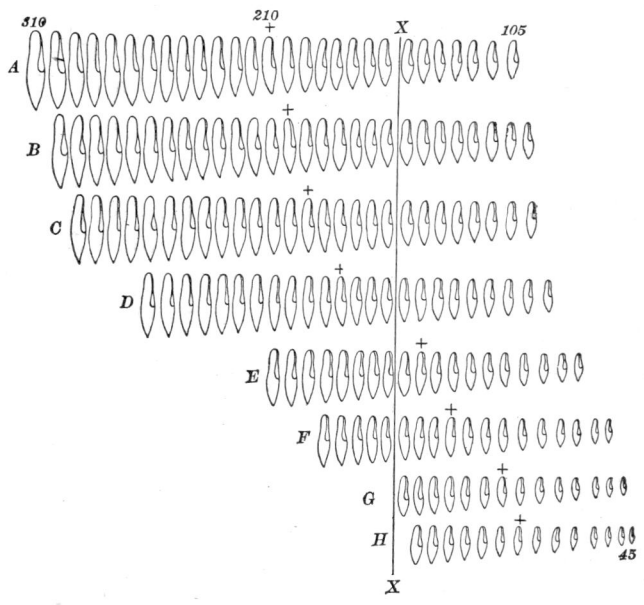

Fig. 22. Diagram of Eight Different Races of *Paramecium*, each horizontal row (A-H) representing a single race. The individual showing the mean size in each race is indicated by $+$; the mean of all the races is shown by the line X-X. The numbers are the lengths in micra (thousandths of a millimeter), $\times$ 43. (After Jennings.)

of one and the same thing, namely organization. For all morphological characters there are functional correlatives, for functional characters morphological expressions, and if the one is inherited so is the other. But there are certain characters in which the physiological aspect is more striking than the morphological one.

Longevity.—For example, longevity is a physiological character which is undoubtedly dependent upon many causes, but in the case of species which differ greatly in length of life there can be little doubt that we are dealing with an inherited character. The great differences in the length of life of an elephant and a mouse, of a parrot and a pigeon, of a cicada and a squash bug, are as surely the result of inherited causes as are the structural differences between these animals. Within the same species different races or lines show characteristic differences in length of life; in the case of man the average length of life is much greater in some families than in others, and life-insurance companies take account of this fact. Even within the same organism certain organs or cells are short-lived, whereas others are long-lived; some cells and organs live only through the early embryonic period, while others live as long as the general organism.

Other Functional Characters.—Obesity is another physiological characteristic which may be inherited; the members of certain families grow fat in spite of themselves, while members of other families remain thin however well fed they may be. Here also many factors enter into the result, but it seems probable that the differentiating factor is an hereditary one. Baldness affects the male members of certain families when they have reached a given age, while in others neither care, dissipation nor age can rob a man of his bushy top. Hæmophilia, or excessive bleeding after an injury, which is due to a deficiency in the clotting power of the blood, is strongly inherited in the male line in certain families. Fecundity and a tendency to bear twins or triplets, left-handedness, a peculiar lack of resistance to certain diseases, and many other physiological peculiarities are probably inherited.

PHENOMENA OF INHERITANCE

(c) *Pathological Peculiarities* are really only unusual or abnormal anatomical or physiological characters, but they are of such interest and importance as to deserve special mention. Many such abnormalities are undoubtedly inherited, among which are the following: polydactylism, in which more than the normal number of digits are present (Fig. 37); syndactylism, or a condition of webbed fingers and toes; brachydactylism, in which fingers are short and stumpy and usually contain less than the normal number of joints (Fig. 38); achondroplasy, or short and crooked limbs, such as occur in certain breeds of dogs and sheep and in certain human dwarfs; myopia, in which the eyeball is elongated; glaucoma, or pressure within the eyeball; coloboma, or open suture of the iris; otosclerosis, or rigidity of tympanum and ossicles, causing "hardness of hearing"; some forms of deaf-mutism, due to certain defects in the inner ear; and many other characters too numerous to mention here. On the other hand many abnormal or monstrous conditions are due to abnormal environment and are not inherited.

Are Diseases Inherited.—The question of the inheritance of diseases may be briefly considered here. If a disease is due to some defect in the hereditary constitution, it is inherited; otherwise, according to our definition of heredity, it is not. Of course no disease develops without extrinsic cause but when one individual takes a disease while another under the same conditions does not, the differential cause may be an inherited one, or it may be due to differences in the previous conditions of life. There is no doubt that certain diseases run in families and have the appearance of being inherited, but in this case as in many others it is extremely difficult in the absence of experiments to distinguish between effects due to intrinsic causes and those due to extrinsic ones. Where the specific cause of disease is some micro-organism the individual must have been infected at some time or other, almost invariably after birth. In few instances is the oosperm itself infected, and even when it is, this is not, strictly speaking, a case of inheritance, but rather one of early infection. Leo Loeb has shown that cancer is inherited in mice and Little finds that there is inheritance of a predisposition to cancer in man. Pearson has found that there is a marked correlation (represented by the number .55 when complete correlation is 1.) between tuberculous parents and tuberculous children, but there is very little evidence that the child is ever infected before birth. What is inherited in this case is probably slight resistance to the tubercle bacillus. There is evidence that almost all adult persons have been infected at one time or another by this bacillus, but it has not developed far in all of them because some have superior powers of resistance. Such greater or smaller resistance, stronger or weaker build, is inherited, and while diminished resistance is not the direct cause of tuberculosis it is a predisposing cause. The same is probably true of many other diseases, the immediate causes of which are extrinsic, while only the more remote, or predisposing causes, are hereditary.

(d) *Psychological Characters* appear to be inherited in the same way that anatomical and physiological traits are; indeed all that has been said regarding the correlation of morphological and physiological characters applies also to psychological ones. No one doubts that particular instincts, aptitudes and capacities are inherited among both animals and men, nor that different races and species differ hereditarily in psychological characteristics.

Certain breeds of dogs such as the mastiff, the bull dog, the terrier, the collie and many others are characterized by peculiarities of temperament, affection, intelligence and disposition. No one who has much studied the subject can doubt that different human races and families show characteristic differences in these same respects. It is quite futile to argue that exceptional individuals may be found in one race with the mental characteristics of another race; the same could be said of different breeds of dogs, or of the sizes of different races of beans or of *Paramecia* (Fig. 22). The fact is that racial characteristics are not determined by exceptional and extreme individuals but by the average or mean qualities of the race; and measured in this way there is no doubt that certain types of mind and disposition are characteristic of certain families.

There is no longer any question that some kinds of feeble-mindedness, epilepsy and insanity are inherited, and that there is often an hereditary basis for nervous and phlegmatic temperaments, for emotional, judicial and calculating dispositions. Nor can it be denied that strength or weakness of will, a tendency to moral obliquity or rectitude, capacity or incapacity for the highest intellectual pursuits, occur frequently in certain families and appear to be inherited. In spite of certain noteworthy exceptions, which may perhaps be due to remarkable variations, statistics collected by Galton show that genius runs in certain families; while the work of some recent investigators, particularly Goddard, Davenport and Weeks, proves that feeble-mindedness and epilepsy are also inherited; and the careful work of Mott and of Rosanoff indicates that certain types of insanity are hereditary. On the other hand, Cotton maintains that mental disorders are not directly inherited, but that "there is probably a constitutional lack of resistance to various toxins and poisons, and not an inherited mental instability, which causes the mind to break down under mental stress and strain." It frequently happens that families in which hereditary insanity occurs also have other members afflicted with epilepsy, hysteria, alcoholism, etc., which seem to indicate that the thing inherited is an unstable condition of the nervous system which may take various forms under slightly different conditions. Indeed there is a good deal of evidence that extraordinary ability, or genius is frequently associated with an unstable nervous organization which sometimes takes the form of insanity or epilepsy or alcoholism. There is perhaps more truth than poetry in Dryden's lines:

> "Great wits are sure to madness near allied,
> And thin partitions do their bounds divide."

Woods has collected data concerning "Heredity in Royalty" which seem to show that very high or low grades of intellect and morality may be traced through the royal families of Europe for several generations. Extensive study of certain families in which an extraordinary number of feeble-minded, degenerate, and criminal individuals have appeared, seems to demonstrate that moral and social qualities are also inherited. One recalls in this connection the famous, or rather infamous, "Jukes", "Kalikaks", "Nams", and "Ishmaels",—these names being pseudonyms for notoriously bad families whose traits have been followed through several generations.

The general tendency of recent work on heredity is unmistakable, whether it concerns man or lower animals. The entire organism, consisting of structures and functions, body and mind, develops out of the germ, and the organization of the germ determines all the *possibilities* of development of the mind no less than of the body, though the actual realization of any possibility is dependent also upon environmental stimuli.

II. HEREDITARY DIFFERENCES

THERE are many exceptions to the general rule that children resemble their parents; indeed no child is ever exactly like a parent and the points in which they differ are known generally as variations. These variations are of two kinds, those which are caused by a different germinal constitution and are therefore inherited and those due to environmental differences which are not inherited. Sometimes inherited variations are due to new combinations of ancestral characters, sometimes they are actually new characters not present so far as known in any of the ancestors, though even such new characters must arise from new combinations of the *elements* of old characters, as we shall see later.

1. *New Combinations of Characters.*—In all cases of sexually produced organisms new combinations of ancestral characters are evident. Usually a child inherits some traits from one parent

PHENOMENA OF INHERITANCE

and other traits from the other parent, so that it is a kind of mosaic of ancestral traits. Such inheritance, bit by bit, of this character from one progenitor and that from another was described by Galton as "particulate" (Fig. 23) and is known today as "Mendelian." As we shall see later (p. 108 *et seq.*) this is probably the only kind of inheritance. On the other hand Galton supposed that in some instances a child might inherit all or nearly all of his traits from one parent, a thing which is most improbable; such inheritance he called "alternative"[1] (Fig. 23).

In other cases the traits of the parents appear to blend in the offspring, as for example, in the skin color of mulattoes; such

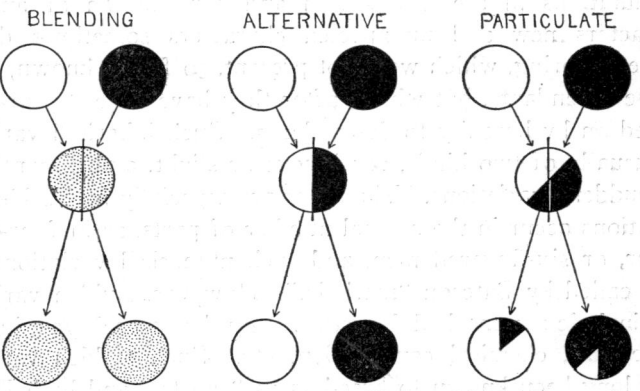

FIG. 23. DIAGRAM TO ILLUSTRATE THREE KINDS OF INHERITANCE described by Galton. Only the last of these (particulate) really occurs. (After Walter.)

cases were called by Galton "blending" inheritance (Fig. 23). Such cases of blending inheritance are now known to be the result of particulate inheritance of many factors (p. 109). Sometimes characters appear in offspring which were "latent" in the parents but were "patent" in one or more of the grandparents;

[1] It is necessary to distinguish between alternative inheritance of a single character (Mendel) and this supposed alternative inheritance of all characters (Galton).

such skipping of a generation, during which a character remains "latent," has long been known as "atavism." At other times characters which were present in distant ancestors, but which have since dropped out of sight or have remained "latent," reappear in descendants; such cases are known as "reversions."

In still other cases certain characters appear only in the male sex, others only in the female, this being called "sex-limited" inheritance; while in some instances characters are transmitted from fathers through daughters to grandsons or from mothers to sons, all such cases being known as "sex-linked" inheritance.[1]

2. *New Characters or Mutations.*—But in addition to these permutations in the distribution and combination of ancestral characters new and unexpected characters sometimes develop in the offspring, which were not present, so far as known, in any of the ascendants, but which, after they have once appeared, are passed on by heredity to descendants. Such inherited variations are usually of two kinds, continuous or slight, and discontinuous or "sudden" variations. The latter are especially noticeable when variations occur in the normal number of parts, as in four-leaved clover, or six-fingered men, and such numerical variations have been called by Bateson "meristic." However, sudden variations may include any marked departure from the normal type, in color, shape, size, chemical composition, etc. Such sudden variations have long been known to breeders as "sports," and both Darwin and Galton pointed out the fact that such sports have sometimes given rise to new races or breeds, though Darwin was not inclined to assign much importance to them in the general process of evolution. Galton, on the other hand, maintained that variations, or what would now be called "continuous variations," cannot be of much significance in the process of evolution, but that the case is quite different with "sports" ("Hereditary Genius," prefatory chapter).

More recently the entire biological world has been greatly influ-

[1] See p. 185.

PHENOMENA OF INHERITANCE

enced by the "Mutation Theory" of deVries, which has placed a new emphasis upon the importance of sudden variations in the process of evolution. At first deVries was inclined to emphasize the degree of difference, that is the discontinuity, in these variations, but in later works this distinction is given a minor place as compared with the question whether variations are inherited or not. Inherited variations, whether large or small, are called by deVries "mutations," whereas non-inherited variations are known as "fluctuations." The former are caused by changes in germinal constitution, the latter by alterations in environmental conditions; the former represent changes in heredity, the latter changes in development.

3. *Mutations and Fluctuations.*—This clear cut distinction between mutations and fluctuations marks one of the most important advances ever made in the study of development and evolution. Thousands of fluctuations occur which are purely somatic in character and which do not affect the germ cells, for every single mutation or change in the hereditary constitution; and yet only the latter are of significance in heredity and evolution. This distinction between variations due to environment (fluctuations) and those due to hereditary causes (mutations) was recognized by Weismann and many of his followers, but the actual demonstration on a large scale of the importance of this distinction is due mainly to deVries.

All hereditary variations, whether due to new combinations of old characters or to the appearance of actually new characters, whether small and continuous or large and discontinuous, have their causes in the organization of the germ cells, just as do inherited resemblances. Heredity is not to be contrasted with variation, nor are hereditary likeness and unlikeness due to conflicting principles; both are the results of germinal organization and both are phenomena of heredity.

4. *Every Individual Unique.*—As a result of the permutations of ancestral characters, the appearance of mutations, and the

fluctuations of organisms due to environmental changes, it happens that in all cases offspring differ more or less from their parents and from one another. No two children of the same family are ever exactly alike (except in the case of identical twins which have come from the same oosperm).[1] Every living being appears on careful examination to be the first and last of its identical kind. This is one of the most remarkable peculiarities of living things. The elements of chemistry are constant, and even the compounds fall into definite categories which have constant characteristics. But the individuals of biology are apparently never twice the same. This may be due to the immense complexity of living units as contrasted with chemical ones,—indeed lack of constancy is evidence in itself of lack of analysis into real elements or of lack of uniform conditions,—but whatever its cause the extraordinary fact remains that every living being appears to be unique. "Reproduction is the generation of unique beings that are, on the average, more like their kind than like anything else" (Brooks).

There seems to be no reason to doubt that all the extraordinary differences which organisms show, as well as all of their resemblances, are due to differences or resemblances in the hereditary and environmental factors which have been operative in their development. But in view of this universal variability of organisms it is not surprising that inheritance has seemed capricious and uncertain,—"a sort of maze in which science loses itself."

B. STATISTICAL STUDY OF INHERITANCE

Francis Galton was one of the first who attempted to reduce the mass of conflicting observations on heredity and variation to some system and to establish certain principles as a result of statistical study. He was the real founder of the scientific study of inheritance; he studied characters singly and he introduced quantitative measures. Galton's researches, which were published

[1] See p. 227.

in several volumes, consisted chiefly in a study of certain families with regard to several selected traits, viz., genius or marked intellectual capacity, artistic faculty, stature, eye color and disease. As a result of his very extensive studies two main principles appeared to be established:

1. *The Law of Ancestral Inheritance* which he stated as follows:

The two parents contribute between them on the average one-half of each inherited faculty, each of them contributing one-quarter of it. The four grandparents contribute between them one-quarter, or each of them one-sixteenth; and so on, the sum of the series $1/2 + 1/4 + 1/8 + 1/16 \ldots$ being equal to 1, as it should be. It is a property of this infinite series that each term is equal to the sum of all those that follow: thus $1/2 = 1/4 + 1/8 + 1/16 + \ldots$, $1/4 = 1/8 + 1/16 + \ldots$, and so on. The prepotencies of particular ancestors in any given pedigree are eliminated by a law which deals only with average contributions, and the various prepotencies of sex with respect to different qualities are also presumably eliminated.

The average contribution of each ancestor was thus stated definitely, the contribution diminishing with the remoteness of the ancestor. This Law of Ancestral Inheritance is represented graphically in the accompanying diagram (Fig. 24). Pearson has proposed a *Law of Reversion* according to which the average reversion of offspring to each ascending generation of ancestors is represented by the series .3, .15, .075, .0375, etc.

Ancestors and Contributors.—Theoretically the number of ancestors doubles in each ascending generation; there are two parents, four grandparents, eight great-grandparents, etc. If this continued to be true indefinitely the number of ancestors in any ascending generation would be $(2)^n$, in which n represents the number of generations. There have been about 57 generations since the beginning of the Christian Era, and if this rule held true indefinitely each of us would have had at the time of the birth of Christ a number of ancestors represented by $(2)^{57}$ or about

120 quadrillions,—a number far greater than the entire human population of the globe since that time. As a matter of fact owing to the intermarriage of cousins of various degrees the actual number of ancestors is much smaller than the theoretical number. For example, Plate says that the late Emperor of Germany, Wilhelm II, had only 162 ancestors in the 10th ascending generation, instead of 512, the theoretical number. Nevertheless this calculation will serve to show how widespread our ancestral lines are, and how nearly related are all people of the same race.

Davenport concludes that no people of English descent are more distantly related than 30th cousins, while most people are much more closely related than that. If we allow three generations to a century, and calculate that the degree of cousinship is determined by the number of generations less two, since first cousins appear only in the third generation, the first being that of the parents and the second that of the sons and daughters, we find that 30th cousins at the present time would have had a common ancestor about one thousand years ago or approximately at

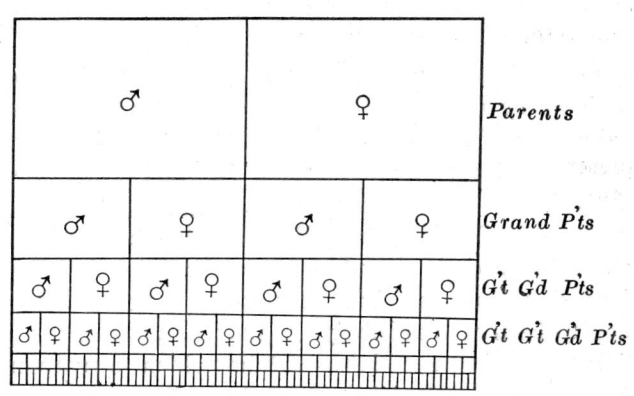

FIG. 24. DIAGRAM OF GALTON'S "LAW OF ANCESTRAL INHERITANCE." The whole heritage is represented by the entire rectangle; that derived from each progenitor by the smaller squares; the number of the latter doubles in each ascending generation while its area is halved. (After Thompson.)

the time of William the Conqueror. As a matter of fact most persons of the same race are much more closely related than this, and certainly we need not go back to Adam nor even to Shem, Ham, or Japheth to find our common ancestor.

On the other hand we now know that we do not inherit equally from all our ancestors; on the average we inherit about as many

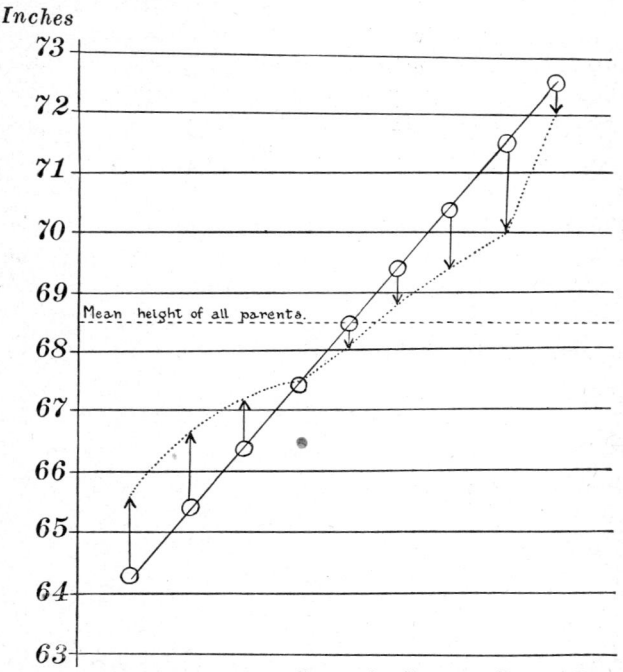

Fig. 25. Scheme to Illustrate Galton's "Law of Filial Regression" as shown in the stature of parents and children. The mean height of all parents is shown by the dotted line between 68 and 69 inches. The circles through which the diagonal line runs represent the heights of graded groups of parents and the arrow heads indicate the average heights of their children. The offspring of undersized parents are taller and of oversized parents are shorter than their respective parents. (After Walter.)

traits from our fathers as from our mothers, but inheritance from the four grand-parents is usually unequal and the farther back we go the more ancestors we find who have contributed nothing to our inheritance. Of all the thousands or even millions of ancestors that each of us has had, only a relatively small number have contributed anything to our inheritance; although we are descended from all the others we are not related to them biologically and have received none of their traits. Those who have contributed to our inheritance may be called "contributing ancestors" or merely "contributors"[1] to distinguish them from non-contributing ones, and the fact that ancestors do not contribute equally to heredity disproves Galton's "law of ancestral inheritance."

2. *The Law of Filial Regression* is the second principle which Galton deduced from his statistical studies, or it may be called the tendency to mediocrity. He found that, on the average, extreme peculiarities of parents were less extreme in children. Thus, "the stature of adult offspring must on the whole be more mediocre than the stature of their parents, that is to say more near to the mean or mid of the general population"; and again, "the more bountifully a parent is gifted by nature, the more rare will be his good fortune if he begets a son who is as richly endowed as himself." This so-called law of filial regression is represented graphically in Fig. 25 in which the actual stature of individual parents is shown by the oblique line, the stature of children by the dotted curve, and the mean stature of the race in the horizontal dotted line.

Statistical vs. Physiological Methods.—One of the chief aims and results of statistical studies is to eliminate individual peculiarities and to obtain general and average results. Such work may be of great importance in the study of heredity, especially where questions of the occurrence or distribution of particular phenomena

[1] I have adopted this term proposed by Dr. H. H. Laughlin in preference to "transmitters" which I had previously used.

are concerned; but the causes of heredity are individual and physiological, and averages are of less value in finding the causes of such phenomena than is the intensive study of individual families.

By observation alone it is usually impossible to distinguish between inherited and environmental resemblances and differences, and yet this distinction is essential to any study of inheritance. If all sorts of likenesses or unlikenesses are lumped together, whether inherited or not, our study of inheritance can only end in confusion. The value of statistics depends upon a proper classification of the things measured and enumerated, and if things which are not commensurable are grouped together the results may be quite misleading and worthless.

Statistical Studies Insufficient.—Unfortunately Galton and Pearson, as well as some of their followers, have not always carefully distinguished between hereditary and environmental characters. Furthermore much of their material was drawn from a general population in which were many different families and lines not closely related genetically. Consequently their statistical studies are of little value in discovering the physiological principles or laws of heredity. Jennings (1910) well says, "Galton's laws of regression and of ancestral inheritance are the product mainly of a lack of distinction between two absolutely diverse things, between non-inheritable fluctuations on the one hand, and permanent genotypic differentiations on the other." In the case of man we have few certain tests to determine whether the differential cause of any character is hereditary or environmental, but in the case of animals and plants, where experiments may be performed on a large scale, it is possible to make such tests by (1) experiments in which the environment is kept as uniform as possible while the hereditary factors differ, and (2) experiments in which, in a series of cases, the hereditary factors are fairly constant while the environment differs. In this way the differential cause or causes of any character may be located in heredity, in environment or in both.

The observational and statistical study of inheritance helped to outline the problem but did little to solve it. Certain phenomena of hereditary resemblances between ascendants and descendants were made intelligible, but there were many peculiar and apparently irregular or lawless phenomena which could not be predicted before they occurred nor explained afterward. For example when Darwin crossed different breeds of domestic pigeons, no one of which had a trace of blue in its plumage, he sometimes obtained offspring with more or less of the blue color and markings of the wild rock pigeon from which domestic pigeons are presumably descended. He described many cases of dogs, cattle and swine, as well as many cultivated plants, in which offspring resembled distant ancestors and differed from nearer ones; such cases had long been known and were spoken of as "reversions." He observed many cases in which certain characters of one parent prevailed over corresponding characters of the other parent in the offspring, this being known as "prepotency"; but there was no satisfactory explanation of these curious phenomena. They did not come under either of Galton's "laws," and their occurrence was apparently so irregular that every such case seemed to be a law unto itself.

C. EXPERIMENTAL STUDY OF INHERITANCE

1. MENDELISM

THE year 1900 marks the beginning of a new era in the study of inheritance. In the spring of that year three botanists, deVries, Correns, and Tschermak, discovered independently an important principle of heredity and at the same time brought to light a long neglected and forgotten work on "Experiments in Plant Hybridization" by Gregor Mendel, in which this same principle was set forth in detail. This principle is now generally known as "Mendel's Law." Mendel, who was a monk, and later abbot, of the *Königskloster,* an Augustinian monastery in Brünn, Moravia, published the results of his experiments on hybridization

PHENOMENA OF INHERITANCE

in the Proceedings of the Natural History Society of Brünn in 1866. The paper attracted but little attention at the time although it contained some of the most important discoveries regarding inheritance which had ever been made, and it remained buried and practically unknown for thirty-five years. Plant hybridization had been studied extensively before Mendel began his work, but he carried on his observations of the hybrids and of their progeny for a longer time and with greater analytical ability than any previous investigator had done. The methods and results of his work are so well known through the writings of Bateson, Punnett, and many others that it is unnecessary to dwell at length upon them here. In brief Mendel's method consisted in crossing two forms having distinct characters, and then in counting the number of offspring in successive generations showing one or the other of these characters.

Mendel's Experiments on Peas.—During the eight years preceding the publication of his paper in 1866 Mendel hybridized some twenty-two varieties of garden peas. This group of plants was chosen because the different varieties could be cross-fertilized or self-fertilized and were easily protected from the influence of foreign pollen; because the hybrids and their offspring remained fertile through successive generations; and because the different varieties are distinguished by constant differentiating characters. Mendel devoted his attention to seven of these contrasting characters, which he followed through several generations of hybrids, viz.,

(1) Differences in the form of the ripe seeds, whether round or wrinkled.

(2) Differences in the color of the food material within the seeds, whether pale yellow, orange or green.

(3) Differences in the color of the seed coats (and in some cases of the flowers also), whether white, gray, gray brown, leather brown, with or without violet spots.

(4) Differences in the form of the ripe pods, whether simply inflated or constricted between the seeds.

(5) Differences in the color of the unripe pods, whether light to dark green, or vividly yellow.

(6) Differences in the positions of the flowers, whether axial, that is, distributed along the stem, or terminal, that is, bunched at the top of the stem.

(7) Differences in the length of the stem, whether tall or short.

1. *Results of Crossing Individuals with one Pair of Contrasting Characters.*—Having determined that these characters were constant for certain varieties Mendel then proceeded to cross one variety with another, by carefully removing the unripe stamens, with their pollen, from the flowers of one variety and dusting upon the stigmas of such flowers the pollen of a different variety. In this way he crossed varieties of peas which differed from each other in some one of the characters mentioned above, and then studied the offspring of several successive generations with respect to this character.

Dominant and Recessive Characters.—In every case he discovered that the plants that developed from such a cross showed only one of the two contrasting characters of the parent plants, *i.e.,* all were round-seeded, yellow-seeded, or tall, etc., although one of the parents had wrinkled seeds, green seeds, or short stem, etc. "Those characters which are transmitted entire or almost unchanged in the hybridization are termed *dominant,* and those which become latent in the process, *recessive.*"

Ratio of Dominants to Recessives.—These hybrids[1] when self-

---

[1] Bateson introduced the term "homo-zygote" for *pure-bred* individuals resulting from the union of gametes which are hereditarily similar, and "hetero-zygote" for *hybrids* resulting from the union of hereditarily dissimilar gametes. The gametes formed from a homo-zygote are all of the same hereditary type, those formed from a hetero-zygote are of two different types for every unit difference of the parents. The members of a pair of contrasting characters are called "allelomorphs"; each member of such a pair is "allelomorphic" to the other member.

PHENOMENA OF INHERITANCE 85

fertilized gave rise to a second filial generation of individuals some of which showed the dominant character and others the recessive, the relative numbers of the two being approximately three to one. Thus the hybrids produced by crossing yellow-seeded and green-seeded peas yielded when self-fertilized 6,022

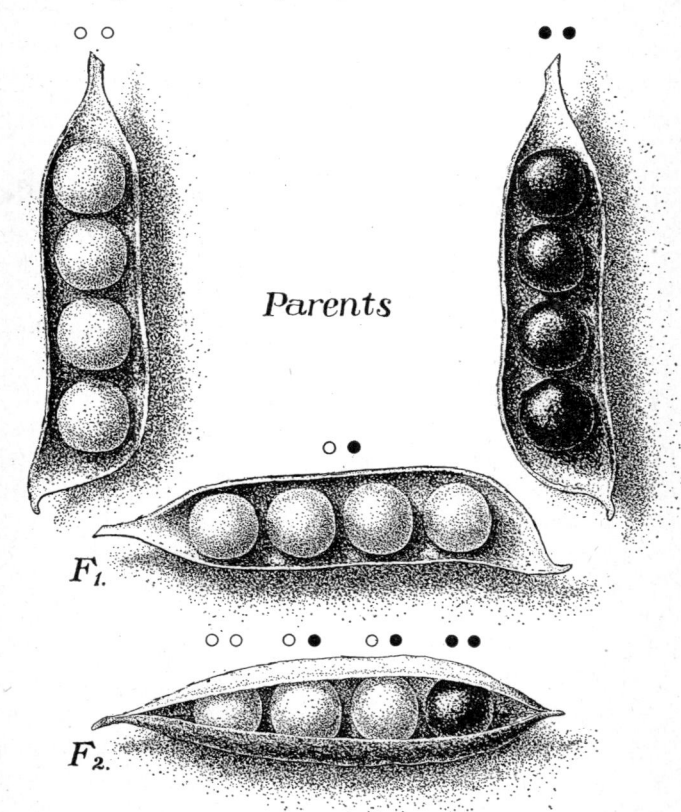

FIG. 26. DIAGRAM SHOWING THE RESULTS OF CROSSING YELLOW-SEEDED (LIGHTER COLORED) AND GREEN-SEEDED (DARKER COLORED) PEAS. The gametes carrying yellow are indicated by small unshaded circles, those carrying green by black circles. (After Thompson.)

yellow seeds and 2,001 green seeds, or very nearly three yellow to one green (Fig. 26). The hybrids produced by crossing round and wrinkled seeded varieties yielded in the second filial generation 5,474 round and 1,850 wrinkled seeds, or approximately three round to one wrinkled (Fig. 30). The hybrids from tall-stemmed and short-stemmed parents produced in the second filial generation 787 long-stemmed and 277 short-stemmed plants, or again approximately three tall to one short. And in every other case Mendel found that the ratio of dominants to recessives in the second filial generation was approximately three to one.

"Extracted" Dominants or Recessives.—These recessives derived from hybrid parents are pure and are known as "extracted" recessives; when self-fertilized they produce only recessives. One-third of the dominants are also pure homozygotes, or "extracted" dominants, and when self-fertilized produce only pure dominants. On the other hand two-thirds of the dominants are heterozygotes and when self-fertilized give rise in the next generation to pure dominants, dominant-recessives and pure recessives in the proportion of 1 : 2 : 1.

These general results are summarized in the accompanying

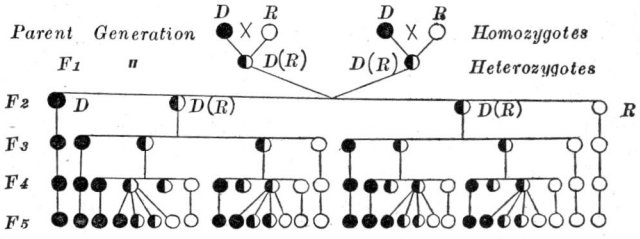

FIG. 27. DIAGRAM SHOWING RESULTS OF MENDELIAN SPLITTING where the parents are pure dominants and pure recessives (homozygotes). All pure dominants are represented by black circles, all pure recessives by white ones, while dominant-recessives (heterozygotes) are represented by circles half white and half black. Successive generations are marked P, F_1, F_2, F_3, etc. For the sake of simplicity all generations, except P and F_1, are represented as self-fertilized.

diagram (Fig. 27) in which dominant characters are indicated by the letter D, recessive characters by R, and dominant-recessives, with the recessive character unexpressed, by $D(R)$; while DD or RR indicate extracted dominants or recessives, that is, pure dominants or recessives which have separated out from dominant-recessives, $D(R)$. The parental generation is usually indicated by the letter P, and the successive filial generations by F_1, F_2, F_3, etc.

Incomplete Dominance.—In the case of the peas studied by Mendel the hybrids of the F_1 generation show only the domi-

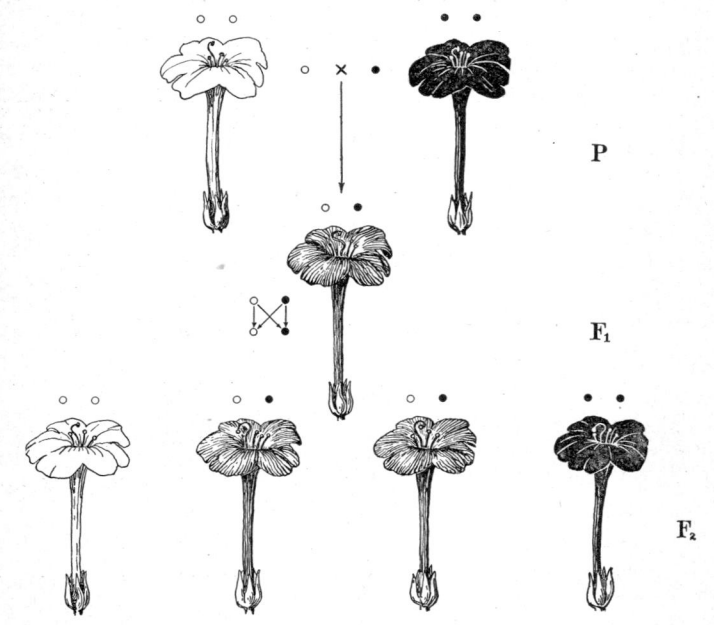

Fig. 28. Results of Crossing White-flowered and Red-flowered races of *Mirabilis Jalapa* ("four o'clocks") giving a pink hybrid in F_1, which when inbred gives in F_2 1 white, 2 pink, 1 red. The gametes bearing white or red are indicated by white or black circles. (From Woodruff after Correns.)

nant character, the contrasted recessive character being present but not expressed. However in certain cases it has been found that the hybrids differ from either parent and in successive generations split up into both parental types and into the hybrid type; thus Correns found that when a white-flowered variety of *Mirabilis,* the "four o'clock," was crossed with a red-flowered variety all of the hybrids in the F_1 generation had pink flowers and from these in the F_2 generation there came white-flowered, pink-flowered and red-flowered forms in the proportion of 1 white: 2 pink: 1 red, as shown in Fig. 28. This is a better illustration of Mendel's principle of splitting than is offered by the peas, since in this case the heterozygotes $D(R)$ are always distinguishable from the pure dominants DD.

Results in Later Generations.—In the F_2 generation and in all subsequent ones the pure dominants and the pure recessives always breed true when self-fertilized, whereas the heterozygotes continue to split up in each successive generation into pure dominants, heterozygotes and pure recessive in the proportion of 1 : 2 : 1. The result of this is that with continued self-fertilization the relative number of dominants and recessives increases in successive generations, whereas the relative number of heterozygotes decreases, and in a few generations a hybrid race will revert in large part to its parental types if continued hybridization is prevented. On the other hand there is no tendency for the relative number of dominants to increase and of recessives to decrease in successive generations; an equal number of pure dominants and pure recessives is produced in each generation (Fig. 27).

"Purity" of Germ Cells.—With remarkable insight Mendel recognized that the real explanation of the splitting of pure recessives and pure dominants from hybrid parents must be found in the composition of the male and female sex cells. Since such extracted dominants and recessives breed true, just as pure species do, it must be that their germ cells are pure.

In the cross between pure races of white-flowered and red-flowered *Mirabilis* the germ cells which unite in fertilization must be pure with respect to white and red, though the individual which develops from this cross is a pink hybrid. But the fact that one-quarter of the progeny of this hybrid are pure white, and another quarter pure red, and that these thereafter breed true, proves that the hybrid produces germ cells which are pure with respect to red and white. Furthermore the fact that one-half the progeny of this hybrid are themselves hybrid may be explained by assuming that they were produced by the union of germ cells carrying pure white and pure red, as in the parental generation.

Mendel therefore concluded that individual germ cells are always pure with respect to any pair of contrasting characters, even though those germ cells have come from hybrids in which the contrasting characters are mixed. A single germ cell can carry the factor for red flowers *or* white flowers, for green seeds *or* yellow seeds, for tall stem *or* short stem, etc., but not for both pairs of these contrasting characters. The hybrids formed by crossing white and red "four o'clocks" carry the factors for both white and red, but the individual germ cells formed by such a hybrid carry the factors for white *or* red, but not for both; these factors segregate or separate in the formation of the germ cells so that one-half of all the germ cells formed carry the factor for white and the other half that for red.

This is the most important part of Mendel's Law,—the central doctrine from which all other conclusions of his radiate. It explains not only the segregation of dominant and recessive characters from a hybrid in which both are present, but also the relative numbers of pure dominants, pure recessives and heterozygotes in each generation. For if all germ cells are pure with respect to any particular character the hybrid offspring of any two parents with contrasting characters will produce in equal numbers two classes of germ cells, one bearing the dominant and the other the recessive factor, and the chance combination of these two classes of male and female gametes will yield on the average one union of dominant with dominant, two unions of dominant with recessive and one union of recessive with recessive, thus producing the typical Mendelian ratio, $1DD:2D(R):1RR$, as shown in the accompanying diagram and in Fig. 29 b.

$$♀ \text{ germ cells } D \qquad R$$
$$♂ \text{ germ cells } D \qquad R$$

Possible combinations $\quad 1\,DD:2\,D(R):1\,RR.$

Other Mendelian Ratios.—When a pure dominant is crossed with a hybrid dominant-recessive (*Fig. 29 c*) all of the offspring show the dominant character, though one-half are pure dominants and the other half dominant-recessives. Thus if a pure round-seeded variety of pea is crossed with a hybrid between a round-seeded and a wrinkled-seeded one, all the progeny are round-seeded, though one-half of them carry the factor for wrinkled

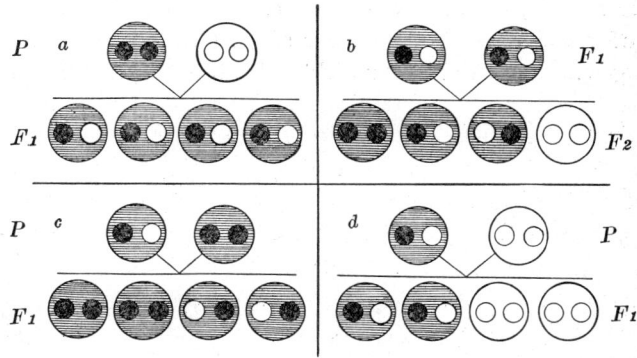

FIG. 29. DIAGRAM OF MENDELIAN INHERITANCE, in which the individual is represented by the large circle, the germ cells by the small ones, dominants being shaded and recessives white. *a*, Pure dominant × pure recessive = (yields) all dominant-recessives; *b*, Dominant-recessive × dominant-recessive = 1 pure dominant : 2 dominant-recessives : 1 pure recessive; *c*, Dominant-recessive × pure dominant = 2 pure dominant : 2 dominant-recessive; *d*, Dominant-recessive × pure recessive = 2 dominant-recessive : 2 pure recessive.

PHENOMENA OF INHERITANCE

seed; this may be graphically represented as follows, R representing the factor for round seed and W that for wrinkled seed:

$$\begin{array}{c} \text{♀ germ cells } R \diagdown R \\ \phantom{\text{♀ germ cells }} \mid \times \mid \\ \text{♂ germ cells } R \diagup W \end{array}$$

Possible combinations $2\,RR : 2\,R(W)$

In subsequent generations the progeny of the pure round (RR) breed true and produce only round-seeded peas, whereas the progeny of the hybrid round-wrinkled (RW) split up into pure round, hybrid round-wrinkled, and pure wrinkled in the regular Mendelian ratio of $1\,RR : 2\,R(W) : 1\,WW$ (Fig. 30).

When a pure recessive is crossed with a hybrid dominant-recessive (Fig. 29, *d*) another typical ratio results. Thus if a wrinkled-seeded variety of pea is crossed with a hybrid between a round-seeded and wrinkled-seeded one, round-seeded and wrinkled-seeded peas are produced in the proportion of 1 : 1 This is due to the fact that the hybrid produces two kinds of germ cells, the pure-bred but one, and the possible combinations of these are as follows:

$$\begin{array}{c} \text{♀ germ cells } W W \\ \phantom{\text{♀ germ cells }} \mid \times \mid \\ \text{♂ germ cells } R W \end{array}$$

Possible combinations $2\,R(W) : 2\,WW$.

This ratio of 2 : 2 or 1 : 1 is approximately the ratio of the two sexes in many animals and plants, and there is good reason to believe that sex is a Mendelian character of this sort, in which one parent is heterozygous for sex and the other homozygous (See p. 164).

2. *Results of Crossings where there is more than one Pair of Contrasting Characters.*—It rarely happens that two individuals differ in a single character only; more frequently they differ in many characters, and this leads to a great increase in the number of types of offspring in the F_2 generation. But however many

pairs of contrasting characters the parents may show each pair may be considered by itself as if it were the only contrasting pair, and when this is done all the offspring may be classified according to the regular Mendelian formula given above.

When the parents differ in one unit character only, the offspring formed by their crossing are called mono-hybrids, when there are two contrasting characters in the parents the offspring are di-hybrids, when three, tri-hybrids, and when the parents differ in more than three characters the offspring are called poly-hybrids. There are certainly few cases in which parents actually differ in only a single character, but since each contrasting character may be dealt with separately, as if it were the only one, and since the number of types of offspring increases greatly when more than one or two characters are considered at the same time, it is cus-

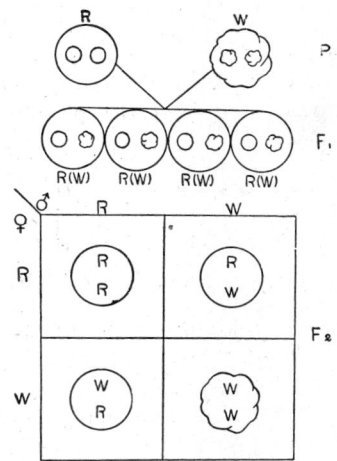

FIG. 30. MONOHYBRID DIAGRAM SHOWING RESULTS OF CROSSING ROUND- (R) SEEDED WITH WRINKLED- (W) SEEDED PEAS. Large circles represent zygotes, small ones, or single letters, gametes. In F_1 all individuals are round but contain round and wrinkled gametes. In F_2 the ♂ gametes are placed above the square, the ♀ ones to the left, and the possible combinations of ♂ and ♀ gametes are shown in the small squares, the relative number of different genotypes being 1 RR : 2 $R(W)$: 1 WW.

PHENOMENA OF INHERITANCE

omary to deal simultaneously with only one or two characters of hybrids, even though the parents may have differed in many characters. The different types of developed organisms are called by Johannsen "phenotypes" whereas the different hereditary types whether patent or latent are called "genotypes." See p. 127.

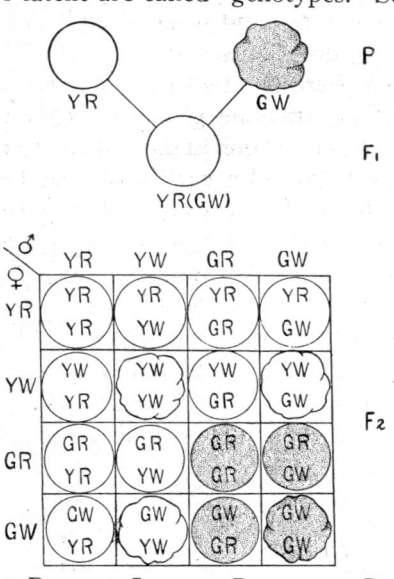

FIG. 31. DIHYBRID DIAGRAM SHOWING RESULTS OF CROSSING PEAS HAVING YELLOW ROUND (YR) SEEDS WITH OTHERS HAVING GREEN WRINKLED (GW) ONES. The hybrids of the first filial generation (F_1) are all yellow and round since these characters are dominant while green and wrinkled are recessive, $YR(GW)$. Four types of germ cells are formed by such a hybrid, viz., YR, YW, GR, GW, and the 16 possible combinations of these ♂ and ♀ gametes are shown in the small squares. Of these 16 combinations 7 contain the same letters (factors) so that there are only 9 different genotypes, and since recessive characters do not appear when mated with dominant ones these 9 genotypes produce only 4 phenotypes in the following relative numbers: 9 YR : 3 YW : 3 GR : 1 GW. There is 1 pure dominant (upper left corner), 1 pure recessive (lower right corner); homozygotes in the diagonal line between these corners, and 12 heterozygotes.

94 HEREDITY AND ENVIRONMENT

Dihybrids.—When two or more contrasting characters of the parents are followed to the F_2 generation all possible permutations of these characters occur, thus giving rise to a larger number of types of individuals than when a single pair of characters is concerned. When there is only one pair of contrasting characters there are three genotypes and usually but two phenotypes in the F_2 generation, viz., dominants and recessives in the ratio of 3:1 (Fig. 30); where there are two pairs of contrasting characters in the parents there are nine genotypes (3^2) and usually four phenotypes in the F_2 generation in the ratio of $(3:1)^2 = 9:3:3:1$. Thus when Mendel crossed a variety of peas bearing round and yellow seeds with another variety having wrinkled and green seeds all the offspring of the F_1 generation bore round and yellow seeds, round being dominant to wrinkled, and yellow to green. But the plants raised from these seeds, when self-fertilized, yielded seeds of four types, yellow and round (YR), yellow and wrinkled (YW), green and round (GR), and green and wrinkled (GW) in the proportion of 9:3:3:1 as shown in Fig. 31.

In this case also this ratio may be explained by assuming that the germ cells are pure with respect to each of the contrasting characters, round or wrinkled, yellow or green, and therefore any combination of these may occur in a germ cell except the combinations RW and YG. Accordingly there are four possible combinations of these characters in both male and female cells as follows:

$$\begin{matrix} Y & & G \\ | & \times & | \\ R & & W \end{matrix} \quad \text{i.e. } YR,\ YW,\ GR,\ GW.$$

Each of these four kinds of male cells may fertilize any one of the same four kinds of female cells, thus giving rise to sixteen combinations, as shown in Fig. 31. The dominant characters are in this case round and yellow, and only when one of these is absent can its contrasting character, wrinkled or green, develop. Accordingly the sixteen possible combinations yield seeds of four different appearances and in the following proportions: 9 YR : 3 GR : 3 YW : 1 GW. Only one individual in each of these four classes is pure (homozygous) and continues to breed true in successive generations; in Fig. 31 these are found in the diagonal from the upper left to the lower right corner. All other individuals are heterozygous and show Mendelian splitting in the next generation.

Trihybrids.—When parents differ in three contrasting characters there are twenty-seven genotypes (3^3) and eight phenotypes (2^3) in the F_2 generation in the proportion of $(3:1)^3 = 27:9:9:9:3:3:3:1$. Thus if a pea with round (R) and yellow (Y) seeds and with tall (T) stem is crossed with one having wrinkled (W) and green (G) seeds and dwarf (D) stem all the progeny of the F_1 generation have round and yellow seeds and tall stem, R, Y, and T being dominant over W, G, and D. In the F_2 generation there are 64 possible combinations (27 genotypes) of these six characters (Fig. 32); but since a recessive character does not develop if its contrasting dominant character is present there are only eight phenotypes which come to expression and in the following ratios: 27 RYT : 9 RYD : 9 RGT : 9 WYT : 3 RGD : 3 WYD : 3 WGT : 1 WGD. Of these sixty-four combinations only eight are homozygous and breed true (those lying in the diagonal between upper left and lower right corners in Fig. 32), while only one is a pure dominant and one a pure recessive (the ones in the upper left and lower right corners of Fig. 32).

3. *Inheritance Formulae.*—Mendel represented the hereditary constitution of the plants used in his experiments by letters employed as symbols, dominant characters being represented by capitals and recessives by small letters. The seven contrasting characters of his peas could be represented as follows:

Seeds, round (A), or wrinkled (a); yellow (B), or green (b); with gray seed coats (C), or white seed coats (c).

Pods, green (D), or yellow (d); inflated (E), or constricted (e).

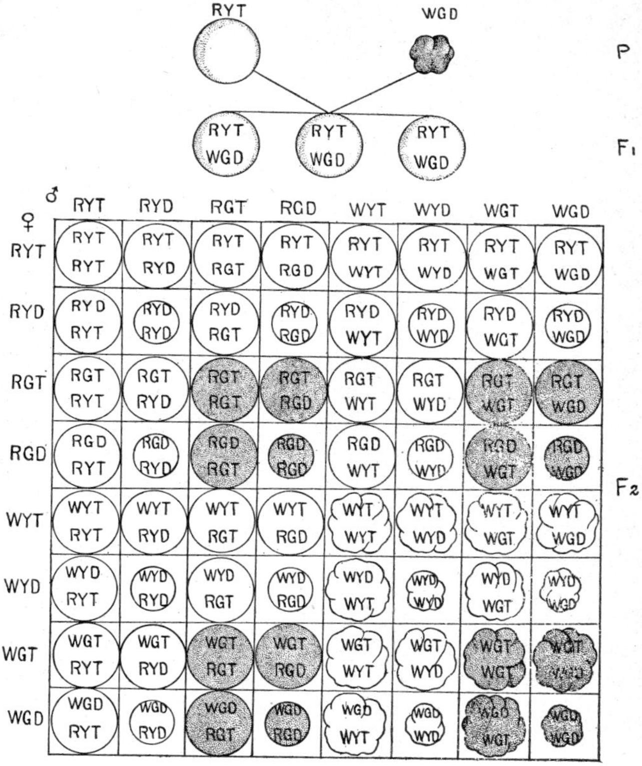

FIG. 32. TRIHYBRID DIAGRAM SHOWING RESULTS OF CROSSING PEAS HAVING ROUND YELLOW SEEDS AND TALL STEM (RYT) WITH PEAS HAVING WRINKLED GREEN SEEDS AND DWARF STEM (WGD). Eight types of germ cells are formed by the F_1 hybrid, as shown in the ♂ gametes above the square and the ♀ ones to the left of it, and the possible combinations of these ♂ and ♀ gametes are shown in the 64 small squares of which only 1 is pure dominant (upper left corner), 1 pure recessive (lower right corner) and 8 homozygotes (in diagonal line between these corners). There are 27 different genotypes, all combinations below this diagonal being homologous with the corresponding ones above and all in the other diagonal being of the same genotype, while 12 other combinations on each side of the first diagonal constitute only 6 genotypes. There are 8 phenotypes, resembling the 8 homozygotes, and their relative numbers are 27 RYT : 9 RYD : 9 RGT : 9 WYT : 3 RGD : 3 WYD : 3 WGT : 1 WGD.

Habit, tall (F), or dwarf (f).

Flowers, axial (G), or terminal (g).

It is possible for one plant to have all of these dominant characters or all of the recessive ones, or part of one kind and part of the other. The gametic formula of a plant having all seven of the dominant characters is $ABCDEFG$; of one having all of the recessive characters $abcdefg$. When two such plants are crossed the zygotic formula of the hybrid is $AaBbCcDdEeFfGg$, and since the dominant and recessive characters (or rather determiners of characters) represented by these seven pairs of letters separate in the formation of the gametes, and since each separate determiner may be associated with either member of the six other pairs, the number of possible combinations of these determiners in the gametes is $(2)^7$ or 128. That is, in this case 128 kinds of germ cells may be produced, each having a different inheritance formula; and since each of these 128 kinds of male germ cells may unite with any one of the 128 kinds of female germ cells the number of combinations of these characters which are possible in the F_2 generation is $(128)^2$ or 16,384, while the number of different genotypes is $(3)^7$ or 2187. Every one of these more than two thousand genotypes may be represented by various combinations of the letters $ABCDEFG$ and $abcdefg$.

When many characters are concerned it is difficult to remember what each letter stands for, and consequently it is customary in such cases to designate characters by the initial letter in the name of that character. By this form of shorthand one can show in a graphic way the possible segregations and combinations of hereditary units in gametes and zygotes through successive generations, and as a result many modern works on Mendelian inheritance look like pages of algebraic formulæ.

4. *Presence-Absence Hypothesis.*—Mendel spoke of the presence of *contrasting* or *differentiating characters* in the plants which he crossed, such as round or wrinkled seeds, tall or short stems, etc. Many others have regarded these contrasting charac-

ters as due to the *presence* or *absence* of single factors: thus round seeds are due to the presence of a factor for roundness (A) while wrinkled seeds were said to be due to the absence of that factor (a). Round seeds were spoken of as wrinkled seeds plus the factor for roundness. But it is practically certain that recessive characters are not due to the absence of factors for dominant characters; there are many genetical and philosophical objections to such a view, which leads logically to some strange conclusions, such as Bateson's speculations on evolution (p. 282). Morgan and his associates have found that a given dominant character may have several different kinds of recessive contrasting characters or allelomorphs; thus the dominant eye color of the wild pomace fly, *Drosophila melanogaster,* is red, but, instead of a single contrasting recessive character, eleven have been found, *viz.,* apricot, blood, buff, cherry, coral, écru, eosin, ivory, tinged, wine, and white. Such a condition is known as "multiple allelomorphism." If the red color is due to the presence of a certain factor, all these other allelomorphic colors cannot be due to its absence, since there can be only one kind of absence. Each of these recessive colors must be due to the presence of a differential factor. Therefore the presence-absence hypothesis must be abandoned.

When both gametes carry similar dominant factors the zygote has a "double dose" of such factors and is said to be *duplex*: when only one of the gametes carries such a factor the zygote has a "single dose" and is *simplex,* when neither gamete carries a positive factor or factors, the zygote receives only negative factors and is said to be *nulliplex*. Thus the union of gametes AB (♀) and AB (♂) yields zygote $AABB$, which is duplex in constitution; gametes Ab (♀) and aB (♂) yield zygote $AaBb$, which is simplex; gametes ab (♀) and ab (♂) yield zygote $aabb$, which is nulliplex.

In some instances a character comes to full expression only when it is derived from both parents, that is, when it is duplex; if derived from one parent only, that is, if simplex, it is diluted appearance and is intermediate between the two parents. For example, when white-flowered "four o'clocks" which are nulliplex are crossed with red-flowered ones which are duplex the progeny, which are simplex, bear pink flowers; in this case red flowers are produced only when the factor for red is derived from both parents, pink flowers when it is derived from one parent, white flowers when the factor for red is derived from neither parent (Fig. 28).

5. *Summary of Mendelian Principles.*—Since the rediscovery in 1900 of Mendel's work many investigators have carried out similar experiments on many species of animals and plants and have greatly extended our knowledge of the principles of inheritance discovered by Mendel, but in the main Mendel's conclusions have been confirmed again and again, so that there is no doubt that they constitute an important rule of inheritance among all sexually produced organisms.

In brief the "Mendelian Law of Alternative Inheritance" or of hereditary "splitting" consists of the following principles:

(a) *The Principle of Unit Characters.*—The heritage of an organism may be analyzed into a number of characters which are not further divisible; these are the so-called "unit characters" (deVries).

(b) *The Principle of Dominance.*—When contrasting unit characters are present in the parents they do not as a rule blend in the offspring, but one is dominant and usually appears fully developed, while the other is recessive and temporarily drops out of sight.

(c) *The Principle of Segregation.*—Every individual germ cell is "pure" with respect to any given unit character, even though it comes from an "impure" or hybrid parent. In the germ cells of hybrids there is a separation of the determiners of contrasting characters so that different kinds of germ cells are produced, each of which is pure with regard to any given unit character. This is the principle of segregation of unit characters, or of the "purity" of the germ cells. Every sexually produced individual

is a double being, double in every cell, one-half of its determine[rs]
having been derived from the male and the other half from t[he]
female sex cell. This double set of determiners again becom[es]
single in the formation of the germ cells only once more to b[e]
come double when the germ cells unite in fertilization.

11. Modifications and Extensions of Mendelian Principles

It is a common experience that natural phenomena are foun[d]
to be more complex the more thoroughly they are investigate[d.]
Nature is always greater than our theories, and with few exce[p]tions hypotheses which were satisfactory at one stage of knowl[l]edge have to be extended, modified or abandoned as knowledg[e]
increases. This observation is well illustrated in the case of th[e]
Mendelian theory. The principles proposed by Mendel were rela[-]
tively simple, but in attempting to apply them to the many phe[-]
nomena of inheritance now known it has become necessary t[o]
modify or extend them in many ways. And yet the general an[d]
fundamental truth of these principles has been established in [a]
surprisingly large number of cases, and they have been extende[d]
to forms of inheritance where at first it was supposed that the[y]
could not apply.

1. *The Principle of Unit Characters and of Inheritance Fac[-]
tors.*—There has been much criticism on the part of some biolo[o]gists of the principle of unit characters. It is said that unit char[-]
acters cannot be independent and discrete things; the organis[m]
itself is a unity and every one of its parts, every one of its char[-]
acters, must influence more or less every other part and ever[y]
other character. Certainly unit characters cannot be absolutel[y]
independent of one another; the various parts and organs of th[e]
body, and even the organism as a whole, are not absolutely inde[-]
pendent, and yet there are varying degrees of independence i[n]
organisms, organs, cells, parts of cells, hereditary units and char[-]
acters which make it possible for purposes of analysis to dea[l]
with these things as if they were really independent though w[e]

how they are not. But the most serious objection to the doctrine of unit characters is not against their independence but against their unity. Every character is complex, many factors enter into its development, and since the combination of these factors is variable the character itself cannot be constant. *Strictly speaking, characters are not units, and while the conception of "unit characters" has served a useful purpose it cannot any longer be regarded as wholly accurate.*

Inheritance Factors are Differential Causes.—Of course characters of adult individuals do not exist as such in germ cells, but there is no escape from the conclusion that in the case of inherited differences between mature organisms there must have been differences in the constitution of the germ cells from which they developed. For every inherited character there must have been a germinal cause in the fertilized egg. This germinal cause, whatever it may be, is often spoken of as a *determiner* of a character. But the character in question is not to be thought of as the result of a single cause nor as the product of the development of a single determiner; undoubtedly many causes are involved in the development of every character, but the *differential* cause or combination of causes is that which is peculiar to the development of each particular character. Of course Mendelian factors are not the only factors of development but merely the *differential* factors which cause, for example, one guinea-pig to be white and its brother to be black. Very many factors are involved in the production of white or black color but there is at least one *differential* factor for every unit character and this alone is the Mendelian factor.

Factors Are Not Undeveloped Characters.—Again it is not necessary to suppose that every developed character is represented in the germ by a distinct determiner, or inheritance unit, just as it is not necessary to suppose that every chemical compound contains a peculiar chemical element; but it is necessary to suppose that each hereditary character is caused by some particular combination of inheritance units and that each compound is produced by some particular combination of chemical elements. An enormous number of chemical compounds exists as the result of various combinations of some ninety different elements, and an almost endless number of words and combinations of words—indeed whole literatures—may be made with the twenty-six letters of the alphabet. It is quite probable that the kinds of inheritance units are few in number as compared with the multitudes of adult characters, and that different combinations of the units give rise to different adult characters; but it is certain that inherited differences in adult organization must have had some differential cause or factor in germinal organization.

Mendel did not speculate about the nature of hereditary units, though he evidently conceived that there was something in the germ which corresponded to each character of the plant. Weismann postulated a determinant in the germ for every character which is independently heritable, and many recent students of heredity hold a similar view. But it is evident that there is not an exact one to one correspondence of inheritance units and adult characters. Many different characters may be determined by a single unit or factor; for example, all the numerous secondary sexual characters which distinguish males from females may be determined by the original factor which determines whether the germ cells shall be ova or spermatozoa.

Multiple Factors.—On the other hand two or more factors may be concerned in the production of a single character. In many cases among both plants and animals the development of color appears to depend upon the presence in the germ cells and the cooperation in development of at least two factors, viz. (1) a pigment factor for each particular color, and (2) a color developer. When both of these factors are present color develops, when either one is absent no color appears.

Such cases have been described for mice, guinea-pigs, and rabbits as well as for several species of plants. Bateson and Punnett found two varieties of white sweet-peas which were apparently alike in every respect except the shape of their pollen grains, one of them having long and the other round pollen. But when these were crossed a remarkable thing occurred for the progeny "instead of being white were purple like the wild Sicilian plant from which our cultivated sweet-peas are descended." This is apparently a typical case of reversion and its cause was found in the fact that at least two factors are necessary in this case for the production of color, a pigment factor R and a color developer C. One of these was lacking in each of the white parents, their gametic formulæ being Cr and cR respectively, but when these two factors came together in the offspring a purple-flowered type was produced with the zygotic formula $CcRr$. These F_1 plants produced colored and white F_2 plants in the proportion of 9 colored to 7 white and the colored forms were of six different kinds (Fig. 33). For the production of these six colored forms five different factors must be present in the gametes, according to Punnett, viz.: (1) a color base R, (2) a color developer C, (3) a purple factor P, (4) a light wing factor L, (5) a factor for intense color I. When all of these factors are present the result is the purple wild form with blue wings, while the omission of one or more of these factors leads to the production of six forms of colored and various types of white-flowered plants of the F_2 generation.

Castle found that eight different factors may be involved in producing the coat colors of rabbits: these are:

C a common color factor necessary to produce any color.

B a factor acting on C to produce *black*.

Br a factor acting on C to produce *brown*.

Y a factor acting on C to produce *yellow*.

I a factor which determines *intensity* of color.

U a factor which determines *uniformity* of color.

A a factor for *agouti*, or wild gray pattern, in which the tip of every hair is black, its middle yellow, its basal part gray.

E a factor for the *extension* of black or brown but not of yellow.

Plate found that all of these factors except the last, E, are also

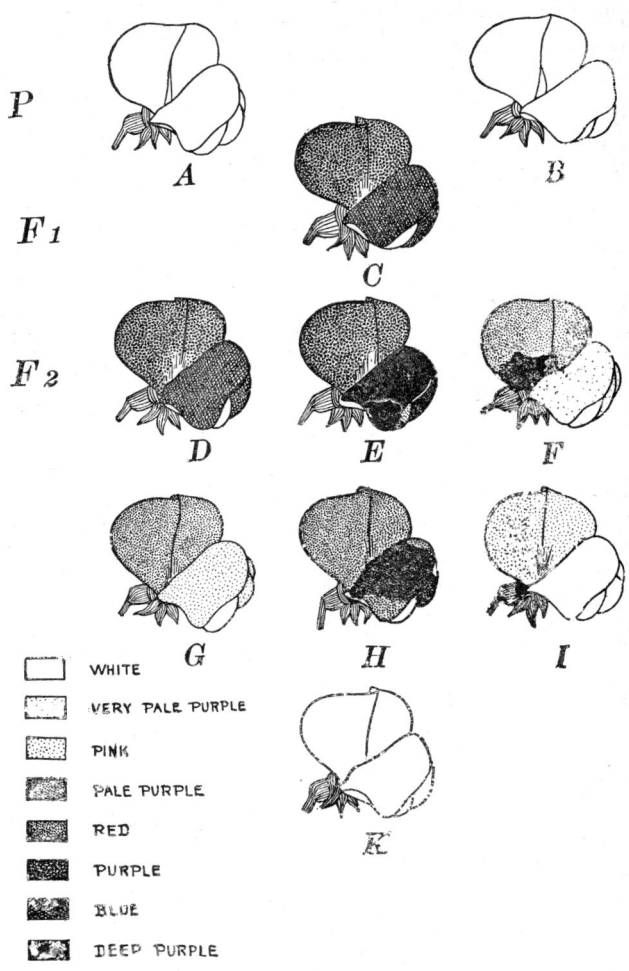

FIG. 33. RESULTS OF CROSSING TWO DIFFERENT RACES (*A* AND *B*) OF WHITE SWEET PEAS; all the F_1 hybrids (*C*) are purple with blue wings like the wild ancestral stock; in F_2 six colored varieties are formed ranging from purple with blue wings (*D*) to tinged white (*I*) and several kinds (genotypes) of white varieties (*K*). (After Punnett).

involved in the production of the coat colors of mice. Baur has recognized more than twenty different factors for the color and form of flowers in the snapdragon, *Antirrhinum.*

Modifying Factors.—Morgan and Bridges have found that the effects of many factors may be modified by other factors. Thus the eye color of *Drosophila* known as "eosin" may be modified by six or seven different factors, occupying different *loci* in the chromosomes, one of which intensifies "eosin" while the others dilute it. These modifying factors are undoubtedly like other Mendelian factors in their behavior and they show that an adult character may be the result of several different inheritance factors. Indeed Morgan says "that an overstatement that each factor may affect the entire body is less likely to do harm than to state that each factor affects only a particular character." And again he says, "It cannot too insistently be urged that when we say a character is the product of a particular factor we mean no more than that it is the most conspicuous effect of the factor" (Morgan, 1916, p. 117).

Lethal Factors.—Morgan and his associates have also demonstrated the existence of a considerable number of lethal factors in *Drosophila* that cause the early death of those gametes or zygotes in which this factor is not balanced by a normal one. This phenomenon greatly modifies expected Mendelian ratios for only heterozygotes survive, and all individuals that are homozygous for a lethal factor usually die so early that they are never seen. Nevertheless their existence can be determined by indirect methods that will be mentioned in the next chapter under "linkage." Such lethal factors greatly complicate the study of genetics but they do not destroy its fundamental principles.

What are factors?—Inheritance factors are probably complex chemical substances which preserve their individuality in various combinations, just as groups of atoms or radicals do in chemical reactions; they may be dropped out or added, substituted or transposed, just as chemical radicals may be in chemical compounds. To this extent they maintain continuity and independence, but they are not absolutely independent for they react upon one another as well as to environmental changes, so that the characters of the developed organism are the resultants of all these reactions and interactions.

Some progress has been made, in identifying certain structures of the germ cells with certain hereditary units, but quite irrespective of what these units may be and where they may be located it is possible, by means of the Mendelian theory of segregation of units in the germ cells and of chance combinations of these in fertilization to predict the number of genotypes and phenotypes which may be expected as the result of a given cross.

2. *Modifications of the Principle of Dominance. Incomplete Dominance.*—A large number of animal and plant hybrids show one contrasting character completely dominant over the other one as Mendel observed in the case of his peas. But in a considerable number of cases this dominance is incomplete or imperfect. When white-flowered strains of "four o'clocks" are crossed with red-flowered ones the F_1 plants bear neither white nor red flowers but pink ones, and the F_2 plants are white-flowered, red-flowered or pink-flowered. The whites and reds are always homozygous, the pinks heterozygous; pure white and pure red are produced only when their factors are duplex (WW), (RR); when they are simplex (WR) pink is produced. In this case red is not completely dominant over white, but the hybrid is more or less intermediate between the two parents (Fig. 28).

It has long been known that the race of fowls called Blue Andalusian does not breed true, but in each generation produces a certain number of blacks and whites as well as blues. Bateson found that the blues are really hybrids between blacks and whites in which neither of the latter is completely dominant. Black and white appear only when they are pure (homozygous), blue only when both black and white are present (heterozygous).

Again a cross of red and white cattle produces roan offspring, but the latter when interbred give rise to reds, roans and whites in the proportion of 1:2:1, showing that the roans are heterozygotes in which red is not completely dominant over white, while the reds and whites are homozygotes and consequently breed true.

Lang found that when snails with uniformly colored shells were crossed with snails having bands of color on the shells the hybrids were faintly banded, thus being more or less intermediate between the two parents; but when these hybrids were interbred they produced banded, faintly banded and uniformly colored snails in the ratio of 1:2:1, thus proving that Mendelian segregation takes place in the F_2 generation, and that dominance is incomplete in the heterozygotes. Many other similar cases of incomplete dominance are known.

Sometimes dominance is incomplete in early stages of development but becomes complete in adult stages. Davenport found that when white and black fowls are crossed the chicks, especially the females, are speckled white and black, but in the adult fowl dominance is complete and the plumage is white. Similar conditions of delayed dominance are well known in the color of hair and eyes of children, though dominance may become complete when they have reached adult life.

Reversible Dominance.—In a few instances a character may be dominant at one time and recessive at another. Thus Davenport found that an extra toe in fowls is dominant under certain circumstances and recessive under others. Tennent found that characters which are usually dominant in hybrid echinoderms may be made recessive if the chemical or physical nature of the sea water is changed. Such cases seem to show that dominance may depend sometimes upon environmental conditions, sometimes upon a particular combination of hereditary units.

Dominance Not Fundamental.—In all cases dominance means merely the development in offspring of certain characters of one parent, while contrasting characters of the other parent remain undeveloped. The appearance of any developed character in an organism depends upon many complicated reactions of germinal units to one another and to the environment. Under certain conditions of the germ or of the environment some characters may develop in hybrids to the exclusion of their opposites whereas under other conditions these results may be reversed or the characters may be intermediate. *The principle of dominance is not a fundamental part of Mendelian inheritance.* Even when the characters of hybrids are intermediate between those of their parents, if the parental types reappear in the F_2 generation we may be certain that we are dealing with cases of Mendelian inheritance.

3. *The Principle of Segregation.*—The individuality of inheritance units, and their segregation or separation in the sex cells and recombination in the zygote are fundamental principles of the Mendelian doctrine. Indeed the evidence for the individuality and continuity of inheritance units is based entirely upon such segregation and recombination, so that *the entire Mendelian theory may be said to rest upon the principle of segregation.* If there are cases in which such segregation does not take place they belong to other forms of inheritance than the Mendelian; if segregation occurs in every instance there is no other type of inheritance than that discovered by Mendel. Are there cases which do not segregate according to Mendelian expectation?

When the Mendelian theory was new it was generally supposed that there were forms of inheritance which differed materially from the Mendelian type; indeed it was supposed that the latter was one of the less common forms of heredity and that blending of parental traits and not segregation was the rule. All cases in which the characters of the parents appeared to blend in the offspring or in which there was not a clear segregation of the parental types in the F_2 generation or in which the ratio of dominants to recessives differed from the well known 3 to 1 ratio were supposed to be non-Mendelian.

Unusual Ratios.—However further work has shown that most of these cases are really Mendelian. Sometimes offspring are intermediate between their parents owing to incompleteness of dominance, rather than to incompleteness of segregation; in such cases the parental types reappear in the F_2 generation as in the cross between red and white "four o'clocks." Sometimes departures from the 3 to 1 ratio are caused by the fact that two or more factors of the same sort are involved in the production of a single character. Nilsson-Ehle found that when oats with black glumes were crossed with varieties having white glumes the ratio of 3 white to 1 black was usually found in the second generation; but one variety of black oats when crossed with white gave in the second generation approximately 15 blacks to 1 white which is the dihybrid ratio. From this and other evidence he concludes that in this variety of oats two hereditarily separable factors are involved in the production of black. In crosses between red-grained and white-grained wheat he usually got in the second generation the monohybrid ratio of 3 red to 1 white, but three strains gave the dihybrid ratio of 15 to 1 and two gave the trihybrid ratio of 63 to 1 and in subsequent generations each of these strains continued to give the same ratios. Consequently he concludes that while the red color of wheat grains is usually due to one factor for red, it may in some cases be due to two or even to three factors; notable departures from expected ratios may thus be explained. Other departures from regular Mendelian ratios are caused by the early death of certain gametes or zygotes due to lethal factors, as explained on page 105.

Blending of Color in Mulatto.—Perhaps the most serious objections which can be presented against the universality of the Mendelian doctrine are found in phenomena of "blending" inheritance. In some instances contrasting characters of parents appear to blend in offspring and even in the F_2 and in subsequent generations the descendants remain more or less intermediate between the parents. One of the best known illustrations of this is found in the skin color of the mulatto which is intermediate between the white parent and the black one, and even in the F_2 and in subsequent generations mulattoes do not usually produce pure white or pure black children, though the children of mulattoes show considerable variation in color. Hence there seems to be a failure of the Mendelian principle of segregation.

But white skin is not really white nor is black skin ever perfectly black. Davenport has shown that there is a mixture of black, yellow and red pigments in both white and black skins, though the amount of each of these pigments varies greatly in negroes and whites. The relative amounts of these pigments in any given case may be determined by means of a rotating color disk. A white person may have a skin color composed of black (b) 8 per cent, yellow (y) 9 per cent, red (r) 50 per cent, and absence of pigment or white (w) 33 per cent. On the other hand a very black negro may have b 68 per cent, y 2 per cent, r 26 per cent, w 4 per cent. The nine children of two mulattoes, the father having 13 per cent of black and the mother 45 per cent, ranged all the way from 46 per cent to 6 per cent of black, the latter so far as skin color is concerned being virtually white. On the other hand where both parents have about the same degree of pigmentation the children are more nearly uniform in color; thus seven children of two mulattoes, the father having 36 per cent and the mother 30 per cent of black, ranged only from 27 per cent to 39 per cent. of black.[1]

Such variations in color in the F_2 and in subsequent generations are exactly what one would expect in a Mendelian character in which more than one factor is involved, as for example in the case of the color of the sweet peas shown in Fig. 33. Davenport, who has made an extensive study of this case, concludes that "there are two double factors (AA, BB) for black pigmentation in the full blooded negro of the west coast of Africa, and these are separably inheritable." These factors are lacking in white persons (this being indicated by the formula aa, bb). Since the germ cells carry only single factors and not double ones the cross between negro and white would have only one set of these fac-

[1] In another family shown in Fig. 35 the father has 18 per cent black pigment, the mother 38 per cent and the children range from 17 per cent to 54 per cent.

PHENOMENA OF INHERITANCE

♀ \ ♂	AB	Ab	aB	ab
AB	AB / AB	AB / Ab	aB / AB	ab / AB
Ab	AB / Ab	Ab / Ab	aB / Ab	ab / Ab
aB	AB / aB	Ab / aB	aB / aB	ab / aB
ab	AB / ab	Ab / ab	aB / ab	ab / ab

FIG. 34. CHECKERBOARD DIAGRAM SHOWING RESULTS OF CROSSING TWO MULATTOES, each having color factors *ABab*. Types of male germ cells are above the square, of female cells on the left and the possible combinations of these are shown in the 16 small squares. Homozygotes are found only along the diagonal. The color of the children varies all the way from black (upper left corner) to white (lower right corner).

FIG. 35. MULLATTO HUSBAND AND WIFE AND THEIR SEVEN CHILDREN ranging in color from the one on left who "passes for white" to the youngest who is typically black. (From Davenport.)

tors for black color, as shown by the formula $AB \times ab = ABab$ hence the color of the F_1 generation is intermediate between that of the two parents. In the F_2 generation there should be a variety of colors ranging all the way from white to black (Fig 34), though pure white ($ab\ ab$) or pure black ($AB\ AB$) would be expected in only 1 out of 16 of the offspring. As a matter of fact it is known that the children of mulattoes vary considerably in color, and in some cases a child may be darker or lighter than either parent which would indicate that segregation does actually occur. It is very probable that this classical case of "blending" inheritance is really Mendelian inheritance in which two or more factors for skin color are involved.

Blending of Size.—Similar "blending" inheritance is found in certain other cases where the parents differ in form or size. Thus Castle found that when long-eared rabbits were crossed with short-eared ones the offspring have ears of intermediate length and in all subsequent generations the ear length remained intermediate between that of the parents. He found the same thing true of length and breadth of the skull (Fig. 36) and of the size of other portions of the skeleton, and he concluded that such quantitative characters are not inherited in Mendelian fashion.

More recently MacDowell, working on the inheritance of size in rabbits, concludes that this character as well as other quantitative differences between parents which appear to blend in the offspring, such as Castle's case of ear length in rabbits, is not due to a single factor, as in the case of Mendel's tall and dwarf peas but to several factors. Consequently in the formation of the germ cells there is not a clean segregation of all the factors for tallness or large size or long ears in half the germ cells and their total absence in the other half of those cells, but some of these factors go into certain cells and others into others, as in the case of dihybrids, trihybrids or polyhybrids. As a result offspring appear more or less intermediate in size between their parents.

Thus it is possible to explain even "blending" inheritance as due not to the real fusion or blending of inheritance factors but to varying combinations of numerous or multiple factors, according to the Mendelian rules. The Mendelian principle of segregation has been found to be of such general occurrence that there is a strong probability that it is universal, and that all cases of "blending" inheritance are due to incomplete dominance or to multiple factors.

Maternal Inheritance.—Another case which seems at first sight to be non-Mendelian is what may be called "maternal inheritance" since certain characters are invariably derived from the mother and not from the father. Among these are the polarity, symmetry and pattern of the egg and of the adult animal which is derived from it (see p. 196). These characters are of such a general sort that they may not be recognized as phenomena of inheritance at all, and yet they form the background and framework for all the other characters. They do not come equally

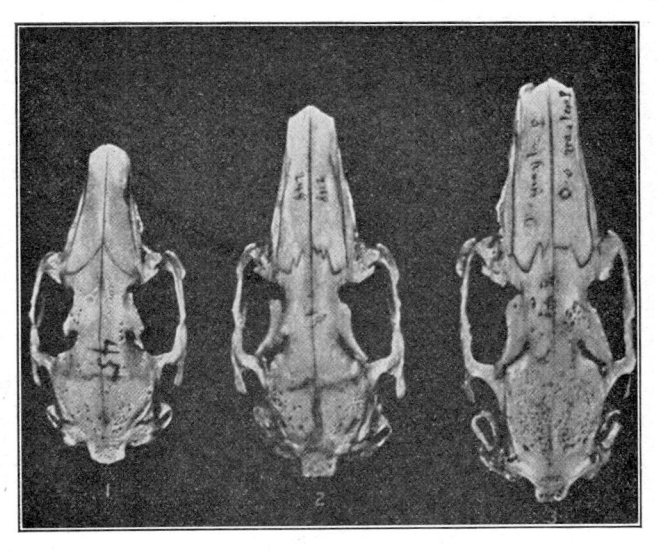

FIG. 36. INHERITANCE OF SIZE IN RABBITS. The skulls of two parents are shown in 1 and 3, of their intermediate offspring in 2. (From Castle.)

114   HEREDITY AND ENVIRONMENT

from the egg and sperm, and they do not undergo segregation i
the formation of the gametes, but are apparently derived from th
egg cytoplasm. Among characters of this sort are the norma
and inverse symmetry of snails, and of many other animals
including man, which are referred to on pages 196-204. Suc
characters are undoubtedly inherited, though they differ fror
other characters not only in the fact that they are transmitte
through the egg only, but also because they are of the same kin
in the egg and in the developed organism; they are in a measur
preformed in the egg; they are differentiated characters carrie
over from a previous generation rather than inheritance factor:
These egg characters probably appeared in the course of oogenesi
under the influence of paternal as well as of maternal factors
if so this is a case of Mendelian inheritance in the previou
generation or what may be called "Pre-inheritance." Simila
phenomena have been described by McCracken and by Toyam
in silk-worms where several egg characters seem to be non
Mendelian, but Toyama has shown that they are in reality Men

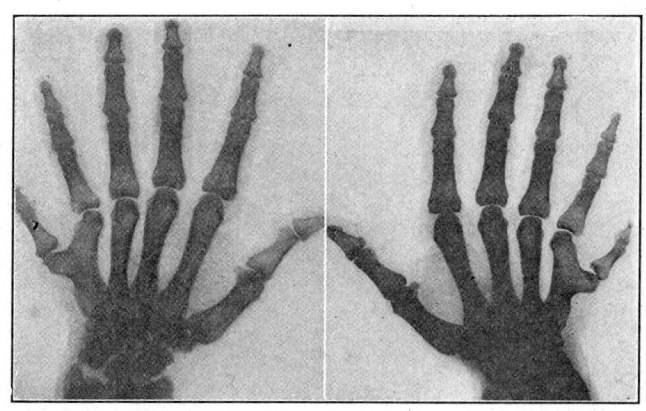

FIG. 37. X-RAY PICTURE OF RIGHT AND LEFT HANDS EACH WITH SI
FINGERS (*polydactyly*) caused by splitting of the little fingers at an earl
stage. (From *Journal of Heredity*.)

delian in the previous generation, this also being a case of pre-inheritance.

It has been found by Correns, Baur, and Shull that the leaf colors of certain plants are not inherited in Mendelian fashion, but the chromoplasts, which produce the chromatophores (chloroplasts), are transmitted from one generation to the next in the cytoplasm of the egg cell and only rarely through the male sex cell. If chromoplasts are integral parts of a plant and undergo differentiation or development this may be a case of pre-inheritance; if they are symbiotic organisms it is an instance of the inclusion of foreign bodies in the cytoplasm and not inheritance at all.

Other forms of transmission are known in which substances are carried over from one generation to the next through the egg, but they are probably not cases of true inheritance. Among these are the occasional transmission of immunity through the mother but never through the father, the carrying over of particular chemical substances such as fat dyes through the egg but not through the sperm, and the transport of symbiotic or parasitic organisms, such as algae, bacteria, etc., through the female sex cell but not through the male cell. These substances, or micro-

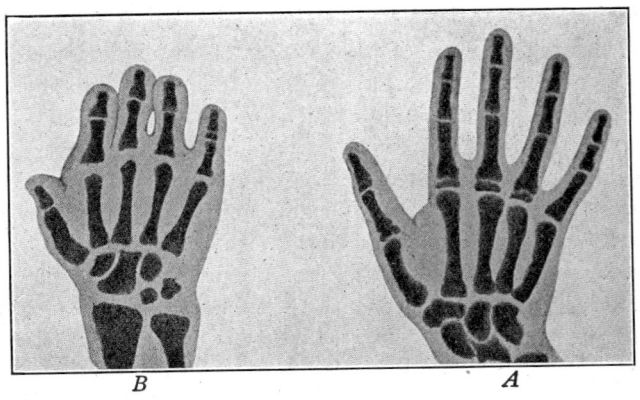

FIG. 38. X-RAY PICTURE, *A* OF A NORMAL, *B* OF A SHORT-FINGERED (*brachydactyl*) hand. (From Bateson.)

organisms, are to be regarded as inclusions in the egg rather than as any permanent part of the germinal organization; consequently they are not inherited in the strict sense of that term.

III. Mendelian Inheritance in Man

The study of inheritance in man must always be less satisfactory and the results less secure than in the case of lower animals and for the following reasons: In the first place there are no "pure lines" but the most complicated intermixture of different lines. In the second place experiments are out of the question and one must rely upon observation and statistics. In the third place man is a slow breeding animal; there have been less than sixty generations of men since the beginning of the Christian era, whereas Jennings gets as many generations of *Paramecium* within two months and Morgan almost as many generations of *Drosophila* within two years. Finally the number of offspring are so few in human families that it is impossible to determine what all the hereditary possibilities of a family may be. Bearing in mind these serious handicaps to an exact study of inheritance it is not surprising that the method of inheritance of many human characters is still uncertain.

Davenport and Plate have catalogued more than sixty human traits which seem to be inherited in Mendelian fashion. About fifty of these represent pathological or teratological conditions while only a relatively small number are normal characters. This does not signify that the method of inheritance differs in the case of normal and abnormal characters, but rather that abnormal characters are more striking, more easily followed from generation to generation, and consequently statistics are more complete with regard to them than in the case of normal characters. In many cases statistics are not sufficiently complete to determine with certainty whether the character in question is dominant or recessive, and it must be understood that in some instances the classification in this respect is tentative. A partial list of these characters is given herewith:

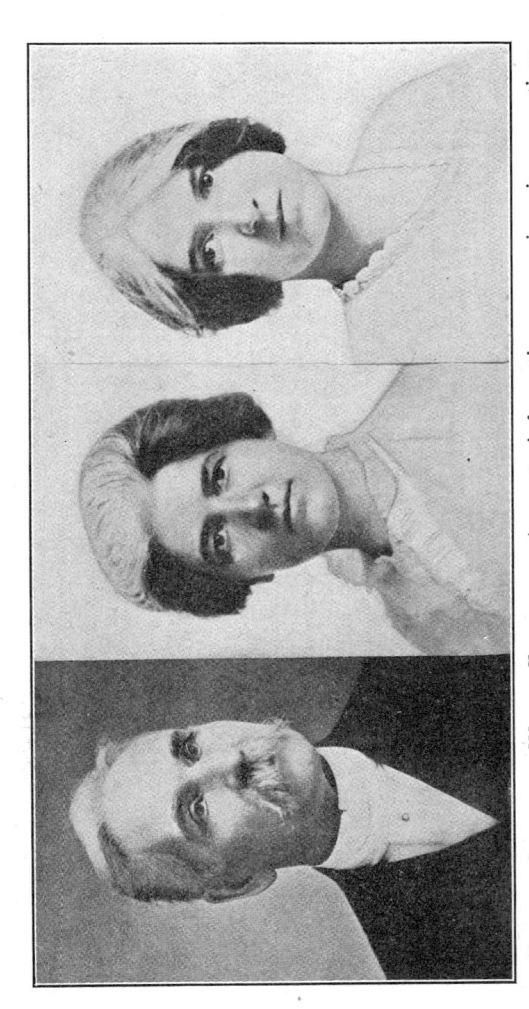

FIG. 39. INHERITED WHITE FORE-LOCK. Accurate information concerning six generations shows that it is a simple Mendelian dominant. (From Miller in *Journal of Heredity*, Vol. 6, No. 4.)

HEREDITY AND ENVIRONMENT

Mendelian Inheritance in Man

NORMAL CHARACTERS

Dominant	*Recessive*
Hair:	
Curly	Straight
Dark	Light to red
Eye Color:	
Brown	Blue
Skin Color:	
Dark	Light
Normal Pigmentation	Albinism
Countenance:	
Hapsburg Type (Thick lower lip and prominent chin)	Normal
Temperament:	
Nervous	Phlegmatic
Intellectual Capacity:	
Average	Very great
Average	Very small

TERATOLOGICAL AND PATHOLOGICAL CHARACTERS

General Size:	
Achondroplasy (Dwarfs with short stout limbs but with bodies and heads of normal size)	Normal
Normal size	True Dwarfs (With all parts of the body reduced in proportion)
Hands and Feet:	
Brachydactyly (Short fingers and toes)	Normal (Fig. 38)
Syndactyly (Webbed fingers and toes)	Normal
Polydactyly (Supernumerary digits)	Normal (Fig. 37)
Skin:	
Keratosis (Thickening of Epidermis)	Normal
Epidermolysis (Excessive formation of blisters)	Normal
Hypotrichosis (Hairlessness associated with lack of teeth)	Normal

MENDELIAN INHERITANCE IN MAN (*Continued*)

TERATOLOGICAL AND PATHOLOGICAL CHARACTERS

Dominant	*Recessive*
Kidneys:	
Diabetes insipidus	Normal
Diabetes mellitus	Normal
Normal	Alkaptonuria (Urine dark after oxidation)
Nervous System:	
Normal Condition	General Neuropathy, *e.g.* Hereditary Epilepsy; Hereditary Feeble-mindedness; Hereditary Insanity; Hereditary Alcoholism; Hereditary Criminality; Hereditary Hysteria
Nervous System:	
Normal	Multiple Sclerosis (Diffuse degeneration of nerve tissue)
Normal	Friedrich's Disease (Degeneration of upper part of spinal cord)
Normal	Meniere's Disease (Dizziness and roaring in ears)
Normal	Chorea (St. Vitus Dance)
Normal	Thomsen's Disease (Lack of muscular tone)
Huntington's Chorea	Normal
Muscular Atrophy	Normal
Eyes:	
Hereditary Cataract	Normal
Pigmentary Degeneration of Retina	Normal
Glaucoma (Internal pressure and swelling of eyeball)	Normal
Coloboma (Open suture in iris)	Normal
Displaced Lens	Normal
Ears:	
Normal	Deaf-mutism
Normal	Otosclerosis (Rigidity of tympanum, etc., with hardness of hearing)

SEX-LINKED CHARACTERS[1]

Recessive characters, appearing in male when simplex, in female when duplex.

Dominant	Recessive
Normal	Gower's Muscular Atrophy
Normal	Hæmophilia (Slow clotting of blood)
Normal	Color Blindness (Daltonism; inability to distinguish red from green)
Normal	Night Blindness (Inability to see by faint light)
Normal	Neuritis Optica (Progressive atrophy of optic nerve)

Summary

The principles of heredity established by Mendel are almost as important for biology as the atomic theory of Dalton is for chemistry. By means of these principles particular dissociations and recombinations of characters can be made with almost the same certainty as particular dissociations and recombinations of atoms can be made in chemical reactions. By means of these principles the hereditary constitution of organisms can be analyzed and the real resemblances and differences of various organisms determined. By means of these principles the once mysterious and apparently capricious phenomena of prepotency, atavism and reversion find a satisfactory explanation.

Before the establishment of Mendel's principles, heredity was, as Balzac said, "a maze in which science loses itself." Much still remains to be discovered about inheritance, but the principles of Mendel have served as the thread of Ariadne to guide science through this maze of apparent contradictions and exceptions in which it was formerly lost.

[1] See page 185.

CHAPTER III

THE CELLULAR BASIS OF HEREDITY AND DEVELOPMENT

CHAPTER III

THE CELLULAR BASIS OF HEREDITY AND DEVELOPMENT

A. INTRODUCTORY

HEREDITY is today the central problem of biology. This problem may be approached from many sides, that of the observer, the statistician, the practical breeder, the experimenter, the embryologist, the cytologist; but these different aspects of the subject may be reduced to three general methods of study, (1) the observational and statistical, (2) the experimental, (3) the cytological and embryological. We have dealt with the first and second of these in the preceding chapter and before taking up the third it is important that we should have clear definitions of the terms employed and a fairly accurate conception of the processes involved.

1. *Confused Ideas of Heredity.*—Heredity originally meant the transmission of property from parents to children, and in the field of biology it has been defined erroneously as "the transmission of qualities or characteristics, mental or physical, from parents to offspring."[1] The colloquial meaning of the word has led to much confusion in biology, for it carries with it the idea of the transmission from one generation to the next of ownership in property. A son may inherit a house from his father and a farm from his mother, the house and farm remaining the same though the ownership has passed from parents to son. And when it is said that a son inherits his stature from his father and his complexion from his mother, the stature and complexion are usually thought of only in their developed condition, while the great fact of development is temporarily forgotten. Of course there are no "qualities" or "characteristics" which are "trans-

[1] *Century Dictionary.*

mitted" as such from one generation to the next. Such terms are not without fault when used merely as figures of speech, but when interpreted literally, as they frequently are, they are altogether misleading; they are the result of reasoning about names rather than facts, of getting far from phenomena and philosophising about them. The comparison of heredity to the transmission of property from parents to children has produced confusion in the scientific as well as in the popular mind. It is only necessary to recall the most elementary facts about development to recognize that in a literal sense developed characteristics of parents are never transmitted to children.

2. *The Transmission Hypothesis.*—And yet the idea that the characteristics of adult persons are transmitted from one generation to the next is a very ancient one and was universally held until the most recent times. Before the details of development were known it was natural to suppose, as Hippocrates did, that white-flowered plants gave rise to white-flowered seeds and that blue-eyed parents produced blue-eyed germs, without attempting to define what was meant by white-flowered seeds or blue-eyed germs. And even after the facts of development were fairly well known it was generally held that the germ cells were made by the adult animal or plant and that the characteristics of the adult were in some way carried over to the germ cells (Fig. 41); but the manner in which this supposed transmission took place remained undefined until Darwin attempted to explain it by his "provisional hypothesis of pangenesis." Darwin assumed that minute particles or "gemmules" were given off by every cell of the body, at every stage of development, and that these gemmules then collected in the germ cells which thus became storehouses of little germs from all parts of the body. Afterward, in the development of the embryo, the gemmules, or little germs, developed into cells and organs similar to those from which they originally came.

3. *Germinal Continuity and Somatic Discontinuity.*—Many

THE CELLULAR BASIS

ingenious hypotheses have been devised to explain things which are not real, and this is one of them. The doctrine that adult organisms manufacture germ cells and transmit their characters to them is now known to be erroneous. Neither germ cells nor any other kind of cells are formed by the body as a whole, but every cell in the body comes from a preceding cell by a process of division, and germ cells are formed, not by contributions from all parts of the body, but by division of preceding cells which are derived ultimately from the fertilized egg (Fig. 40). The hen does not produce the egg, but the egg produces the hen and also other eggs. (Fig. 41). Individual traits are not transmitted from the hen to the egg, but they develop out of germinal factors which are carried along from cell to cell, and from generation to generation.

Germ Cells and Body Cells.—There is a continuity of germinal substance, and usually of germinal cells, from one generation to the next. In some animals the germ cells are set apart at a very early stage of development, sometimes in the early cleavage stages of the egg. In other cases the germ cells are first recognizable at later stages, but in practically every case they arise from germinal or embryonic cells which have not differentiated into somatic tissues. In general then germ cells do not come from differentiated body cells, but only from undifferentiated germinal cells, and if in a few doubtful cases differentiated cells may reverse the process of development and become embryonic cells and even germ cells it does not destroy this general principle of germinal continuity and somatic discontinuity of successive generations.

Thus the problem which faces the student of heredity and development has been cut in two; he no longer inquires how the body produces the germ cells, for this does not happen, but merely how the latter produce the body and other germ cells. The germ is the undeveloped organism which forms the bond between successive generations; the body is the developed organism which arises from the germ under the influence of environmental conditions. The body develops and dies in each generation; the

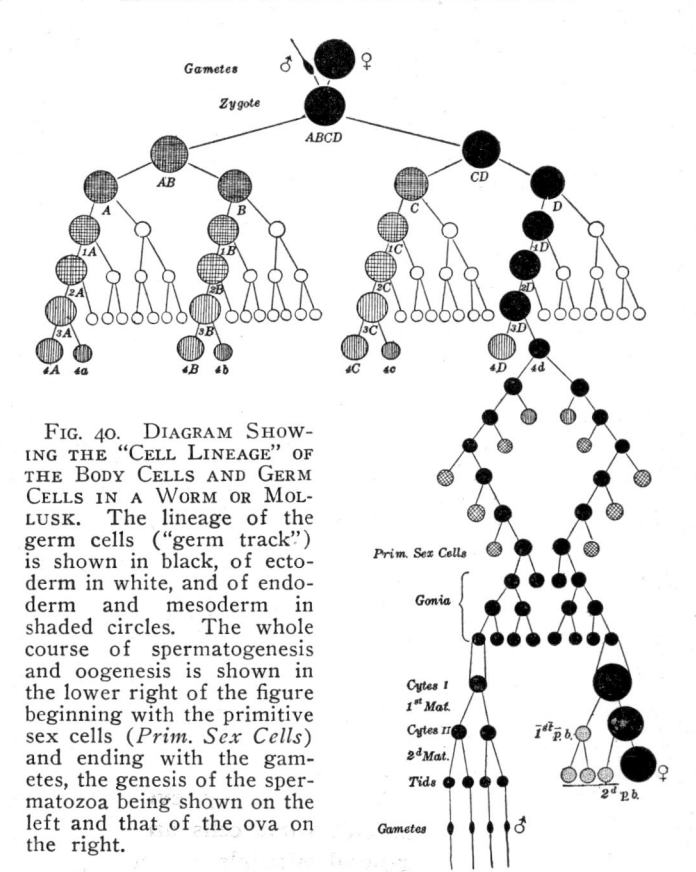

Fig. 40. Diagram Showing the "Cell Lineage" of the Body Cells and Germ Cells in a Worm or Mollusk. The lineage of the germ cells ("germ track") is shown in black, of ectoderm in white, and of endoderm and mesoderm in shaded circles. The whole course of spermatogenesis and oogenesis is shown in the lower right of the figure beginning with the primitive sex cells (*Prim. Sex Cells*) and ending with the gametes, the genesis of the spermatozoa being shown on the left and that of the ova on the right.

germplasm is the continuous stream of living substance which connects all generations. The body nourishes and protects the germ; it is the carrier of the germplasm, the mortal trustee of an immortal substance.

4. *Germplasm and Somatoplasm.*—This contrast between the germ and the body, between the undeveloped and the developed organism, is fundamental in all modern studies of heredity. It was

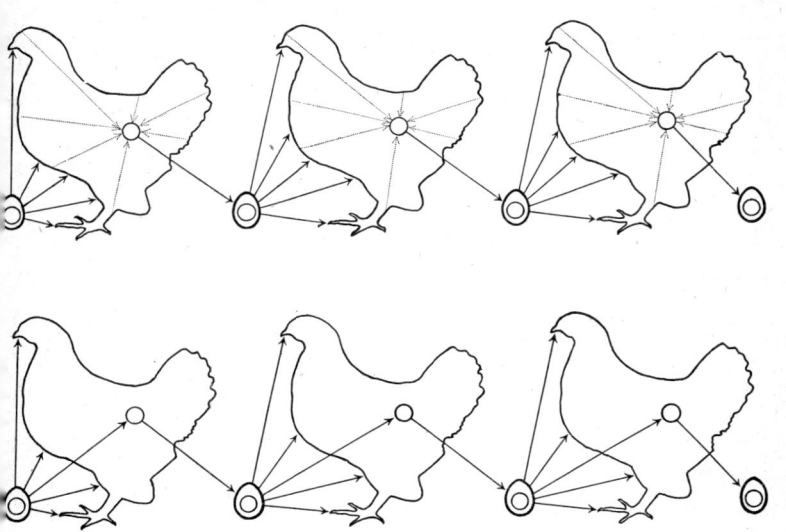

Fig. 41. Diagram Contrasting the Erroneous Hypothesis of Pangenesis (Upper Row) and the True View of the Continuity of the Germplasm (Lower Row): In the former the egg produces all parts of the hen, and all parts of the hen, in turn, produce the egg; in the latter the egg produces the hen and also other eggs.

especially emphasized by Weismann in his germplasm theory and recently it has been made prominent by Johannsen under the terms "genotype" and "phenotype"; the genotype is the fundamental hereditary constitution of an organism, it is the germinal type; the phenotype is the developed organism with all of its visible characters, it is the somatic type.

But important as this distinction is between germ and soma it has sometimes been overemphasized. This is one of the chief faults of Weismann's theory. The germ and the soma are generically alike, but specifically different. Both germ cells and somatic cells have come from the same oosperm, but have differentiated in different ways; the tissue cells have lost certain things which the germ cells retain and have developed other things which remain undeveloped in the germ cells. But the germ cells do not remain undifferentiated; both egg and sperm are differentiated, the former for receiving the sperm and for the nourishment of the embryo, the latter for locomotion and for penetration into the egg. But while the differentiations of tissue cells are usually irreversible, so that they do not again become germinal cells, the differentiations of the sex cells are reversible, so that these cells, after their union, again become germinal cells. The ovum loses its power to form yolk and during the early development it gradually loses all the yolk which it had stored up; the spermatozoon loses its highly differentiated tail or locomotor apparatus and its small compact nucleus absorbs substance from the cytoplasm of the egg and becomes a large germinal nucleus.

Chromatin is Germplasm, Cytoplasm is Somatoplasm.—In many theories of heredity it is assumed that there is a specific "inheritance material," distinct from the general protoplasm, the function of which is the "transmission" of hereditary properties from generation to generation, and the chief characteristics of which are independence of the general protoplasm, continuity from generation to generation and extreme stability in organization. This is the idioplasm of Nägeli, the germplasm of Weismann. Such a substance is no mere fiction or logical abstraction, as many writers have affirmed, for there is in the nucleus of every cell a substance which fulfills all of these conditions, namely, the *chromatin*. It is relatively independent of the surrounding cytoplasm, it is self-propagating and consequently continuous from cell to cell, and from generation to generation and it is relatively stable in organization so that it is but little influenced by environmental conditions. There are many important reasons for believing that the chromatin is the germplasm, or at least that it contains the inheritance units, as we shall see later. It is present not only in germ cells but in every cell of the organism, though in highly differentiated tissue cells it may undergo certain secondary modifications. On the other hand the cyto-

THE CELLULAR BASIS

plasm surrounding the nucleus, undergoes many marked differentiations in the course of development and it constitutes in the main the body plasm or somatoplasm. Germplasm and somatoplasm are not, therefore, vague generalizations, but they are definite cell substances which may be seen under the microscope.

5. *The Units of Living Matter.*—The entire cell, nucleus and cytoplasm, is the smallest unit of living matter which is capable of independent existence. Neither the nucleus nor the cytoplasm can for long live independently of each other, but the entire cell can perform all the fundamental vital processes. It transforms food into its own living material, it grows and divides, it is capable of responding to many kinds of stimuli. But while the parts of a cell are not capable of independent existence they may be differentiated to perform different functions.

Panmerism.—Not only is the cell as a whole capable of assimilation, growth and division, but every visible part of the cell has this power. The nucleus builds foreign substances into its own substance, and after it has grown to a certain size it divides into two; the cytoplasm does the same, and this process of assimilation, growth and division occurs in many parts of the nucleus and cytoplasm, such as the chromosomes, chromomeres, centrosomes, etc. In all cases cells come from cells, nuclei from nuclei, chromosomes from chromosomes, centrosomes from centrosomes, etc.

Indeed, the manner in which all living matter grows indicates that every minute particle of protoplasm has this power of taking in food substance and of dividing into two particles when it has grown to maximum size; this is known as panmerism. Presumably this power of assimilation, growth and division is possessed by particles of protoplasm which are invisible with the highest powers of our microscopes, though it is probable that these particles are much larger than the largest molecules known to chemistry. The smallest particle which can be seen with the most powerful microscope in ordinary light is about 250 $\mu\mu$ (millionths of a millimeter) in diameter. The largest molecules are probably about 10 $\mu\mu$ in diameter. Between these molecules and the just visible particles of protoplasm there may be other units of organization. These hypothetical particles of protoplasm have been supposed by many authors to be the ultimate units of assimilation, growth and division, and in so far as these units are supposed to be the differential causes of hereditary characters, they are known as inheritance units.

Inheritance Units.—It is assumed in practically all theories of heredity that the "inheritance material," or the germinal protoplasm, is composed of ultra-microscopical inheritance units which have the power of individual growth and division and which are capable of undergoing many combinations and dissociations during the course of development, by which combinations and dissociations they are transformed into the structures of the adult. Various names have been given to these units by different authors; they are the "physiological units" of Herbert Spencer, the "gemmules" of Darwin, the "plastidules" of Elsberg and Haeckel, the "pangenes" of de Vries, the "plasomes" of Wiesner, the "idioblasts" of Hertwig, the "biophores" and "determinants" of Weismann.

With the publication of Weismann's work on the germplasm in 1892 speculation with regard to these ultra-microscopic units of life and of heredity reached a climax and began to decline, owing to the highly speculative character of the evidence as to the existence, nature and activities of such units. But with the rediscovery of Mendel's principles of heredity the necessity of assuming the existence of inheritance units of some kind once more became evident, and, without being able to define just what such units are or just how they behave, modern students of heredity assume their existence. They are now called *determiners* or *factors* or *genes,* and they are usually thought of as units in the germ cells which condition the characters of the developed organism, and which are in a measure independent of one another;

though of course neither they nor any other parts of a cell are really independent in the sense that they can exist apart from one another. They are to be thought of as analogous to chemical radicals which are never independent but exist only in combination with other chemical elements in the form of molecules, and yet preserve their identity in many different combinations.

It is certain that Mendelian factors are not to be regarded as gemmules or the germs of particular characters. There is not a separate factor for every character, and factors are not "representatives" or "carriers" of characters. They are the differential causes of particular characters just as in the compounds H_2SO_4 and K_2SO_4 the hydrogen and potassium atoms are the differential causes of the properties manifested by these two substances.

Location of Inheritance Units.—If there are inheritance units, such as determiners or genes, as practically all students of heredity maintain, they must be contained in the germ cells, and it becomes one of the fundamental problems of biology to find out where and what these units are. There are many evidences that these genes are located in the chromatin of the nucleus, that they are arranged in a linear series when the chromatin takes the form of threads, or chromosomes, preparatory to cell division, that in the division of each chromosome every gene which it contains is also divided and that daughter chromosomes and daughter genes are distributed equally to the daughter cells at every typical cell division (Figs. 6, 7, 8). For nearly fifty years this complex process of nuclear division, known as *mitosis* or *karyokinesis,* has been recognized as a mechanism for the equal distribution of the chromosomes to the daughter cells, and for nearly that length of time it has been suggested that the inheritance material or germplasm was located in the chromosomes, but only within recent years has critical experimental evidence been obtained that inheritance units occupy definite positions in these chromosomes. With this advance in our knowledge, which we owe chiefly to Morgan and his associates, it may be said that an important part, at least, of the "mechanism of heredity" has been discovered.

It must be said however that there are biologists who still refuse to believe that heredity is associated with any particular cell substance, while many others who would grant this are not yet ready to admit that there are particular units or genes which are concerned in the production of particular characters. However anyone who will examine at first hand the evidences in favor of this cannot fail to be impressed with its importance, and no one has proposed any other hypothesis that is at all satisfactory. But whether we assume the existence of these units or not we know that the germ cells are exceedingly complex, that they contain many visible units such as chromosomes, chromomeres, plastosomes and microsomes, and that with every great improvement in the microscope and in microscopical technique other structures are made visible which were invisible before, and whether the particular hypothetical units just named are invisible or not seems to be a matter of no great importance, seeing that, so far as the analysis of the microscope is able to go, there are in all protoplasm differentiated units which are combined into a system; in short, there is organization.

6. *Heredity and Development.*—The germ cells are individual organisms and after the fertilization of the egg the new individual thus formed remains distinct from every other one. Furthermore, from its earliest to its latest stage of development it is one and the same organism; the egg is not one being and the embryo another and the adult a third, but the egg of a human being is a human being in the one-celled stage of development, and the characteristics of the adult develop out of the egg and are not in some mysterious way grafted upon it or transmitted to it.

Parents do not transmit their characters to their offspring, but their germ cells in the course of long development give rise to adult characters similar to those of the parents. The thing which persists more or less completely from generation to generation is the organization of the germ cells which differentiate in similar ways in successive generations if the extrinsic factors of development remain similar.

Definitions.—In short, *heredity may be defined as the continuity from generation to generation of certain elements of germinal organization. Heritage is the sum of all those qualities which are determined or caused by this germinal organization. Development is progressive and coordinated differentiation of the oosperm, under the joint influence of heredity and environment, by which it is transformed into the adult organization. Differentiation is the formation and localization of many different kinds of substances out of the germinal substance, of many different structures and functions out of the relatively simple structures and functions of the oosperm.*

This germinal organization influences not merely adult characters but also the characters of every stage from the egg to the adult condition. For every inherited character, whether embryonic or adult, there is some germinal basis. In the last analysis the causes of heredity and development are problems of cell structures and functions, problems of the formation of particular kinds of germ cells, of the fusion of these cells in fertilization, and of the subsequent formation of the various types of somatic cells from the fertilized egg cell.

B. THE GERM CELLS

Observations and experiments on developed animals and plants have furnished us with a knowledge of the finished products of inheritance, but the actual stages and causes of inheritance, the real mechanisms of heredity, are to be found only in a study of the germ cells and their development. Although many phenomena of inheritance have been discovered in the absence of any definite knowledge of the mechanism of heredity, a scientific explanation of these phenomena must wait upon the knowledge of their causes. In the absence of such knowledge it has been necessary to formulate theories of heredity to account for the facts, but these theories are only temporary scaffolding to bridge the gaps in our knowledge, and if we knew all that could be known about the germ cells and their development we should have little need for theories. In the first chapter we looked at the germ cells and their development from the outside, as it were; let us now look inside these cells and study their minuter structures and functions.

Only a beginning has been made in this minute study of the germ cells and of their transformation into the developed animal, and it seems probable that it may engage the attention of many future generations of biologists, but nevertheless we have come far since that day in 1875 when Oscar Hertwig first saw the approach and union of the egg and sperm nuclei within the fertilized egg. Indeed so rapid has been the advance of knowledge in this field that many of the pioneers in this work are still active in research.

1. *Fertilization.* a. *Stimulus to Development.*—The development of the individual may be said to begin with the fertilization of the egg, though it is evident that both egg and sperm must have had a more remote beginning, and that they also have undergone a process of development by which their peculiar characteristics of structure and function have arisen,—a subject to which we shall return later. But the developmental processes which lead to the formation of fully developed ova and spermatozoa come to a full stop before fertilization and they do not usually begin again until a spermatozoon has entered an ovum, or until the latter has been stimulated by some other outside means.

Parthenogenesis.—In some animals and plants, eggs may develop regularly without fertilization, the stimulus to development being supplied by certain external or internal conditions; in other cases, as Loeb discovered, eggs which would never develop if left to themselves may be experimentally stimulated by physical or chemical changes in the environment, so that they undergo reguar development. The development of an egg without previous fertilization is known as parthenogenesis or virgin reproduction;

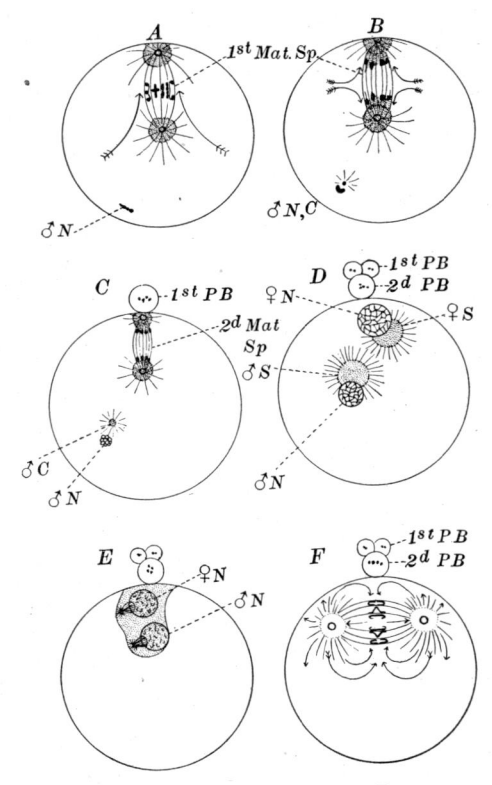

Fig. 42. Diagrams of the Maturation and Fertilization of the Egg of a Mollusk (*Crepidula*). *A, B,* First maturation division (*1st Mat. Sp.*). *C,* Second maturation division (*2d Mat. Sp.*) and first polar body (*1st PB*) resulting from first division. ♂ *N*, Sperm nucleus. ♂ *C*, Sperm centrosome. *D,* Approach of sperm nucleus (♂ *N*) and sphere (♂ *S*) to egg nucleus (♀ *N*) and sphere (♀ *S*); the second polar body (*2d PB*) has been formed and the first has divided (*1st PB*). *E,* Meeting of egg and sperm nuclei and origin of cleavage centrosomes. *F,* First cleavage of egg showing direction of currents in the cell.

if it occurs in nature it is natural parthenogenesis, if in experiments it is artificial parthenogenesis. Natural parthenogenesis is relatively rare and in the vast majority of animals and plants the egg does not begin to develop until a spermatozoon has entered it.

b. *Union of Germplasms.*—But the spermatozoon not only stimulates the egg to develop, as environmental conditions may also do, but it carries into the egg living substances which are of great significance in heredity. Usually only the head of the spermatozoon enters the egg (Fig. 4) and this consists almost entirely of nuclear chromatin (Fig. 4 *D-H*, 42 *A-B*); when the egg has matured and is ready to be fertilized its nucleus also consists of a small mass of chromatin (Fig. 43 *A*). Both of these condensed chromatic nuclei then grow in size and become less chromatic by absorbing from the egg a substance which is not easily stained by dyes and hence is called achromatin (Figs. 4 *I-L*, 42 *D-E*). The chromatin then appears to become scattered through each nucleus in the form of granules or threads which are embedded in the achromatin; this is the condition of a typical "resting" nucleus. It is evident however that these chromatin granules are not scattered broadcast throughout the nucleus, since at the next mitosis they come together into particular chromosomes similar in every way to the chromosomes of the previous mitosis. Probably the chromosomes preserve their identity from one division to the next either in the form of chromosomal vesicles (Fig. 8, p. 20) or as strings of granules. The spermatozoon also brings into the egg a centrosome or division center, around which an aster appears consisting of radiating lines in the protoplasm of the egg (Fig. 4 *F-I*, Fig. 42 *B-E*, Fig. 43 *A-D*).

The moment that the spermatozoon touches the surface of the egg the latter throws out at the point touched a prominence, or reception cone (Fig. 4 *A-E*), and as soon as the head of the sperm has entered this cone some of the superficial protoplasm of the egg flows to this point and then turns into the interior of the egg in a kind of vortex current. Probably as a result of this

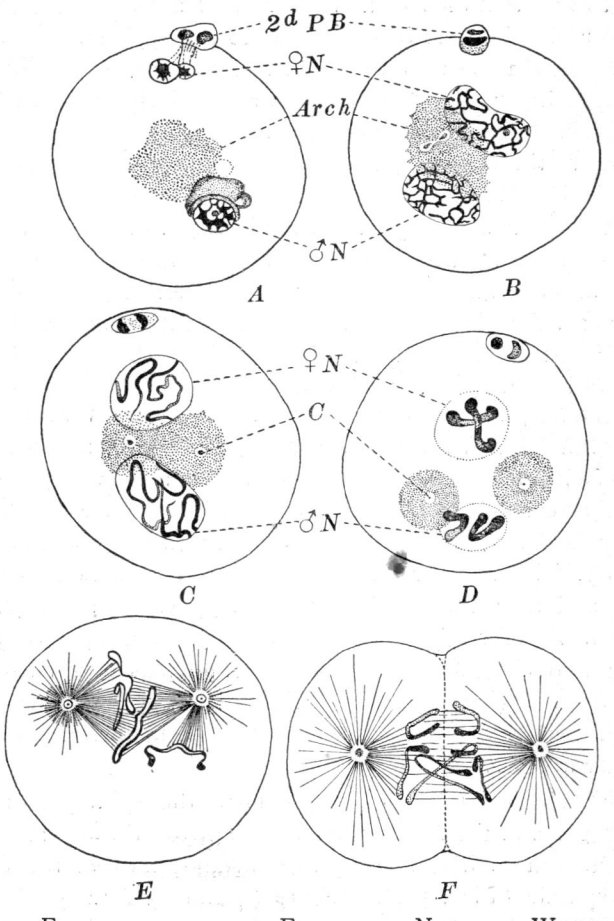

FIG. 43. FERTILIZATION OF THE EGG OF THE NEMATODE WORM *Ascaris megalocephala.* ♀N, Egg nucleus. ♂N, Sperm nucleus. *Arch*, Archiplasm. C, Centrosome. A, B, Approach of germ nuclei. C, D, Formation of two chromosomes in each germ nucleus. E, F, Stages in the division of the chromosomes which are split in E and are separating in F; only three of the four chromosome pairs are shown in F. (From Wilson after Boveri.)

current the sperm nucleus and centrosome are carried deeper into the egg and finally are brought near to the egg nucleus (Fig. 42 *D* and *E*). In the movements of egg and sperm nuclei toward each other it is probable that they are passively carried about by currents in the cytoplasm; the entrance of the sperm serves as a stimulus to the egg cytoplasm which moves according to its pre-established organization.

2. *Cleavage and Differentiation.*—When the sperm nucleus has come close to the egg nucleus the sperm centrosome usually divides into two minute granules, the daughter centrosomes, which move apart forming a spindle with the centrosomes at its poles and with astral radiations running out from these into the cytoplasm (Figs. 4, 42 *F*, 43 *B-E*).

Egg and Sperm Chromosomes.—At the same time the chromatin granules and threads in the egg and sperm nuclei take the form of chromosomes, and at this stage it is sometimes possible to see that each chromosome is composed of a series of granules, like beads on a string; these granules are the chromomeres (Fig. 8). The number of chromosomes is constant for every species and race, though the number may vary in different species. In the thread worm, *Ascaris megalocephala,* there are usually two chromosomes in the egg nucleus and two in the sperm nucleus (Fig. 43 *D*). In the gastropod, *Crepidula* (Figs. 42, 45), there are about thirty chromosomes in each germ nucleus and sixty in the two.

Distribution of Chromosomes.—Then the spindle and asters grow larger and the nuclear membrane grows thinner and finally disappears altogether, leaving the chromosomes in the equator of the spindle (Figs. 5 *A*, 6 *F*, 42 *F*, and 43 *F*). Each of the chromosomes then splits lengthwise into two equal parts, and in the splitting of the chromosomes it is sometimes possible to see that each bead-like chromosome divides through its middle (Fig. 8 *D*). The daughter chromosomes then separate and move to opposite poles of the spindle, where they form the daughter nuclei, and at

THE CELLULAR BASIS

the same time the cell body begins to divide by a constriction which pinches the cell in two in the plane which passes through the equator of the spindle (Figs. 5, 7, 43 *F*, 45 *B*). Finally the chromosomes grow in size by the absorption of achromatin from the cell body forming the chromosomal vesicles in which the chromatin takes

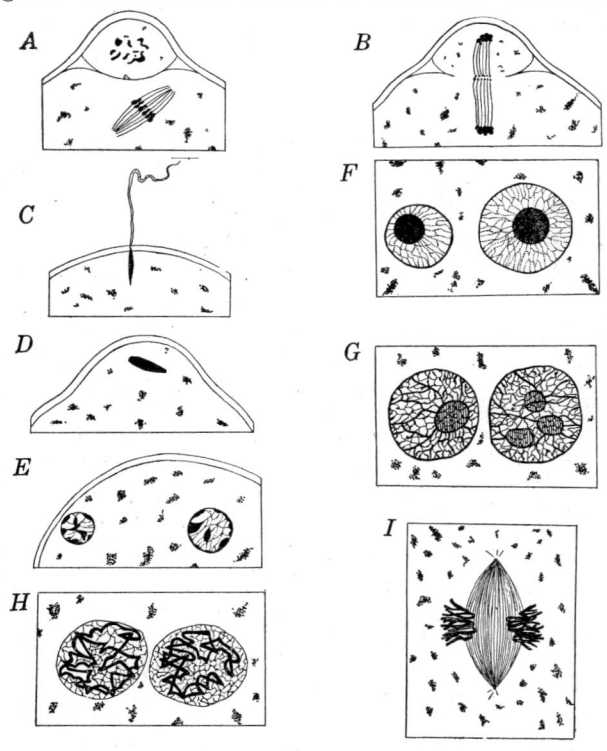

FIG. 44. MATURATION AND FERTILIZATION OF THE EGG OF THE MOUSE. *A*, First polar body and second maturation spindle. *B*, second polar body and maturation spindle. *C*, Entrance of the spermatozoon into the egg. *D-G*, Successive stages in the approach of egg and sperm nuclei. *H*, formation of chromosomes in each germ nucleus. *I*, First cleavage spindle showing chromosomes from egg and sperm on opposite sides of spindle. (After Sobotta.)

the form of threads and granules, the chromosomal vesicles unite to form the daughter nuclei and these nuclei come back to a "resting" stage similar to that with which the division began, thus completing the "division cycle" of the cell (Fig. 8).

Identity of Chromosomes.—During the whole division cycle it is possible in a few instances to distinguish the chromosomes of the egg from those of the sperm, and in every instance where this can be done it is perfectly clear that these chromosomes do not fuse together nor lose their identity, but that every chromosome splits lengthwise and its halves separate and go into the two daughter cells where they form the daughter nuclei. Each of these cells therefore receives half of its chromosomes from the egg and half from the sperm. Even in cases where the individual chromosomes are lost to view in the daughter nuclei those nuclei are sometimes clearly double, one-half of each having come from the egg chromosomes and the other half from the sperm chromosomes (Fig. 45).

At every subsequent cleavage of the egg the chromosomes divide in exactly the same way as has been described for the first cleavage. Every cell of the developing animal receives one-half of its chromosomes from the egg and the other half from the sperm, and if the chromosomes of the egg differ in shape or in size from those of the sperm, as is sometimes the case when different races or species are crossed, these two groups of chromosomes may still be distinguished at advanced stages of development. Where the egg and sperm chromosomes are not thus distinguishable it may still be possible to recognize the half of the nucleus which comes from the egg and the half which comes from the sperm even up to an advanced stage of the cleavage (Fig. 45).

Distribution of Cytoplasm.—At the same time that the maternal and paternal chromosomes are being distributed with such precise equality to all the cells of the developing organism the different substances in the cell body outside of the nucleus may be distributed very unequally to the cleavage cells. The move-

THE CELLULAR BASIS 141

nents of the cytoplasm of the egg, which began with the flowing of the surface layer to the point of entrance of the sperm, and

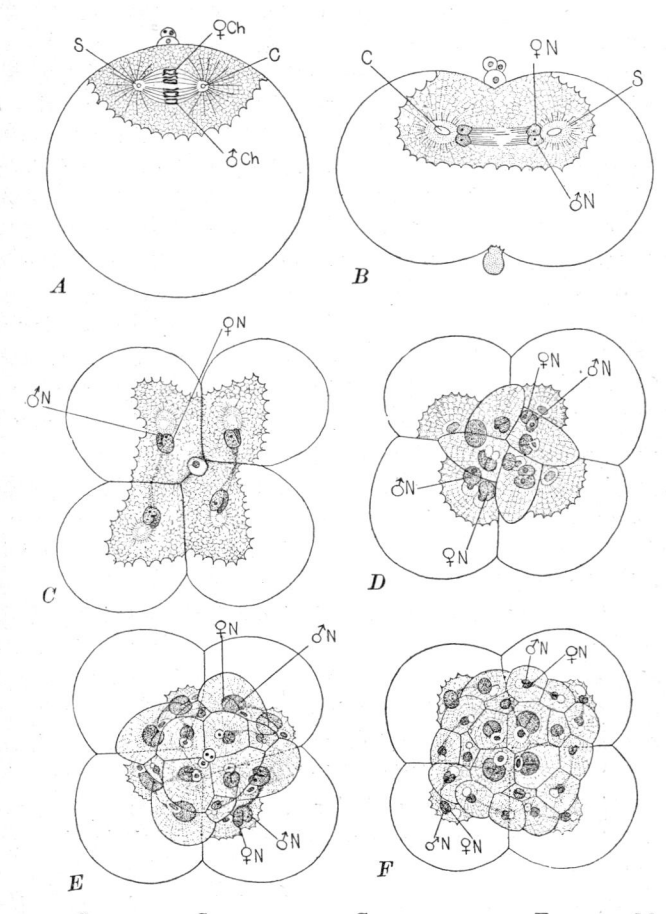

Fig. 45. Successive Stages in the Cleavage of the Egg of a Mollusk (*Crepidula*), showing the separateness of the male and female chromosomes (♂ *ch*, ♀ *ch*) and of the male and female halves of each nucleus (♂ *N*, ♀ *N*).

which continue during every cleavage of the egg, lead to the segregation of different kinds of plasms in different parts of the egg and to the unequal distribution of these substances to different cells (Figs. 10, 46, 47).

One of the most striking cases of this is found in the ascidian *Styela,* in which there are four or five substances in the egg which differ in color, so that their distribution to different regions of the egg and to different cleavage cells may be easily followed and even photographed, while in the living condition. The peripheral layer of protoplasm is yellow and it gathers at the lower pole of the egg, where the sperm enters, forming a yellow cap (Fig. 46, 1, *pl.*). This yellow substance then moves, following the sperm nucleus, up to the equator of the egg on the posterior side and there forms a yellow crescent extending around the posterior side of the egg just below the equator (Fig. 46, 2-4). On the anterior side of the egg a gray crescent is formed in a somewhat similar manner and at the lower pole between these two crescents is a slate blue substance, while at the upper pole is an area of colorless protoplasm. The yellow crescent goes into cleavage cells which become muscles and mesoderm, the gray crescent into cells which become nervous system and notochord, the slate blue substance into endoderm cells and the colorless substance into ectoderm cells. (Figs. 47 and 48; see also Figs. 10 and 11.)

Localization of Substances.—Thus within a few minutes after the fertilization of the egg, and before or immediately after the first cleavage, the anterior and posterior, dorsal and ventral, right and left poles are clearly distinguishable, and the substances which will give rise to ectoderm, endoderm, mesoderm, muscles, notochord and nervous system are plainly visible in their characteristic positions.

At the first cleavage of the egg each of these substances is divided into right and left halves (Fig. 46, 5). The second cleavage cuts off two anterior cells containing the gray crescent from two posterior ones containing the yellow crescent (Fig. 46, 6 and Fig

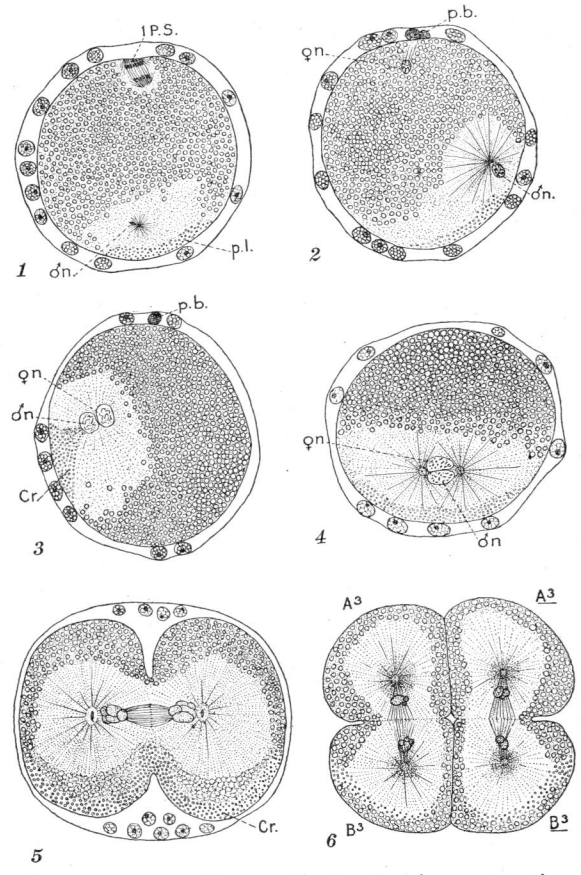

FIG. 46. SECTIONS OF THE EGG OF *Styela*, showing maturation, fertilization and early cleavage. *1 P.S.*, First polar spindle. *p.b.*, Polar bodies; ♂ *N*, sperm nucleus, ♀ *N*, egg nucleus. *p.l.*, peripheral layer of yellow protoplasm. *Cr.*, Crescent of yellow protoplasm. A_3, A_3, Anterior cells, B_3, B_3 Posterior cells of the 4-cell stage. In 1 the sperm nucleus and centrosome are at the lower pole near the point of entrance; in 2 and 3 they have moved up to the equator on the posterior side of the egg; in 4 the egg and sperm nuclei have come together and the sperm centrosome has divided and formed the cleavage spindle; in 5 the egg is dividing into right and left halves; in 6 it is dividing into anterior and posterior quarters. A membrane, the chorion, surrounds the egg inclosing a number of small "test" cells.

47, 1). The third cleavage separates the colorless protoplasm in the upper hemisphere from the slate blue in the lower (Fig. 47, 2). And at every successive cleavage the cytoplasmic substances are segregated and isolated in particular cells, and in this way the cytoplasm of the different cells comes to be unlike (Figs. 47 and 48). When once partition walls have been formed between cells the substances in the different cells are permanently separated so that they can no longer commingle.

What is true of *Styela* in this regard is equally true of many other ascidians, as well as of *Amphioxus* and of the frog (Figs. 9, 10, 11), though the segregation of substances and the differentiation of cells are not so evident in the last named animals because these substances are not so strikingly colored. Indeed the segregation and isolation of different protoplasmic substances in different cleavage cells occurs during the cleavage of the egg in all animals, though such differentiations are much more marked in some cases than in others.

This same type of cell division, with equal division of the chromosomes and more or less unequal division of the cell body, continues long after the cleavage stages, indeed throughout the entire period of embryonic development. Sometimes the division of the cell body is equal, the daughter cells being alike; sometimes it is unequal or differential, but always the division of the chromosomes is equal and non-differential. When once the various tissues have been differentiated the further divisions in these tissue cells are usually non-differential even in the case of the cell bodies.

Significance of Cleavage.—There can be no doubt that this remarkably complicated process of cell division has some deep significance; why should a nucleus divide in this peculiarly indirect manner instead of merely pinching in two, as was once supposed to be the rule? What is the relation of cell division to embryonic differentiation? In this process of mitosis, or indirect cell division, two important things take place: (1) Each chronosome, chromomere and centrosome is divided exactly into two equal parts so that each daughter cell receives precisely similar halves of these structures. (2) Accompanying the formation of radiations, which go out from the centrosomes into the cell body, diffusion currents are set up in the cytoplasm which lead to the localization of different parts of the cytoplasm in definite regions of the cells, and this cytoplasmic localization is sometimes of such a sort that one of the daughter cells may contain one kind of cytoplasm and the other another kind.

Cytoplasm Differentiates, Nuclei Do Not.—Thus while mitosis brings about a scrupulously equal division of the elements of the nucleus, it may lead to a very unequal and dissimilar division of the cytoplasm. In this is found the significance of mitosis, and it suggests at once that the nucleus contains non-differentiating material, viz., the idioplasm or germplasm, which is characteristic of the race and is carried on from cell to cell and from generation to generation; whereas the cell body contains the differentiating substance, the personal plasm or somatoplasm, which gives rise to all the differentiations of cells, tissues and organs in the course of ontogeny.

Weismann supposed that the mitotic division of the chromosomes during development was of a differential character, the daughter chromosomes differing from each other at every differential division in some constant and characteristic way, and that these differentiations of the chromosomes produced the characteristic differentiations of the cytoplasm which occur during development. But there is not a particle of evidence that the ordinary division of chromosomes is differential; on the contrary, there is the most complete evidence that their division is remarkably equal both quantitatively and qualitatively. If daughter chromosomes and nuclei ever become unlike, as they sometimes do, this unlikeness occurs long after division and is probably the result of the action of different kinds of cytoplasm upon the nuclei, as is true for example, in the differentiation of the chromosomes in the somatic cells as contrasted with the germ cells of *Ascaris* (Fig. 49). In this case Boveri has shown that the nuclei and chromosomes of germ cells and of somatic cells are at first

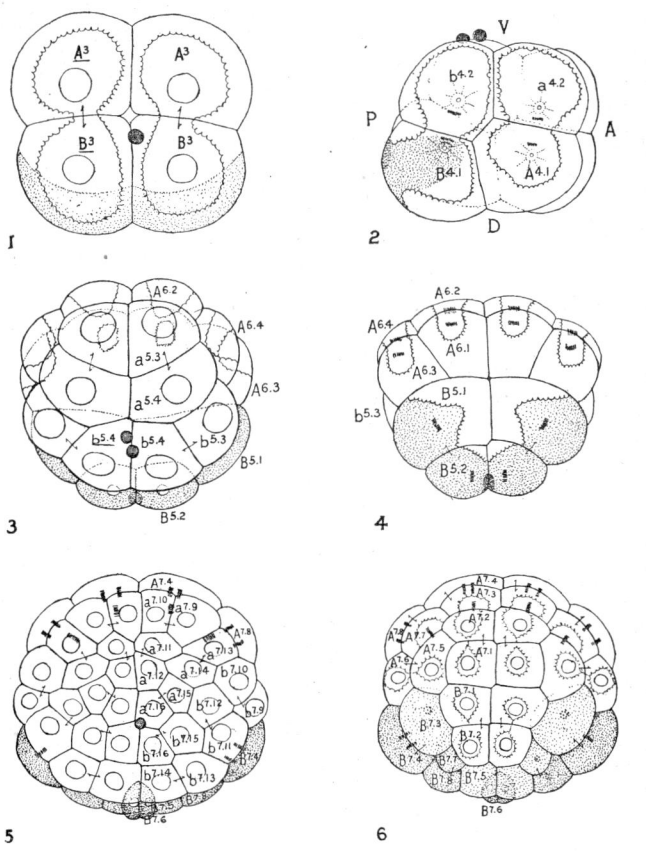

FIG. 47. CLEAVAGE OF THE EGG OF *Styela*, showing distribution of the yellow protoplasm (stippled) and of the clear protoplasm (designated by lower-case letters) and of the gray protoplasm (designated by capital letters) to the various cells, each of which bears a definite letter and number and gives rise to a definite part of the embryo.

THE CELLULAR BASIS

alike whereas the cytoplasm in these cells is unlike; later the nuclei and chromosomes of these two kinds of cells become unlike,

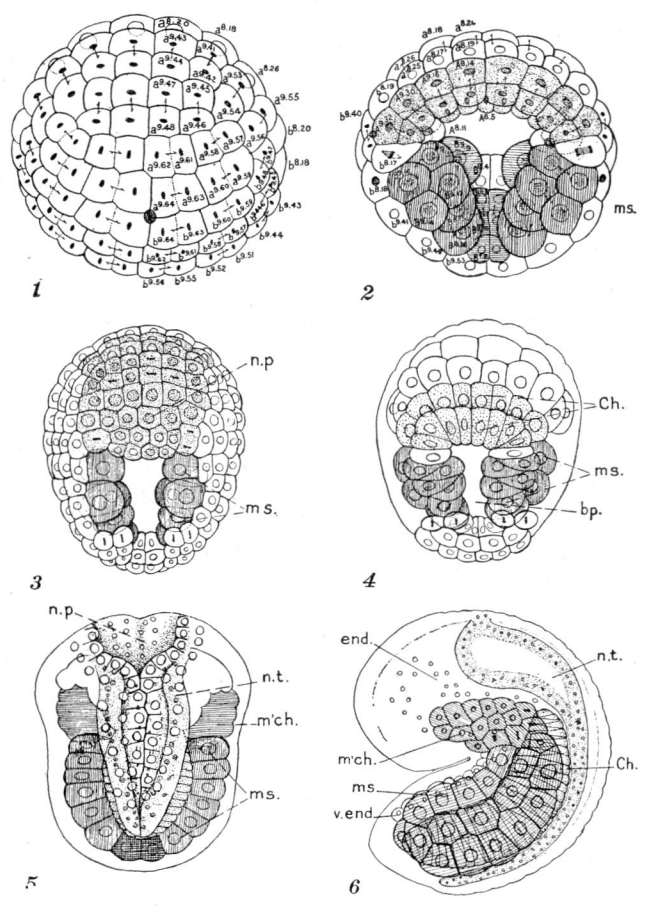

FIG. 48. GASTRULA AND LARVA OF *Styela*, showing the cell lineage of various organs, and the distribution of the different kinds of protoplasm to these organs. Muscle cells are shaded by vertical lines, mesenchyme by horizontal lines, nervous system and chorda by stipples.

owing probably to the peculiarities of the cytoplasm of these cells. These nuclear differentiations are caused by differential divisions of the cytoplasm and not of the nucleus. But while the chromosomes themselves divide equally, other portions of the nucleus may not do so. Nuclear achromatin and oxychromatin, like the cytoplasm, may divide unequally and differentially, and this is probably a prime factor in development.

On the other hand, the differential division of the cytoplasm is a regular and characteristic feature of ontogeny; indeed, the segregation and isolation of different kinds of cytoplasm in different cells is one of the most important functions of cell division during development. *Thus we find in the division apparatus of the cell a mechanism for the preservation in unaltered form of the species plasm or germplasm of the nucleus, and for the progressive differentiation of the personal plasm or somatoplasm of the cell body.*

3. *The Origin of the Sex Cells.*—The sex cells are among the latest of all cells of a developing animal to reach maturity, and yet they may be among the earliest to make their appearance. Every sex cell, like every other type of cell, is a lineal descendant of the fertilized egg (Fig. 40), but the period at which the sex cells become visibly different from other cells varies from the first cleavage of the egg in some species to a relatively advanced stage of development in others.

(a) *The Division Period. Oogonia and Spermatogonia.*—When the primitive sex cells are first distinguishable they differ from other cells only in the fact that they are less differentiated; they have relatively larger nuclei and smaller cell bodies, a condition which is indicative of little differentiation of the cell body since the products of differentiation such as fibres, secretions, etc., swell the size of the cell body but do not contribute to the growth of the nucleus. These primitive sex cells or gonia divide repeatedly, but the oogonia grow more rapidly and divide less frequently than the spermatogonia. As a result of this difference in the

THE CELLULAR BASIS 149

rate of growth and division the spermatogonia become much smaller and immensely more numerous than the oogonia. This period in the genesis of the sex cells is known as the division period (Fig. 40).

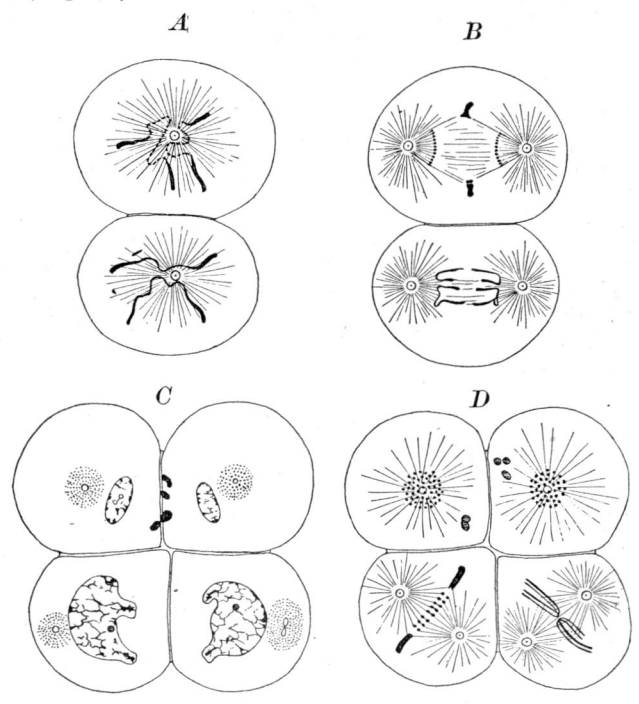

FIG. 49. DIFFERENTIATION OF GERM CELLS AND SOMATIC CELLS IN THE EGG OF *Ascaris*. *A* and *B*, Second cleavage division showing that the chromosomes remain entire in the lower cell, which is in the line of descent of the sex cells ("germ track"), but that they throw off their ends and break up into small granules in the upper cells, which become somatic cells. *C*, 4-cell stage, the nuclei in the upper (somatic) cells being small and the ends of the chromosomes remaining as chromatic masses in the cell body outside of the nuclei, while the nuclei in the lower cells are much larger and contain all of their chromatin. *D*, Third nuclear division, showing the somatic differentiation of the chromosomes in all the cells except the lower right one, which alone is in the germ track and will ultimately give rise to sex cells. (After Boveri.)

150   HEREDITY AND ENVIRONMENT

(b) *The Growth Period. Oocytes and Spermatocytes.*—This period of rapid cell division is followed by a period of growth without division during which the developing sex cells are called primary oocytes or spermatocytes. This growth period may be very long in the case of the oocytes, lasting, for example, in the human female from the time of birth to the end of the reproductive period; during this long time the oocytes in the ovary rarely if ever divide, probably there are as many of them at birth as at any later time; during this period of growth the ovarian egg becomes relatively large,—in some animals, *e.g.,* birds, the largest of all

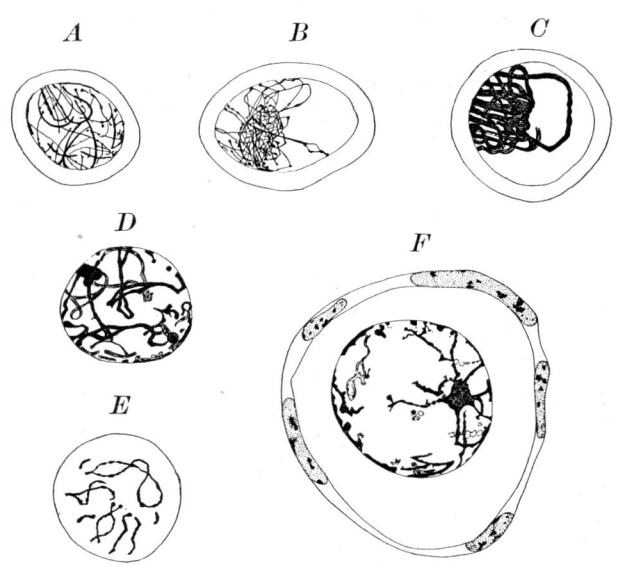

FIG. 50. DIFFERENT STAGES IN THE DEVELOPMENT OF THE EGG OF THE RABBIT. *A,* At the beginning of the growth period showing slender chromatic threads in the nucleus. *B,* Later stage in which these threads ball up and parallel threads conjugate forming the shorter, thicker thread shown in *C* and *D*. *E,* Later stage showing pairs of chromosomes due to conjugation. *F,* Later oocyte, surrounded by follicle cells, in which the distinctness of the chromosomes is temporarily lost. (After Winiwarter.)

cells. The growth period of a spermatocyte lasts for a briefer time than does that of an oocyte so that the former remains relatively small (Fig. 40).

Synapsis.—All of the cell divisions which take place during the division period are of the usual kind, in which every chromosome splits lengthwise into two and the two halves then separate and move to opposite poles of the spindle where they swell up into chromosomal vesicles and form the daughter nuclei, as is shown in Figs. 5-8. But during the growth period of the oocytes and spermatocytes the chromosomes form a closely wound coil of long chromatin threads (Fig. 50 *A* and *B*), and when these threads uncoil later it is seen that the chromosomes have united in pairs (Figs. 50 *D* and *E*, 51 *A-C*, 52 *B*, 53 *B*) this process is known as synapsis, or the conjugation of the chromosomes, and there is evidence that one member of each synaptic pair is derived from the father, and the other from the mother. The union of these chromosomes is a temporary one and is not so close that they lose their identity. By this union of the chromosomes into pairs the number of separate chromosomes is reduced to half the normal number; if there are usually 4 chromosomes, as in *Ascaris,* they are reduced to 2 pairs; if 48 chromosomes, as in man, there are 24 of these pairs.

Conjugation of Homologous Chromosomes.—In the conjugation of the chromosomes it is plain that, generally speaking, those chromosomes unite which are similar in shape and size; big chromosomes unite with big ones, little ones with little ones, and those of peculiar shape with others of similar shape (Figs. 51, 52 *B*, 53 *B*, 55, 59). It is probable that the two members of a pair of conjugating chromosomes are homologous not merely in shape and size but also in function, though this homology does not amount to identity. These homologous chromosomes may be compared to the fingers of the two hands; each digit differs from every other one, but the thumb, index finger and other fingers of the right hand are homologous, but not identical, with the correspond-

152   HEREDITY AND ENVIRONMENT

ing digits of the left hand, and the conjugation of homologous chromosomes may be compared to the placing together of the two hands so that homologous digits come together.

In some instances it can be proved that one member of each conjugating pair of chromosomes comes from one parent and the other from the other parent, and it is probable that this is always

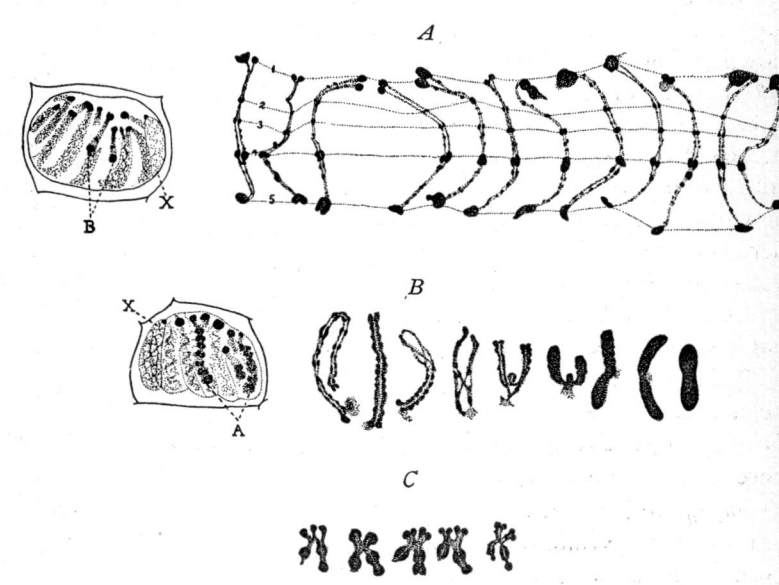

FIG. 51. SYNAPSIS (CONJUGATION) OF CHROMOSOMES in the grasshopper *Phrynotettix*. *A*. At the left, Telophase of Spermatogonium showing chromosomes in nucleus, among them chromosome X and a pair B. At the right, are 12 pairs of B chromosomes in synapsis, each pair from a different animal, and one member of each pair from the father, the other from the mother. Homologous chromomeres (granules 1, 2, 3, 4, 5) are shown in each chromosome. *B* similar stages in chromosome pair *A*. Some of the chromosomes in the middle of the row show a "secondary" longitudinal split and a "crossing over" (?) of the halves. *C*. "Tetrads" (conjugated chromosomes) of pair *B* formed by shortening and thickening of the chromosome pairs and by the appearance of the "secondary" split. (After Wenrich.)

THE CELLULAR BASIS

true. In every cell of every individual which has developed from a fertilized egg there are two full sets of chromosomes, one of which came from the sperm and the other from the egg; but when this individual in its turn produces germ cells homologous chromosomes of each set unite in pairs, side by side, during the growth period. This again may be compared to the union of the two hands, the right digits, for example, representing the paternal and the left the maternal chromosomes which come together in homologous pairs, corresponding joints lying opposite each other, as corresponding genes in the chromosomes lie opposite each other.

These synaptic pairs are the bivalent chromosomes, and in addition to showing the line of junction by which they are united they frequently show a longitudinal split through the middle of each chromosome and at right angles to the line of junction. It thus happens that these bivalent chromosomes are frequently four-parted and such four-parted chromosomes are known as tetrads (Figs. 51 *C*, 52 *B*, 53 *B*, *C*).

(c) *The Maturation Period.*—Finally at the close of the growth period both oocyte and spermatocyte undergo two peculiar divisions, one following immediately after the other, which are unlike any other cell divisions. These are known as the first and second maturation divisions and they are the last divisions which take place in the formation of the egg and sperm.

Reduction Division.—In one or the other of these two maturation divisions the pairs of conjugated chromosomes separate along the line of junction, one member of each pair going to one pole of the spindle and the other to the other pole, so that in each of the daughter cells thus formed only a single set of chromosomes is present (Figs. 51 *C*, 52 *C*); but since the position of the pairs of chromosomes in the spindle is a matter of chance it rarely happens that all the paternal chromosomes go to one pole and all the maternal ones to the other; thus each of the sex cells comes to contain a complete set of chromosomes, though particular indi-

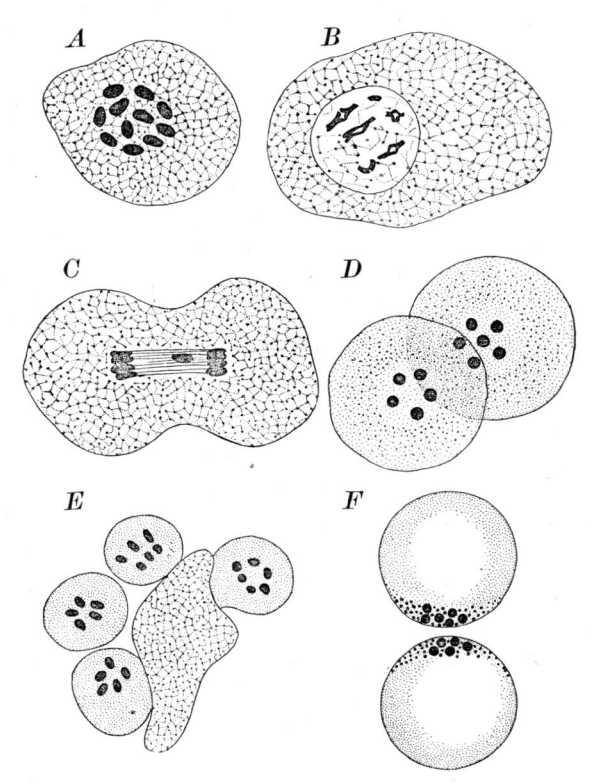

Fig. 52. Spermatogenesis of a Nematode Worm (*Ancyracanthus*). *A*, Chromosomes of sperm mother cell, 11 in number, before their union into pairs. *B*, early stage of first maturation division; 10 of the chromosomes have united into 5 pairs and each of these has split lengthwise; 1 chromosome remains unpaired. *C*, First maturation division after the 5 pairs of chromosomes have pulled apart; the unpaired chromosome is going entire to one pole of the spindle. *D*, Two cells resulting from this division, one containing 5 and the other 6 chromosomes. *E*, Four cells resulting from the division of the two cells like *D*, in which every chromosome has split into two so that two of the cells contain 5 and two contain 6 chromosomes. *F*, two of these cells changing into spermatozoa, one containing 5 and the other 6 chromosomes. (After Mulsow.)

THE CELLULAR BASIS

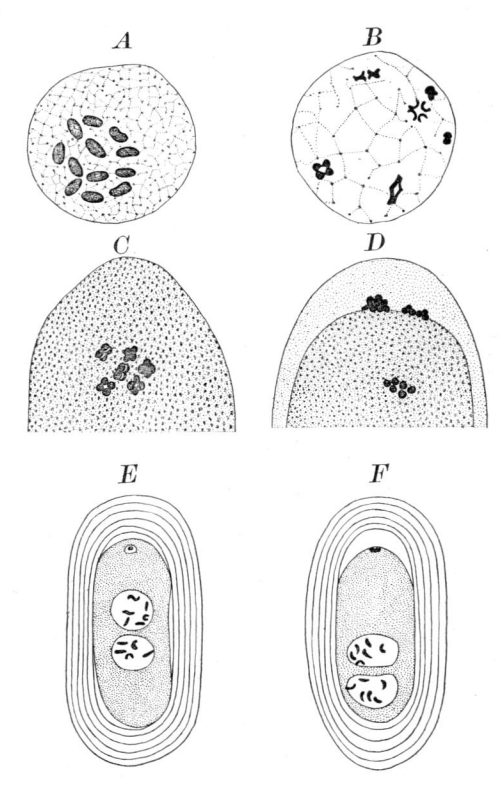

FIG. 53. OOGENESIS OF A NEMATODE WORM (*Ancyracanthus*). *A*, Egg mother cell containing 12 chromosomes before their union into pairs. *B*, Early stage of first maturation division; all the chromosomes have united into 6 pairs, and all but one of these has split in two so that the pairs are really four-parted (tetrads). *C*, The six tetrads in the first maturation division. *D*, Egg containing 6 chromosomes, after both first and second maturation divisions; the eliminated chromosomes are shown as the polar bodies at the margin of the egg. *E* and *F*, Eggs after fertilization; the egg nucleus is above and contains 6 chromosomes, the sperm nucleus is below and contains 5 chromosomes in one case and 6 in the other; in the former case the egg becomes a male with 11 chromosomes, in the latter a female with 12 chromosomes. (After Mulsow.)

vidual chromosomes may have come from the father while others have come from the mother. There is reason to believe that homologous chromosomes show general resemblances but individual differences, and consequently when the members of each pair separate and go into the sex cells these cells differ among themselves because the individual chromosomes in different cells are not the same in hereditary value (Fig. 59).

Again this is comparable to the separation of the digits of the two hands after these have been placed together in homologous pairs, except that all the right digits must go with the right hand, all the left ones with the left hand since they are permanently attached to the hands. But the homologous chromosomes are free and each chromosome of a pair may separate to the right or to the left as the digits might do if they were severed from the hands. Then each digit of a pair could separate either to the right or to the left and it would rarely happen that all the right hand ones would go in one direction and all the left hand ones in the other, but since the members of each pair could separate in two directions and since there are five pairs, the possible number of different combinations after separation would be $(2)^5$ or 32, and yet in each of these combinations the full set of digits from the thumb to the little finger would be present. Finally the fertilization of the egg may be likened to the game of "bean porridge" in which different hands (gametes) each with its full set of digits (chromosomes) are struck together thus making new combinations of digits (chromosomes).

In this way the number of chromosomes in the mature egg or sperm comes to be one-half the number present in other kinds of cells, and when the egg and sperm unite in fertilization the whole number is again restored. The double set of chromosomes is known as the diploid number, the single set as the haploid number, and the maturation division in which this reduction from the double to the single set takes place is the reduction division. It is generally held that this reduction takes place in the first of the

THE CELLULAR BASIS

two maturation divisions (Fig. 52 C, D), and that the second of these divisions is like an ordinary mitosis in that each chromosome splits into two and the halves move apart, such a division being known as an equation division (Fig. 52 E), but it is possible that some chromosome pairs undergo an equation division in the first maturation mitosis and a reduction division in the second, while other chromosome pairs may reverse this order.

It is an interesting fact that long before the reduction of chromosomes had been actually seen Weismann maintained on theoretical grounds that such a reduction must occur, otherwise the number of chromosomes would double in every generation, and he held that this reduction must take place in one of the maturation divisions; this hypothesis of Weismann's is now an established fact.

Mature Egg and Sperm.—As the result of these two maturation divisions four cells are formed from each cell (spermatocyte or oocyte) of the growth period. In the spermatogenesis each of these four cells is transformed into a functional spermatozoon (Figs. 40, 52 F) by the condensation of the nucleus into the sperm head and the outgrowth of the centrosome and cytoplasm to form the tail. In the oogenesis only one of these four cells becomes a functional egg while the other three are small rudimentary eggs which are called polar bodies and which take no further part in development (Figs. 40, 42 C-F). The fertilization of the egg usually takes place coincidently with the formation of the polar bodies, and so we come back once more to the stage of fertilization from which we started (p. 134), thus completing the life cycle.

C. SEX DETERMINATION

In the formation of the sex cells one can distinguish at an early stage differences between the larger oogonia and the smaller and more numerous spermatogonia; this difference is the first visible distinction in the development of the two sexes. In the case of the human embryo this distinction can be made as early as the fifth week, and it is evident that the real causes of this difference must be found at a still earlier period of development.

The cause of sex has been a favorite subject of speculation for thousands of years. Hundreds of hypotheses have been advanced to explain this perennially interesting phenomenon. The causes of sex determination have been ascribed to almost every possible external or internal influence and the world is full of people who think they have discovered by personal experience just how sex is determined. Unfortunately these hypotheses and rules are generally founded upon a few observations of selected cases. Since there are only two sexes the chances are that any hypothesis will be right half the time, and if only one forgets the failure of a rule and remembers the times when it holds good it is possible to believe in the influence of food or temperature or age, or war or peace or education on the relative numbers of the sexes or on almost any other thing. By statistics it has been shown that each of these things influences the sex ratio, and by more extensive statistics it has been proved that they do not.

1. *Chromosomal Determination: XO Type.*—This was the condition regarding the causes of sex determination which prevailed up to the year 1902. Immediately preceding that year it had been found that two kinds of spermatozoa were formed in equal numbers in certain insects; one of these kinds contained a peculiar "accessory" or "odd" chromosome, and the other lacked it. The manner in which these two types of spermatozoa were formed had been carefully worked out by several investigators without any suspicion of the real significance of the facts. It was shown that an uneven number of chromosomes might be present in the spermatogonia of certain insects and that when maternal and paternal chromosomes united in pairs in synapsis one "odd" chromosome was left without a mate (Fig. 52 B). Later, in the reduction division, when the synaptic pairs separated, the odd chromosome went entire into one of the daughter cells, and the spermatozoa formed from this cell contained one

chromosome more than those formed from the other daughter cell (Fig. 52 *C-F*).

Chiefly because these two kinds of spermatozoa occur in equal numbers McClung in 1902 concluded that this accessory chromosome was a sex-determinant. In 1905 Wilson discovered in a number of bugs that while there were two types of spermatozoa, one of which contained and the other lacked the accessory chromosome, there was only one type of egg, since every egg contained the accessory chromosome, and he pointed out that if an egg were fertilized by a sperm containing an accessory, two accessories would be present in the zygote, this being the condition of the female, while if it were fertilized by a sperm without an accessory there would be present in the zygote only the accessory

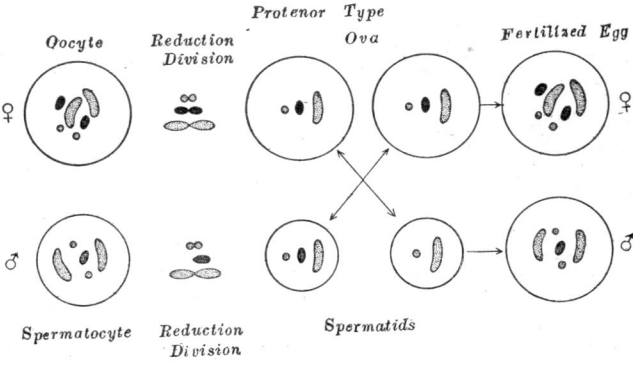

FIG. 54. DIAGRAMS OF SEX DETERMINATION IN THE BUG, *Protenor*. The oocyte contains 6 chromosomes and the spermatocyte 5 chromosomes which are not yet united into synaptic pairs; the "sex" chromosomes are shown in black, two are present in the oocyte, but only one in the spermatocyte. In the reduction division the synaptic pairs separate so that all ova contain 3 chromosomes, one of which is the sex chromosome (black), whereas one-half of the spermatids contain 3 chromosomes and the other half only 2, the sex chromosome being absent. If an egg is fertilized by a sperm without the sex chromosome a male results; if fertilized by a sperm containing the sex chromosome a female results. (After Wilson with modifications.)

derived from the egg (Fig. 53 E and F, Fig. 54). This "accessory" chromosome was therefore called the "sex determining" or merely the "sex" chromosome and was designated by the letter X consequently its double occurrence in the female was indicated by XX; its single occurrence in the male by XO, the O standing for zero or no chromosome.

XY Type.—In other cases Miss Stevens as well as Wilson discovered that two accessory chromosomes, differing in size, might be present in the male whereas in the female they are of equal size (Fig. 55). In such cases two types of spermatozoa are produced in equal numbers, one containing a large and the other a small accessory chromosome, whereas every egg contains one large accessory chromosome. If such an egg is fertilized by a sperm containing a large accessory (the X chromosome) it gives rise to a female with the formula XX, if by a sperm containing

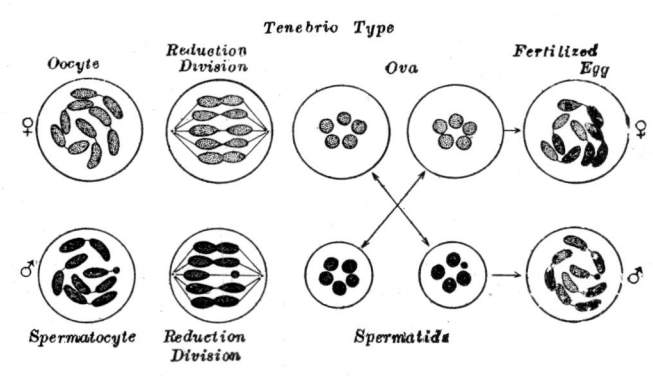

FIG. 55. DIAGRAM OF SEX DETERMINATION IN THE BEETLE, *Tenebrio* showing 5 synaptic pairs of chromosomes (there are actually 10 pairs) in the oocyte the members of each pair are equal in size; in the spermatocyte the members of one pair are unequal. These pairs separate in the reduction division giving rise to two types of spermatozoa and one type of ova; eggs fertilized by one type of sperm give rise to females, those fertilized by the other type give rise to males. (After Stevens with modifications.)

small accessory (the Y chromosome) it gives rise to a male with the formula XY (Fig. 55).

In other animals one may not be able to distinguish separate X or Y chromosomes and yet such structures may be joined to one or two ordinary chromosomes. This is the case in the thread worm, *Ascaris* (Fig. 56), where two such accessory elements are present in the female, each being joined to the end of an ordinary chromosome, whereas in the male only one such element is present. Here also two classes of spermatozoa are found one with and the other without the accessory element, whereas all ova have this element, and in this case also sex is probably determined by the type of spermatozoon which enters the egg (Fig. 56).

Recurring to the comparison of digits and chromosomes, the chromosomal theory of sex determination may be illustrated by assuming that all males have lost a particular digit, say the

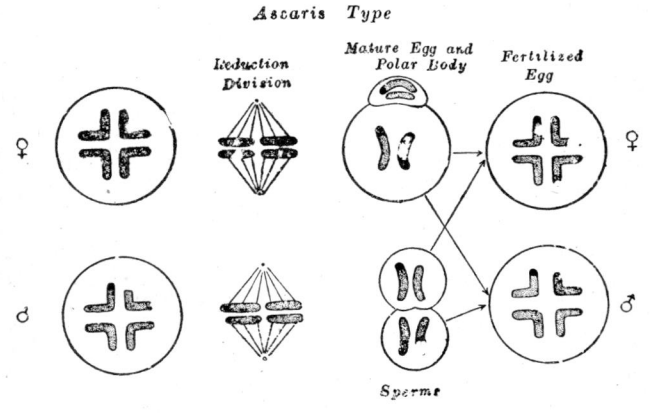

Ascaris Type

FIG. 56. DIAGRAM OF SEX DETERMINATION IN THE THREAD WORM, *Ascaris*. The X chromosomes (black) are here joined to ordinary chromosomes, there being two in the egg mother cell and one in the sperm mother cell. All eggs contain one of these X chromosomes, while half of the spermatozoa have it and half do not. Eggs fertilized by one type of sperm produce females, those fertilized by the other type produce males. (From Wilson.)

thumb, from one hand while all females have the full number on both hands. When the hands (gametes) are struck together as in the game of "bean porridge" there will be an equal number of cases in which a hand with five digits meets one with five (female) and one with four meets one with five (male). In the latter case there will be an "odd" thumb (chromosome) which has no mate.

Sex Determination in Man.—Even in man sex is determined in the same manner, according to several recent investigators. Winiwarter concluded that there are in the spermatogonia of man 47 chromosomes, one of which is the X or accessory chromosome. These unite in synapsis into 23 pairs, leaving the X chromosome unpaired; in the reduction division the pairs separate, while the X chromosome goes entire into one of the daughter cells, which consequently contains $23 + X$ chromosomes, whereas the other daughter cell contains 23 chromosomes. In the female there are probably 48 chromosomes, there being two X chromosomes, one from each parent, and after the reduction divisions every egg contains 24 chromosomes. Winiwarter held that if an egg is fertilized by a sperm containing 24 chromosomes an individual with 48 chromosomes, or a female, is produced; if fertilized by a sperm with 23 chromosomes an individual with 47 chromosomes or a male, results.

It is a curious fact that it has been very difficult to determine the exact number of chromosomes in man. This is probably due to the difficulty of preserving in an unaltered condition the chromosomes of mammals in general, as McClung and his pupils have shown, and also to the peculiar difficulty of obtaining human tissues in a perfectly fresh and normal condition. Thus Guyer and Montgomery found not 47 but about 22 chromosomes in the spermatogonia of man. Since both the latter investigators worked on negroes whereas Winiwarter worked on white men it was suggested by Morgan and Guyer that there may be twice as many chromosomes in the white race as in the black. A

THE CELLULAR BASIS 163

similar condition in which one race has twice as many chromosomes as another race of the same species is found in two races of the thread worm, *Ascaris megalocephala*. However, more recent work has shown that the number of chromosomes in white men and in negroes is the same. Wieman found 24 chromosomes in the spermatogonia (?) of both races and he inferred that in the male there is an XY pair of sex chromosomes, in the female an XX pair. In view of more recent work which proves that there are 48 chromosomes in both sexes of both races it seems probable that Wieman counted the reduced number of chromosomes in the spermatocytes.

The most recent work on the number of chromosomes in man is by Painter, who has studied normal testes of both whites and

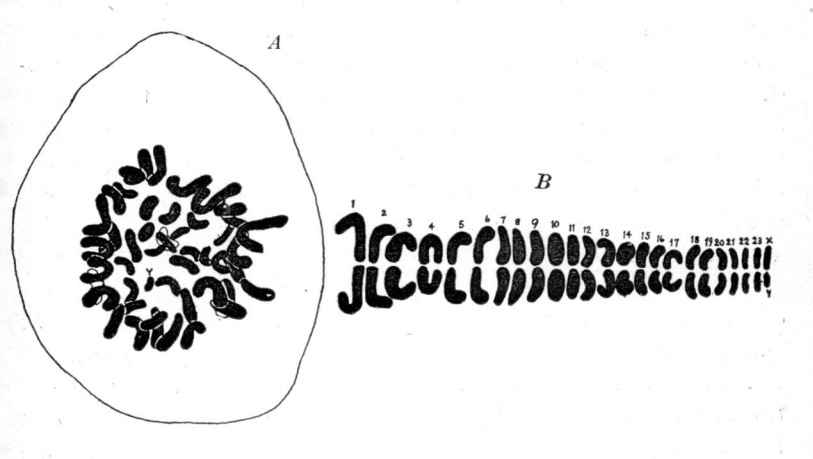

FIG. 57. SPERMATOGNIAL CHROMOSOMES OF NEGRO. *A,* Spermatogonium showing 48 chromosomes, the smallest of which is the sex chromosome Y; the other sex chromosome X, is a small rod-shaped one. As reproduced the figure is magnified about 3000 diameters. From a preparation and drawing by Painter.

B. The same chromosomes spread out and arranged, according to size and shape, in 24 synaptic pairs, the last in the row being the XY pair.

negroes, removed in castration and fixed immediately by the most approved methods. He finds that in both whites and blacks, there are 48 chromosomes in the spermatogonia, or 24 synaptic pairs in the first maturation division; one of these is plainly the XY pair, the X and Y being unequaled in size (Fig. 57). Guyer has also informed me personally that in new and better material he now finds 48 chromosomes in human spermatogonia, one of these being an X and another a Y. These discoveries appear to settle once for all this vexing question and to establish the fact that in man, as well as in many other animals, sex is determined by the chromosomes, the sex chromosomes being XX in the female, and XY in the male.

Sex a Mendelian Character.—Correlations between chromosomes and sex have been observed in more than one hundred species of animals belonging to widely different phyla. In a few classes of animals, particularly Lepidoptera and birds, the evidence while not entirely convincing seems to point to the fact that two types of ova are produced and but one type of spermatozoa; but the general principle that sex is determined by the chance union of male-producing or female-producing gametes is not changed by such cases.

Sex, therefore, appears to be inherited, that is, its factors are present in the germ cells but probably not as particular genes occupying definite loci in a chromosome but rather as a relation of whole chromosomes, such as XX, XY or XO (see p. 170); it is a Mendelian character in which the female is usually homozygous for sex while the male is heterozygous. Consequently in the formation of the gametes every egg cell receives one sex determiner, while only one-half of the spermatozoa receive such a determiner, the other half of them being without it. If then an egg is fertilized by a sperm with one of these determiners a female is produced, if by a sperm without the sex determiner a male results. This is graphically illustrated in Fig. 58 in which X represents the sex-determiner, which is duplex in the female and simplex in the male, and the chance unions of male and female germ cells yields females (*XX*) and males (*XO*) in equal numbers. Of course there is no such thing as a "sex-producing" chromosome, sex being a developed character which is the result of many intrinsic and extrinsic causes. The *X* chromosome is only one factor in the development of sex but if it is a factor which differs in the case of the two sexes, it is a "sex-differential."

2. *Environmental Influence. Alteration of Sex Ratios.*—On the other hand there are many observations which seem to indicate that the sex ratio may be changed by environmental conditions acting before or after fertilization and therefore some

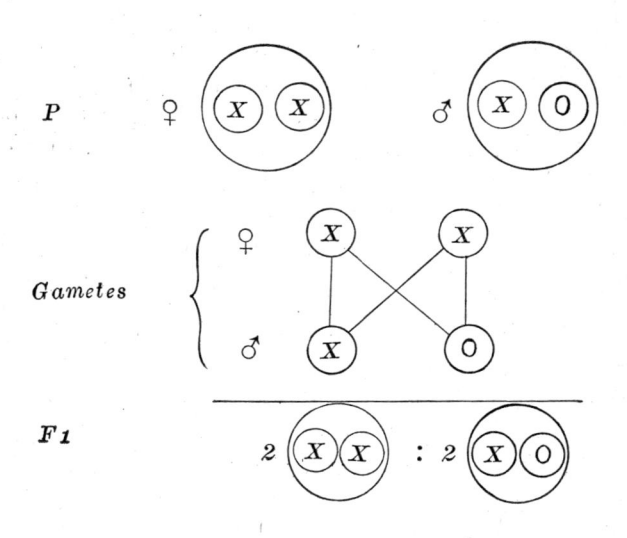

Fig. 58. Diagram Showing Sex as a Mendelian Character, the female being homozygous, the male heterozygous for sex. The female forms gametes all of which contain the *X* chromosome; the male forms two sorts of gametes one-half of which contain the *X* chromosome and the other half lack it. All possible combinations of these gametes give **2:2 or 1:1 ratio of females to males.**

have concluded that sex is determined by extrinsic rather than by intrinsic causes. Many of these observations, as already remarked are now known to be erroneous or misleading, since they do not prove what they were once supposed to demonstrate. But there remain a few cases which cannot at present be explained away in this manner. Perhaps the best attested of these are the observations of R. Hertwig and some of his pupils on the effect of the time of fertilization on the determination of sex. If frog's eggs, which are always fertilized after they are laid, are kept for some hours before spermatozoa are mixed with them, or if the female is prevented for two or three days from laying the eggs after they have entered the oviducts, the proportion of males to females is enormously increased. A wholly similar result has been observed by Pearl and Parshley in the case of cattle, where delayed fertilization of the egg leads to a great preponderance of males. Hertwig attempts to explain his extremely interesting and important observations as due to the relative size of the nucleus and cytoplasm of the egg; but in general this nucleus-plasma ratio may vary greatly irrespective of sex and there is no clear evidence that it is a cause of sex determination.

Miss King also, working on toad's eggs, has increased the proportion of *females* by slightly drying the eggs or by withdrawing water from them by placing them in solutions of salts, acids, sugar, etc., but the manner in which drying increases the proportion of females is wholly unknown. All of these experiments on sex determination in frogs and toads are somewhat complicated by the difficulty of recognizing the sex of tadpoles and young animals.

Whitney has shown that in several species of rotifers a scanty diet produces in the second filial generation only female offspring while a copious diet produces as high as 95 per cent of male offspring. Many earlier investigators had found that food influences sex, though usually it was held that scanty food led to the production of males and abundant food to females, but this older work unlike Whitney's, was generally uncritical. A. F. Shull finds that "the irrevocable event leading to the determination of the sex of any given parthenogenetically produced individual (rotifer) occurs in the maturation of the egg from which that individual's mother develops. . . . Probably a definite chemical change in the proteins of the chromosomes occurs at the time of maturation." The diet may thus affect the chromosomes and through these the sex.

Extensive statistics show that in many animals including man more males are born than females, whereas according to the chromosome theory of sex-determination as many female-producing spermatozoa are formed as male-producing ones. It is possible to explain such departure from the 1:1 ratio of males to females in conformity with the chromosome theory if one class of spermatozoa are more active or have greater vitality than the other class, or if after fertilization one sex is more likely to live than the other. In the human species it is known that mortality is greater in male babies before and after birth than in female babies, but if before fertilization the activity or vitality of male-producing spermatozoa is greater than that of female-producing ones it would offer a possible explanation of the greater number of males than of females at the time of birth. In certain insects it is known that only females develop from fertilized eggs, and in one of these cases, viz., *Phylloxera*, Morgan has discovered that this is due to the fact that all the male-producing spermatozoa degenerate and that only female-producing spermatozoa become functional. Possibly experimental alterations of the sex ratio, such as Hertwig, King, Whitney and others have brought about may be explained as due to a differential action of the modified egg cells or of the environment upon the two types of spermatozoa. In the Drosophila work many lethal mutations have appeared which cause the early death of those zygotes in which the lethal gene is not balanced by a normal allelomorph. Some of these lethals are sex-linked and alterations of the normal sex ratio in certain cases may be explained as the result of these lethal factors. Finally the chromosomal theory of sex determination is so well supported in so many instances where at first it seemed impossible of application, that it ought not to be rejected until unmistakable evidences can be adduced against it.

3. *Hermaphrodites and Intersexes.*—Finally a number of cases have been brought to light which indicate the necessity of distinguishing between the hereditary determination of sex and its ontogenetic development. It has been known for a long time that in bees and ants the workers are imperfect females, while the queens are perfect females, and that the kind or amount of food which is fed to the larvae determines whether they will be workers or queens (see p. 230). Again in many animals the development of male or female characters is dependent upon internal secretions or hormones from the sex glands or other organs (see pp. 232-235). In these cases it is evident that sex was determined at an early stage, probably at fertilization, but the development of those male or female characters, which occurs later, is influenced also by the external or internal environment (p. 216). Breeders of cattle are familiar with the fact that when twin calves are of opposite sex, the male is sexually perfect, but the female usually has many male characters and grows into a steer-like animal which is sterile and is known as a "free martin." In a recent paper Lillie has shown that in all such cases the twins are connected by blood vessels at an early stage *in utero* and that there is a more or less complete circulation of blood from one foetus to the other, and he concludes that "sex hormones," which are probably formed earlier in the male than in the female, are carried from the male to the female twin, thus causing the development of male organs in an animal which would otherwise have been a female. Therefore the chromosomal or "zygotic determination of sex is not irreversible predestination but a quantitative overbalance in the direction of one sex or the other," which may later be changed. Somewhat similar conclusions had previously been reached by Whitman and Riddle regarding the

THE CELLULAR BASIS 169

sex of pigeons, by Shull in the case of *Lychnis,* and especially by Goldschmidt for the gypsy-moth. Goldschmidt supposes that sex is determined by certain enzymes which he calls "andrase" and "gynase"; an excess of the former leads to the development of males, an excess of the latter to females, and varying mixtures of the two to varying intergrades or "intersexes." He assumes that these enzymes are present in the "sex determining" chromosomes at fertilization as well as in later stages and thus he attempts to identify all sex determining factors with these "sex enzymes." The difference between determination by chromosomes and by internal secretions, that is, between heredity and development, is found chiefly in the time at which these enzymes act. Crew and Riddle have found that complete reversal of sex in fowls and pigeons may be caused by disease of the ovary, but Greenwood and Finlay find that even in embryo chicks heterosexual gonod grafts do not completely change the sex.

Morgan has shown that "gynandromorphs" or "sex mosaics" are due to the irregular distribution or loss of certain "sex chromosomes" owing to abnormalities in fertilization or cleavage. In such cases one portion of the body shows male characters, another portion female ones and a study of the chromosomes in these regions shows that in the former the male combination of chromosomes is present, in the latter the female combination. This comes as near to a demonstration of the truth of the chromosomal theory of sex determination as is possible.

Finally, the most notable recent work on this subject is by Bridges. In his studies of the pomace fly, *Drosophila melanogaster,* he found intersexes whose genetical behavior was such as to suggest that they had more or less than the usual number of chromosomes. By means of breeding experiments as well as by microscopical study of their germ cells he has demonstrated that this is true. In the normal fly of this species there are four pairs of chromosomes (Fig. 67); the first pair (Chromosomes I) are the sex chromosomes which are XX in the female and XY in the male; the second and third pairs (Chromosomes II and III) are large and V-shaped; the fourth pair (Chromosomes IV) are very small and round. Through the failure of chromosome pairs to separate in the maturation divisions an egg or sperm may come to contain an abnormal number of any or all of these chromosomes (Fig. 67 C).

As opposed to the sex chromosomes, all others are known collectively as "autosomes." A normal female has two X chromosomes and two of each of the autosomes; a normal male has one X and two of each of the others; but when two X's occur with three of each of the others, or with three of some of them, intersexes result, and these may grade all the way from perfect females to perfect males depending upon the ratio of X chromosomes to autosomes. Sex is therefore determined by a quantitative relation of the X chromosomes to the autosomes, and if one assumes that there are sex enzymes, such as Goldschmidt postulated, it is probable that the X chromosomes produce mainly "gynase" and the autosomes "andrase." Such an explanation harmonizes well not only with all that is known regarding the chromosomal *determination* of sex and of intersexes but also with much that is known concerning the possibility of modifying the *development* of sex by enzymes or hormones from glands of internal secretion.

In either sex many secondary sexual characters of the other sex are present during development and traces of these may persist in the adult; but one set of these characters develops fully in the male and another set in the female, so that they may be called *sex-limited*. The development of the secondary sex characters is usually determined by internal secretions from the ovaries or testes, though in some cases they may develop after these organs have been removed, but in the last analysis both primary and secondary sex characters are dependent upon the sex determiners in the germ cells. Sex and sex-limited inheritance are only special cases of Mendelian inheritance and the full development of the male or female condition is dependent upon the predominance of male-determining, or of female-determining factors, both hereditary and environmental; while the condition of "intersexes" is the result of the lack of such predominance.

On one point there is general agreement, namely every organism is at the beginning of ontogeny so evenly balanced between maleness and femaleness that very slight changes in heredity or environment may cause it to go one way or the other; every organism is potentially both male and female, and even in the fully developed state each sex carries the vestiges of suppressed organs of the other sex.

D. THE MECHANISM OF HEREDITY

The mechanism of heredity, as contrasted with the mechanism of development, consists in the formation of particular kinds of germ cells and in the union of certain of these cells in fertilization. We have briefly traced the origin, maturation and union of male and female sex cells in a number of animals, and in these phenomena we have the mechanism of the hereditary continuity between successive generations. But in addition to these specific facts there are certain general considerations which need to be emphasized.

1. THE SPECIFICITY OF GERM CELLS

The conclusion is inevitable that the germ cells of different species and even those of different individuals are not all alike. Every individual difference between organisms must be due to one or more differentiating causes or factors. Specific results come only from specific causes. These causes may be found in the organization of the germ cells or in environmental stimuli, *i.e.*, they may be intrinsic or extrinsic, but as a matter of fact experience has shown that they are generally intrinsic in the germ. In the same environment one egg becomes a chicken and another a duck; one becomes a frog and another a fish and another a snail; one becomes a black guinea-pig and another a white one; one becomes a male and another a female; one gives rise to a tall man and another to a short man, etc. Since these differences may occur in the same environment they must be due to differences in the germ cells concerned.

Environment Non-specific.—On the other hand different environmental conditions may be associated with similar developmental results. Loeb and others have found that artificial parthenogenesis may be induced by a great variety of environmental stimuli, *viz.*, by salt solutions, by acids and alkalis, by fatty acids and fat solvents, by alkaloids and cyanides, by blood serum and sperm extract, by heat and cold, by agitation and electric current. There is certainly nothing specific in these different stimuli. Similarly Stockard has discovered that cyclopia, or one-eyed monsters, may be produced by magnesium salts, alcohol, chloretone, chloroform, and ether, and to this list McClendon has added various other salts and anæsthetics. In all such cases it is evident that the specific results of such treatment are due to a specific organization of the germ rather than to specific stimuli.

Why does one egg give rise to a chicken and another to a duck, or a fish, or a frog? Why does one egg give rise to a black guinea-pig and another to a white one, though both may be produced by the same parents? Why does one child differ from another in the same family? Why does one cell give rise to a gland and another to a nerve, one to an egg and another to a sperm? If these differences are not due to environmental causes, and the evidence shows that they are not, they must be due to differences in the structures and functions of the cells concerned.

Protoplasm Specific.—Many differences in the material substances of cells are visible, and many more are invisible though still demonstrable. These differences may not be detectable by chemical or physical tests, and yet they may be demonstrated

THE CELLULAR BASIS

physiologically and developmentally. The most delicate of all tests are physiological, as is shown by the Weidal test in typhoid fever, the Wassermann reaction in syphilis, the reactions of immunized animals to different toxins, etc. Lillie has recently shown that egg cells give off a substance which he calls "fertilizin," which can be detected only by the way in which spermatozoa react to it. No chemical or physical test can distinguish between the different eggs or spermatozoa produced by the same individual, but the reactions of these cells in development prove that they are different. Undoubtedly chemical and physical differences are here present but no chemical methods at present available are sufficiently delicate to detect them.

It is one of the marvelous facts of biology that practically every sexually produced individual is unique, the first and last of its identical kind, and although some of these individual differences are due to varying environment, others are evidently due to germinal differences, so that we must conclude that every fertilized egg cell differs in some respects from every other one.

But are there molecules and atoms enough in a tiny germ cell, such as a spermatozoon, to allow for all these differences? Miescher has shown that a molecule of albumin with 40 carbon atoms may have as many as one billion stereo-isomers, and in protoplasm there are many kinds of albumin and other proteins, some with probably more than 700 carbon atoms. In such a complex substance as protoplasm the possible variations in molecular constitution must be well nigh infinite, and it can not be objected on this ground that it is chemically and physically impossible to have as many varieties of germ cells as there are different kinds of individuals in the world.

Permutations of Chromosomes.—Even with regard to morphological elements which may be seen with the microscope it can be shown that an enormous number of permutations is possible (Fig. 59). It seems probable, as Boveri has shown, that different chromosomes of the fertilized egg differ in hereditary potencies,

174   HEREDITY AND ENVIRONMENT

and where the number of chromosomes is fairly large the number of possible combinations of these chromosomes in the germ cells become very great. In every human being, where there are after synapsis, 24 maternal-paternal pairs, the possible number of permutations in the distribution of these chromosomes to the different germ cells would be 2^{24}, or 16,777,216, and the possible number of different combinations of fertilized eggs or oosperms which could be produced by a single pair of parents would be $(16,777,216)^2$, or approximately three hundred thousand billions. But probably other things than chromosomes differ in different germ cells, and it is by no means certain that individual chromosomes are always composed of the same chromomeres, or units of the next smaller order, and in view of these possibilities it may well be that every human germ cell differs morphologically and physiologically from every other one, in short that every oosperm and every individual which develops from it is absolutely unique.

Significance of Sexual Reproduction.—Indeed the production of unique individuals seems to be the chief purpose and result of sexual reproduction. In asexual reproduction the individual variations which occur are chiefly if not entirely due to environment but in sexual reproduction they are also due to new combinations of hereditary elements. The particular germinal organization transmitted from one generation to the next depends upon (a) the organization inherited from ancestors, (b) the particular character of the cell divisions by which the germ cells are formed (c) the particular kinds of egg and sperm cells which combine in fertilization. The inherited organization determines all the general characteristics of race, species, genus, order, phylum. It determines the possibilities and limitations of individual variations. Given a certain inherited organization, the individual peculiarities of the germ cells are determined by the particular character of

[1] Excluding duplications there would be 3^{24} different genotypes and 2^2 different phenotypes, assuming that chromosomes always preserve their identity and that dominance is always complete.

THE CELLULAR BASIS

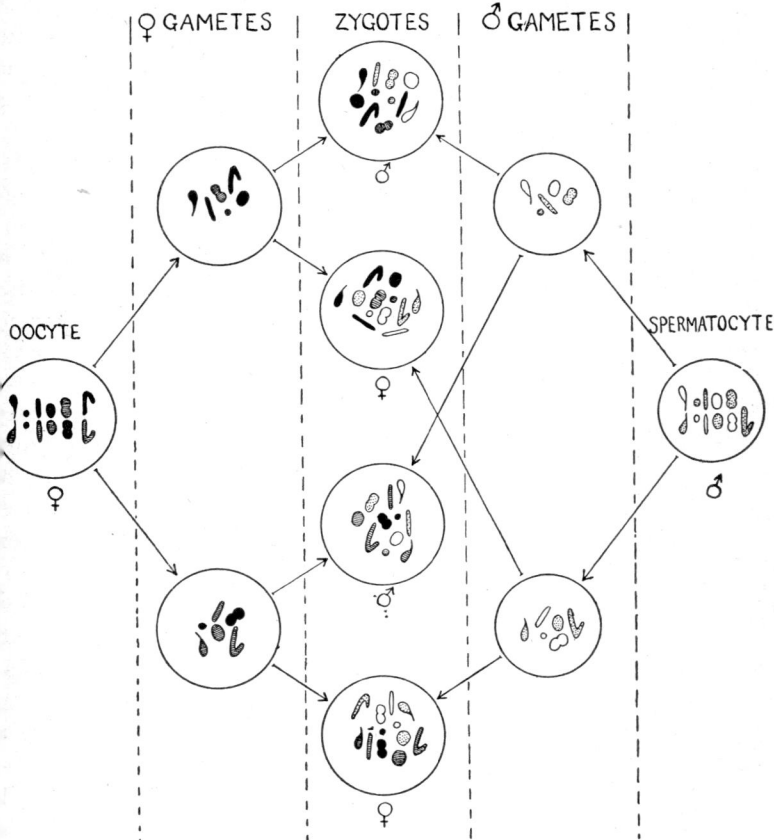

FIG. 59. DIAGRAM SHOWING SOME OF THE POSSIBLE DISTRIBUTIONS OF CHROMOSOMES OF GRANDPARENTS TO GRANDCHILDREN. Chromosomes of maternal grandfather shaded black, of maternal grandmother with cross lines; of paternal grandmother stippled, of paternal grandfather unshaded. "Sex-chromosomes" are J-shaped, a pair being present in the female, a single one in the male. In the maturation of oocyte and of spermatocyte homologous chromosomes unite in pairs and then separate into the gametes. With 6 pairs of chromosomes, each free to separate in 2 directions the largest possible number of combinations in the gametes is $(2)^6$ or 64 and since any sperm may unite with any egg, the number of combinations possible in the zygotes is $(64)^2$ or 4096 and, excluding all duplications, there are $(3)^6$ or 729 genotypes only 4 of which are shown in the diagram in the column marked "Zygotes."

176   HEREDITY AND ENVIRONMENT

cell division by which the germ cells are formed, and the peculiarities of the individuals or persons which develop from these cells are determined in large part by the particular kinds of germ cells which unite in fertilization.

Comparison of Cards and Chromosomes.—The behavior of chromosomes in maturation and fertilization is like the shuffle and deal of cards in a game, and apparently with the same object, namely, never to deal the same hand twice. To make this comparison more complete suppose that kings be discarded from the pack, leaving 48 cards, 24 red and 24 black, which we will compare to the 48 chromosomes, 24 maternal and 24 paternal. The dealing of these cards into two hands, each containing varying numbers of red and black cards, would correspond to what takes place in the maturation of the germ cells, and the putting of any such hand with a corresponding hand *from another pack* would parallel what takes place in fertilization. In no game of cards are half of the cards taken from one pack and half from another at every game, but this is just what happens in this game with the chromosomes. Because of the mixture of chromosomes from different individuals in every generation, each of which has its own peculiar value, the game of heredity becomes vastly more complex than any game of cards. This illustration may serve to make plain the fact that in the process of maturation and fertilization there is this shuffle and deal of the chromosomes, with the result that every oosperm and every individual which develops from it is different from every other one.

Comparison of Letters and Chromosomes.—Perhaps a more instructive illustration may be found in comparing human chromosomes with the letters of the alphabet. If we discard the letter Z and also Y, except when it is used to replace X, we have 24 letters left, corresponding to the 24 chromosomes of the egg or sperm. All eggs contain the chromosomes A to X, whereas half of the spermatozoa have A to X and the other half A to Y, in which Y replaces X. In the fertilized egg and in every cell

THE CELLULAR BASIS

derived from it there are two alphabets, one derived from the egg, the other from the sperm, the female constitution being represented by the formula 2A to 2W+XX and the male by 2A to 2W+XY. Every human being has the same alphabet of chromosomes, but different individuals have different numbers of capitals and lower case letters, and different families and races have different styles of letters, such as Roman, Italic, Gothic, Old English, Script, etc. In the crossing of these families and races we get a mixture of different fonts of type that would be the despair of a printer, though it makes every page and almost every sentence in the book of life different from every other one.

Germ Cells as Specific as Persons.—This conception of the specificity of every germ cell, as well as of every developed individual, sets the whole problem of heredity and development in a clear light. The visible peculiarities of an adult become invisible as development is traced back to the germ, but they do not wholly cease to exist. Similarly the multitudinous complexities of an adult fade out of view as development is traced to its earliest stages, but it is probable that they are not wholly lost. In short the specificity of the germ applies not merely to those things in which it differs from other germs, but also to characters in which it resembles others; in short, to hereditary resemblances no less than to hereditary differences.

The mistake of the doctrine of preformation (see p. 57) was in supposing that germinal parts were of the same kind as adult parts; the mistake of epigenesis was in maintaining the lack of specific parts in the germ. The development of every animal and plant consists in the transformation of the specific characters of the germ into those of the adult. From beginning to end development is a series of morphological and physiological changes but not of new formations or creations except in so far as new structures or functions appear as a result of "creative synthesis" (p. 31). It is only the incompleteness of our knowledge of development which allows us to say that the eye or ear or brain begins to form in this or that stage. They become visible at certain stages, but their real beginnings are indefinitely remote.

11. Correlations Between Germinal and Somatic Organization

All the world knows that the organization of the germ is not the same as that of the developed animal which comes from it, and yet the specificity of the germ indicates that there must be some correlation between the germinal and the developed organization; in short, there is not identity of organization but there is correlation of organization between the germ and the adult. What correlations are known to exist between the oosperm and the developed animal?

Inheritance Factors and Developed Characters.—We have considered in the preceding chapter (pp. 100-105) the evidence that there are specific inheritance factors or genes which are correlated with the development of specific adult characters. These factors are not the characters in miniature nor are they the "representatives" or "carriers" of characters, but they are the differential causes of characters. Every inherited character must have a differential cause in the germ cells, but this cause may not be a peculiar vital corpuscle nor even a peculiar atom or molecule but it must at least be some particular combination of atoms or molecules. It is practically certain that this differential cause of each inherited character is associated with some material substance which occupies some place in the germ cells.

Location of Inheritance Factors.—If it is asked whether there are particular structures in germ cells which correspond to particular inheritance factors it must be admitted that we have no certain knowledge on this subject and that opinions differ with respect to it. On the one hand it is maintained that the entire germ is concerned in the development of every character, and on the other hand that the *differential cause* of any character may be located in some differentiated structure or function of the germ. These views are not mutually exclusive and it may well be that both are true. We know that germ cells are composed of many parts which differ from one another in both structure and function, and it is highly probable that there are enough of these parts to provide a *locus* for every inheritance factor.

There was a time when the cell was the *Ultima Thule* of biological analysis and when the contents of cells were supposed to be "perfectly homogeneous, diaphanous, structureless slime." Then the nucleus was discovered within the cell, then the chromosomes within the nucleus, then the chromomeres within the chromosomes, and there is no reason to suppose that organization ceases with the powers of our present microscopes. With every improvement of the microscope and of microscopical technique, structures have been found in cells which were undreamed of before, and it is not probable that the end has been reached in this regard. We know that cells contain nuclei and chromosomes and chromomeres, centrosomes and plastosomes and microsomes, and we know that some of these parts differ in function as well as structure. And there is no reason to doubt that if we had sufficiently powerful microscopes we should find still smaller and smaller units until we came at last to molecules and atoms.

The fact that inheritance units from the two parents unite in fertilization and later segregate in the formation of gametes, so that the latter are pure with respect to any character, is a familiar part of Mendelian inheritance (Fig. 60). Even if these units be regarded as physiological processes they must be associated with particular structures, since function and structure are inseparable in life processes. What are the units in terms of cell structures and where are they located in the cell?

1. CHROMOSOMAL INHERITANCE THEORY.—We have in the chromosomes, as Wilson especially has emphasized, an apparatus which fulfils all the requirements of carriers of Mendelian factors (Fig. 61). Both factors and chromosomes come in equal numbers from both parents; both maternal and paternal factors and chromosomes pair in the zygote and separate in the gamete

as shown in Figs. 60 and 61; and so far as is known the chromosomes are the only portions of the germ cells which fulfil these conditions. The association, segregation and distribution of Mendelian factors and of maternal and paternal chromosomes are exactly parallel and it is not reasonable to suppose that this remarkable coincidence is without significance.

Furthermore there is much additional evidence that the chromosomes are important factors in heredity and development. Boveri has studied the abnormal distribution of chromosomes to different cleavage cells in doubly fertilized sea urchin eggs and has found evidence that the hereditary value of different chromosomes is different. McClung, Stevens, Wilson and others have discovered that the determination of sex is associated with the presence or absence of a particular chromosome, the X or Y chromosome, in the spermatozoon which fertilizes the egg. If an egg

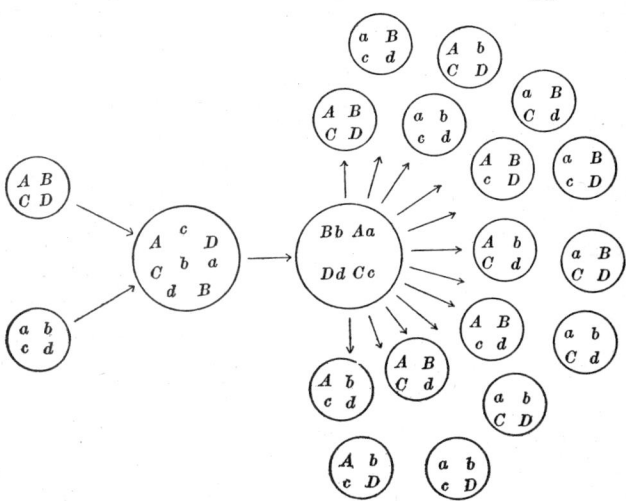

FIG. 60. DIAGRAM SHOWING UNION OF FACTORS IN FERTILIZATION AND THEIR SEGREGATION IN THE FORMATION OF GERM CELLS. With 4 pairs of factors (*Aa, Bb, Cc, Dd*) 16 types of gametes are possible as shown in the two series of small circles at the right. (After Wilson.)

THE CELLULAR BASIS 181

is fertilized by a sperm which lacks the X chromosome a male is produced, if fertilized by the other type a female results. This correlation between the presence or absence of a whole chromosome and of a developed character such as sex, was the first case of the kind that was known and more than anything else it served to prove that the chromosomes contain the Mendelian factors. The absence of a whole chromosome is plainly visible under the microscope, whereas the absence of a single factor or gene from a chromosome would never have been seen. These fortunate cases in which the male lacks a whole chromosome seem almost to have been intended to give ocular proof of the chromosomal theory of heredity; they are to biology what the rings of Saturn are to astronomy,—a visible confirmation of a great theory.

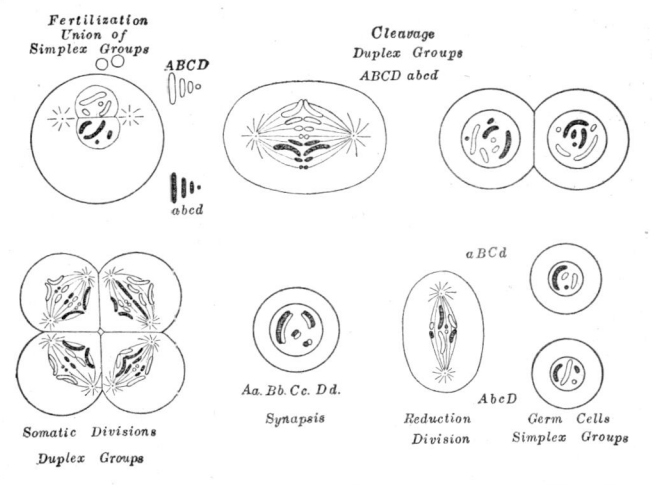

FIG. 61. DIAGRAM OF GERM CELLS CORRESPONDING TO FIG. 60, showing the union of maternal chromosomes ($ABCD$) and paternal ones ($abcd$) in fertilization, their distribution in cleavage, their union into 4 pairs (Aa, Bb, Cc, Dd) in synapsis and the separation of the pairs in the reduction division. Only 2 of the 16 possible types of germ cells are shown at the lower right. (After Wilson.)

We have in these facts a remarkable correlation between the distribution of the chromosomes and the occurrence of certain characters of the adult animal. The association of maternal and paternal chromosomes in fertilization and their segregation in the maturation of the germ cells is parallel to the association of Mendelian characters in the zygote and their segregation in the gametes; if the distribution of chromosomes in cleavage is abnormal the larva shows abnormal characters (Boveri); sex determination is associated with the distribution of a particular chromosome to one-half of the spermatozoa, and the fertilization of the egg by one or the other type of spermatozoa. (Wilson) There are many parts of a germ cell, all of which may be concerned in heredity and development, but the chromosomes appear to be the seat of the differential factors for Mendelian characters.

On the other hand it has been objected by certain investigators, notably Child, Foot and Strobell, *et al*, that chromosomes are not the causes of anything, but that they are the "results of dynamic processes," "the expression rather than the cause of cell activities." This objection seems to confuse the idea of natural cause with that of final cause. Science knows nothing of the latter; any natural cause is only a link in the chain of cause and effect, it is itself the result of antecedent causes and the cause of subsequent results. Undoubtedly the chromosomes are the results of antecedent processes, and yet they may also be the causes of subsequent results. No thoughtful person has ever maintained that chromosomes or any other things in nature are autonomous, absolute, uncaused causes.

Abnormal Distribution of Chromosomes and Factors.—Experimental evidence that the chromosomes are the seat of inheritance factors is found in the correlation between the abnormal distribution of chromosomes and the development of abnormal characters in the embryo or adult. An abnormal distribution of chromosomes to the cleavage cells may be caused in a variety of ways but one of the least injurious methods of accomplishing this is by

THE CELLULAR BASIS

causing two spermatozoa instead of one to enter an egg. In such doubly fertilized eggs Boveri discovered that different cleavage cells receive a different number of chromosomes and in general those cells which receive the largest number develop most typically, while those which receive a small number develop atypically. By a skillful analysis Boveri proved that normal develop-

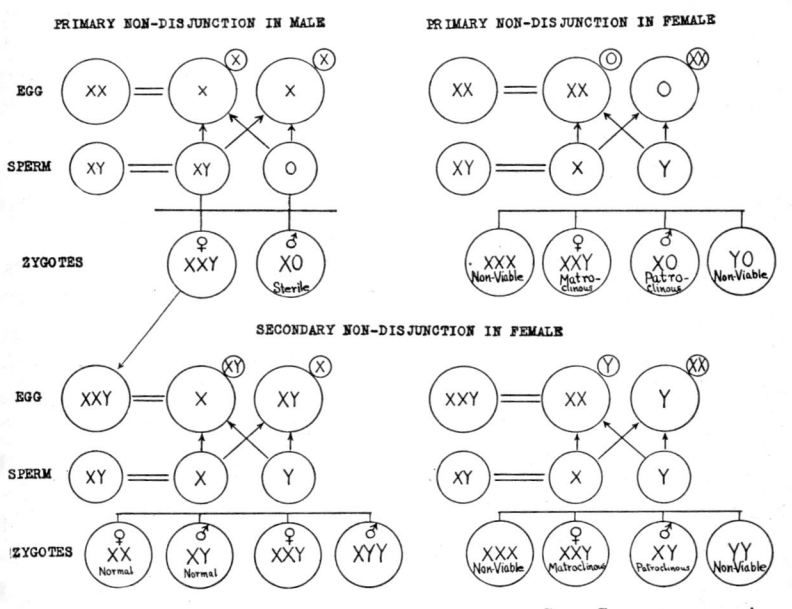

FIG. 62. DIAGRAM OF NON-DISJUNCTION OF THE SEX CHROMOSOMES in the Maturation of the Egg and Sperm of *Drosophila melanogaster,* with the resulting types of Zygotes (According to Bridges). Primary non-disjunction in the sperm leads to the production of XXY females and XO males; primary non-disjunction in the egg leads to four combinations, two of which (XXX and YO) are non-viable; secondary non-disjunction occurs in the maturation of an XXY egg. In *Drosopohila* Y is an "empty chromosome," *i.e.,* it appears to contain no genes, yet it has some influence since a male without it (XO) is sterile. Zygotes are matroclinous or patroclinous depending upon whether they get a larger number of X chromosomes from the egg or from the sperm.

ment depends not so much upon the absolute number of chromosomes in a given cell as upon a complete set of all the different kinds of chromosomes, and when a complete set was not present certain characters failed to develop. By this means he showed that different chromosomes of a set differ in hereditary value, as, for example, the fingers of a hand differ from one another, and that two chromosomes of one kind could not make up for the lack of one of another kind, any more than two thumbs could make up for the loss of a little finger.

A still more detailed correlation between the presence or absence of a particular chromosome and the presence or absence of particular characters in the developed organism has been described by Bridges (1916). In his study of the pomace fly *Drosophila melanogaster,* he found that the occasional appearance (1 in 1700) of a matroclinous daughter or patroclinous son was due to the fact that the members of the XX pair of chromosomes of the oocyte fail to separate in the reduction division so that both XX's are included in the egg (Fig. 67, C) or both are extruded in the polar body, the eggs being accordingly XX or O; or the two chromosomes of the XY pair in the spermatocyte fail to separate in the reduction division so that one sperm may have both X and Y, and another lack both of these chromosomes (Fig. 62). This phenomenon he calls "non-disjunction" and it results in the production of matroclinous daughters or patroclinous sons, and in many other irregularities of inheritance which follow precisely the abnormal distribution of these chromosomes. A patroclinous son is the result of the fertilization of an O egg by an X sperm; such an XO son is sterile whereas the normal XY son is not, thus showing that the chromosome Y has some function though in *Drosophila* it does not contain any of the genes; fertilization of an O egg by a Y sperm produces a combination which is non-viable. Fertilization of an XX egg by a Y sperm produces a matroclinous daughter (XXY), whereas fertilization of an XX egg by an X sperm produces a combination (XXX) which is non-viable.

These relations are shown schematically in the accompanying diagrams (Fig. 62).

2. LINKAGE OF CHARACTERS AND CHROMOSOMAL LOCALIZATION.—Finally the study of characters which are linked together in heredity, joined with the study of the chromosomes and their distribution in the maturation and fertilization of the germ cells, has not only confirmed the chromosomal theory of heredity but has also shown that certain chromosomes carry the genes for certain characters and has even indicated the relative positions of different genes in the chromosomes.

Thanks to the work of Bateson, Morgan, and many others, it is now known that many characters are linked together in inheritance. Darwin had long ago noted that male albino cats with blue eyes are usually deaf and many other cases of the association of peculiar characters had been reported by earlier observers. In 1906, Bateson and Punnett found that sweet peas with purple flowers usually have elongated pollen grains, whereas those with red flowers have round pollen. Since 1910 Morgan and his pupils have discovered about four hundred new characters, or mutations in the pomace fly, *Drosophila melanogaster,* which are usually linked together in four groups.

a. *Sex-linked Inheritance.*—The first cases of such linkage studied by Morgan were in characters which are usually associated with one or the other sex, but which may have nothing to do with reproduction and may affect any part of the body. Such characters are not necessarily limited to one sex or the other, as are many primary and secondary sexual characters, but they may appear in either sex though they are usually transmitted from fathers to daughters, or from mothers to sons ("criss-cross" inheritance) in exactly the way in which the sex chromosomes (X) are transmitted. Morgan has therefore concluded that the factors for these characters are carried by the sex chromosomes and has named them *sex-linked* characters. In the pomace fly, *Drosophila,* he has discovered a large number of such characters which are

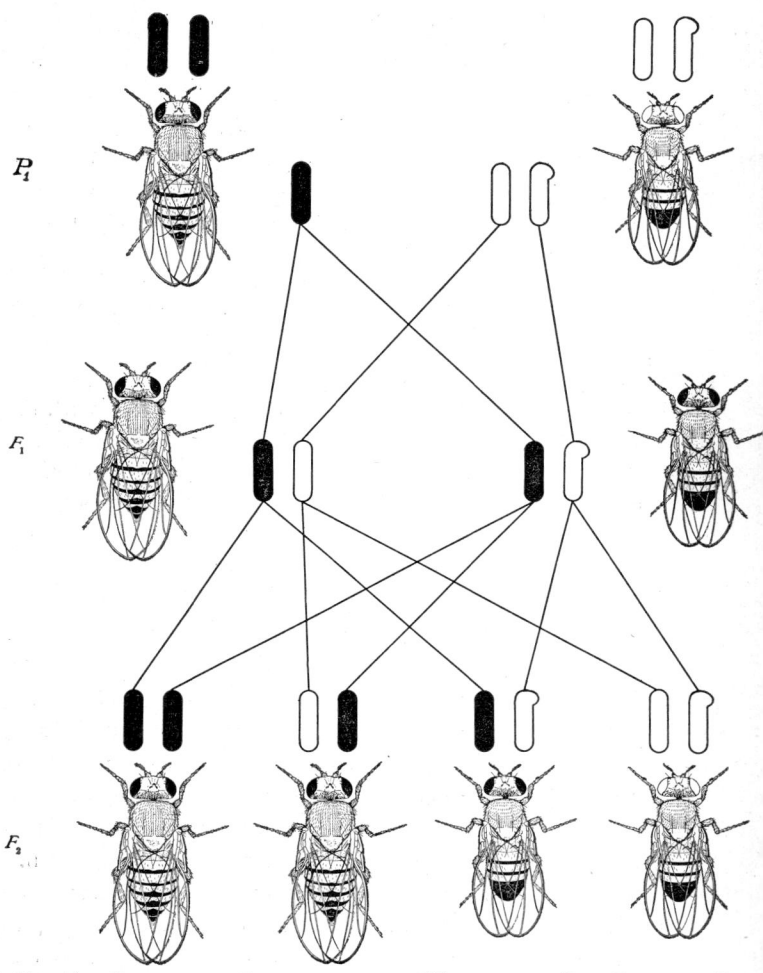

Fig. 63. Sex-linked Inheritance of White and Red Eyes in *Drosophila*. Parents, red-eyed female and white-eyed male (females are larger than males); F_1, red-eyed males and females; F_2, red-eyed females and equal numbers of red-eyed and white-eyed males. The distribution of sex chromosomes is shown through the middle of the figure, the straight rods being X-chromosomes and the hooked rods Y-chromosomes; the X-chromosomes that carry the factor for red eye are black, those that carry the factor for white eye are unshaded; the Y-chromosomes carry no factors for eye color. (From Morgan.)

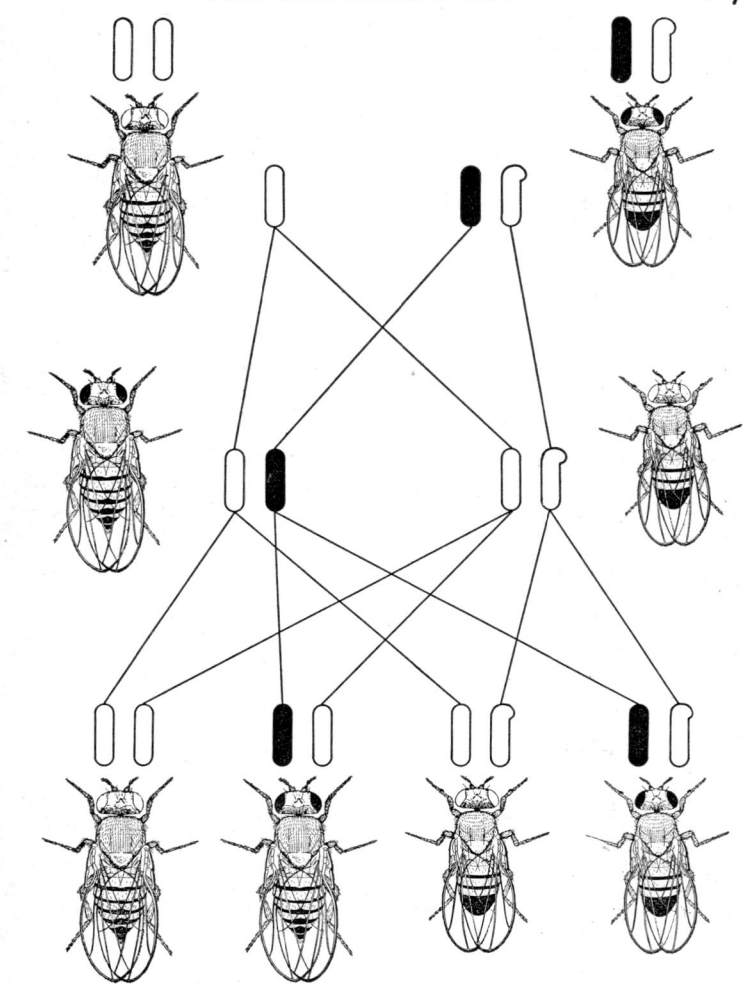

Fig. 64. Reciprocal of Cross shown in Fig. 63. Parents, white-eyed female and red-eyed male; F_1, red-eyed females and white-eyed males ("criss-cross inheritance"); F_2, equal numbers of red-eyed and white-eyed females and males. The distribution of the sex-chromosomes is shown in the middle of the figure as in Fig. 63. (From Morgan.)

linked with sex, such as the color of the eyes and of the body the length of the wings, etc. A typical case is shown in Figs. 63 and 64. The eye color of this fly is normally red, but mutation have arisen in which the eye is white. Such a mutation first appears in males, though it may later be transferred to females, as we shall see. If now a white-eyed male and a red-eyed female are crossed all the F_1's are red-eyed, but if these F_1's are interbred all the females of F_2 have red eyes while half of the males have red eyes and the other half have white eyes (Fig. 63). On the other hand if one of the F_1 females of this cross is bred with a white eyed male (Fig. 63, F_1), half of the females of F_2 are red-eyed and half are white-eyed, and half of the males are red-eyed and half are white-eyed.

If now one of these white-eyed females is bred with a red-eyed male (Fig. 64, P) all the females of the F_1 generation are red eyed and all the males white-eyed ("criss-cross" inheritance) and if these are interbred there are produced in the F_2 generation equal numbers of red-eyed and white-eyed males and females (Fig. 64).

The distribution of the maternal and paternal sex chromosomes exactly parallels this distribution of this sex-linked character, as is shown in Figs. 63 and 64, and this proves that the differential factors for these characters are carried in these sex chromosomes.

By a series of ingenious experiments Foot and Strobell have shown that the differential factors for certain *sex-limited* characters in insects, that is, characters which are limited to one sex are not contained in the "sex chromosomes," and they argue that the differential factors for *sex* and for *sex-linked* characters cannot be located in these chromosomes. Their conclusions apply only to sex-limited and not to sex-linked characters. There is no doubt that the factors for the determination of sex and sex linked characters are distributed in the same way as the "sex chromosomes" are, and this proves that these factors are located in the "sex chromosomes."

THE CELLULAR BASIS

Haemophilia.—Another case of sex-linked inheritance is found in an abnormal condition in man known as *haemophilia*, which is characterized by a deficiency in the clotting power of the blood, and consequently by excessive bleeding after injury. "Bleeders" are almost always males, though the defect is always transmitted to a son from his mother who does not usually show the defect because it appears in females only when both parents were affected. The manner of inheritance of this character is exactly similar to the inheritance of white eyes in *Drosophila* and is in all probability due to similar causes.

Daltonism.—One of the most striking cases of sex-linked inheritance is that form of color blindness known as Daltonism, in which the affected person is unable to distinguish between red and green. It is known that males are more frequently affected than females, and that color blindness is in some way associated with sex. It requires two determiners for color blindness, one from the father, the other from the mother, to produce a color blind female, whereas only a single determiner is necessary to produce a color blind male, just as is true of sex. The accompanying diagrams (Figs. 65, 66) illustrate the method of inheritance of color blindness. 𝕏 represents the normal *X*-chromosome, ⓞ its absence (or rather the *Y*-chromosome, since the sex chromosomes of the human male are XY and not XO) and **X** the *X*-chromosome which carries the factor for color blindness.

It will be seen that a color blind father and a normal mother have only normal children, but the father transmits to his daughters and not to his sons the sex-determiner which carries the factor for color blindness. But since color blindness does not develop in females unless it is duplex (*i.e.* comes from both father and mother) whereas it develops in males if it is simplex (*i.e.* comes from either parent) all the daughters of a color blind father and normal mother will appear normal although carrying one determiner for color blindness, while all the sons will be normal because they carry no determiner for color blindness. But these daughters transmit to one-half of their children the single determiner for color blindness and if any of those receiving this determiner are males they will be color blind. Consequently we have the curious phenomenon of simplex color blindness appearing only in males and being transmitted to them only through apparently normal females.

On the other hand if a female is color blind she has inherited it from both father and mother, *i.e.*, the character in her is duplex and in all of her children by a normal male the character will be simplex; accordingly all of her sons will be color blind and all of her daughters will be normal, though carrying the simplex determiner for color blindness.

Sex-linked Lethal Factors.—One of the most interesting cases of linkage is found where early death is linked with sex. In

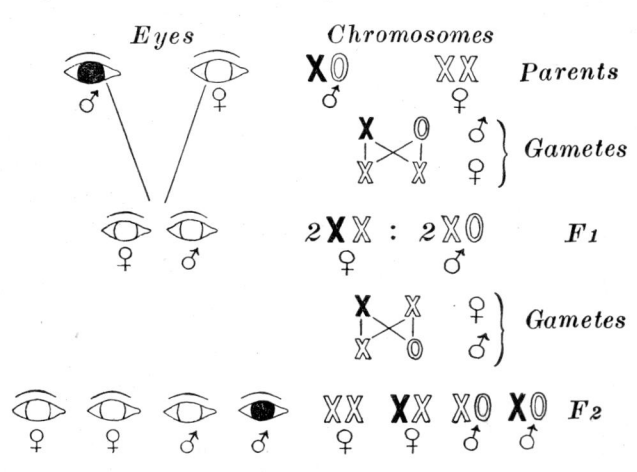

FIG. 65. DIAGRAM OF INHERITANCE OF COLOR BLINDNESS THROUGH THE MALE. A color blind male (here black) transmits his defect to his grandsons only. The corresponding distribution of the sex chromosomes is shown on the right, the one carrying the factor for color blindness being black. Recent work shows that the sex chromosomes of the human male are XY and not XO, as shown in this diagram.

THE CELLULAR BASIS

Drosophila a considerable number of lethal mutant factors have been demonstrated in the X chromosome and all individuals in which such a factor is not balanced by a normal allelomorph die early. All males that receive such a lethal die, since there is only one X chromosome in the male; all homozygous females that have the factor in both X chromosomes die, while only those survive that are heterozygous for this factor. Such a heterozygous female produces in equal numbers eggs with and without the lethal factor and if she is bred to a normal male all of the daughters are viable though half of them carry the lethal factor in one of the X chromosomes, but all of the males that receive the lethal factor are non-viable since the male has only one X chromosome, while all the males that survive lack this factor altogether. Thus the sex ratio in this case is 2 females to 1 male. Other lethal factors have occurred in other chromosomes of

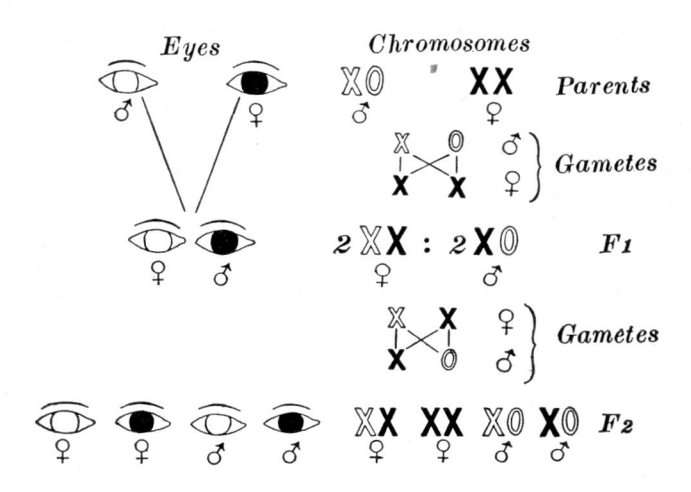

FIG. 66. DIAGRAM OF INHERITANCE OF COLOR BLINDNESS THROUGH THE FEMALE. A color blind female transmits her defects to all her sons, to half of her granddaughters and to half of her grandsons. Corresponding distribution of sex chromosomes on right. (After Morgan.)

192   HEREDITY AND ENVIRONMENT

Drosophila but they were first studied and are most easily demonstrated in the X chromosome.

b. *Other Cases of Linkage.*—In addition to characters which are sex-linked other characters may be bound together in heredity without being linked with sex. Morgan and his associates have found and studied about four hundred mutations of *Drosophila* (see Figs. 103-105), which are inherited in four groups, all the characters of each group usually going together. There have been found in the first group 140 different mutant characters, in the second 125, in the third 120 and in the fourth 3. Or eliminating lethal and modifying factors and those of more doubtful location, there remain 188 mutant genes, 50 of which are in the first group, 70 in the second, 65 in the third, and 3 in the fourth. Corresponding with the number and size of these groups there are four pairs of chromosomes in *Drosophila*, three of which are large and one is very small (Fig. 67). The sex chromosomes (XX in the female, XY in the male) constitute one of the large pairs and the genes of the characters which are sex-linked are located in these chromosomes; the genes of the second and third groups of characters are in the other large chromosomes, while the fourth group of only three characters have their genes in the very small chromosomes (Fig. 67). If this interpretation is correct, linkage is due to the

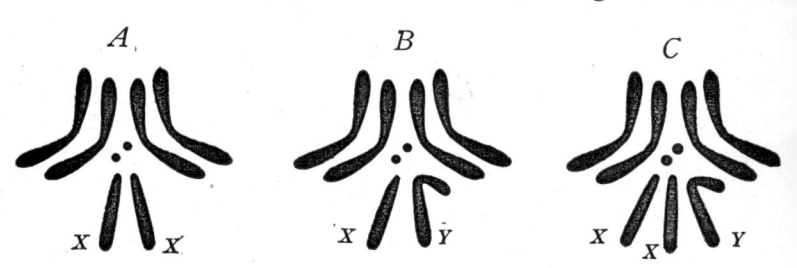

Fig. 67. Chromosomes (Diploid number) of *Drosophila melanogaster*. *A*. Female with 2 X chromosomes; *B*. Male with 1 X and 1 Y; *C*. Matroclinous female (XXY) resulting from non-disjunction of the 2 X chromosomes of the egg. (After Morgan.)

THE CELLULAR BASIS

grouping together of certain genes in certain chromosomes, there are as many groups of characters as there are pairs of chromosomes and as long as the chromosomes preserve their identity the linkage of genes in the chromosomes and of characters in the developed organisms will persist.

c. *"Cross-Overs."*—But linkage of inherited characters is not quite so simple as this statement would indicate for an extensive study of this phenomenon in *Drosophila* has shown that while characters are usually linked in four constant groups this is not always true. For example Morgan has found that when a female fly with white eyes and yellow wings is crossed to a male with red eyes and gray wings, the genes for these characters being linked together in the X chromosomes, all the sons are yellow and have white eyes while all the daughters are gray and have red eyes, gray wings and red eyes being dominant over yellow and white; but when these F_1s are crossed about 99 per cent of the offspring show the same linkage of the colors yellow-white, gray-red, but in 1 per cent the linkage is yellow-red, gray-white. This interchange of characters in the two groups, or "cross-over" as Morgan calls it, may be explained by assuming that there has been an interchange of genes between the two sex chromosomes of the female.[1] When the paired chromosomes lie side by side in synapsis it is known that they sometimes twist around each other and if in their subsequent separation each chromosome should break at the point where the two cross and a portion of one chromosome should be joined to the other one we would have a relatively simple explanation of the interchange or "cross-over" of genes and consequently of the breaking up of the old group of characters and the establishment of a new group (Fig. 68).

Similar interchange of characters takes place in each of the other three groups of *Drosophila,* and can be explained in the

[1] Such "cross-overs" occur only in the female of *Drosophila,* though they may occur in the males of other species. They occur at the stage when the synaptic pairs of chromosomes are long, slender threads.

same way. If chromosomes of a pair are twisted round each other at more than one place and are then broken at these points we get double or multiple crossing over and a corresponding re-grouping of genes and of characters. Unless chromosomes of a pair are very tightly twisted two cross-overs will not occur near together and in general the farther apart points are in a chromosome the more likely is a cross-over to occur. If one per cent of "crossing over" occurs the genes are assumed to be one unit of distance apart; if ten per cent, ten units, etc. On this basis Morgan and his associates have constructed a "map" of each chromosome of *Drosophila* indicating the positions of those genes which

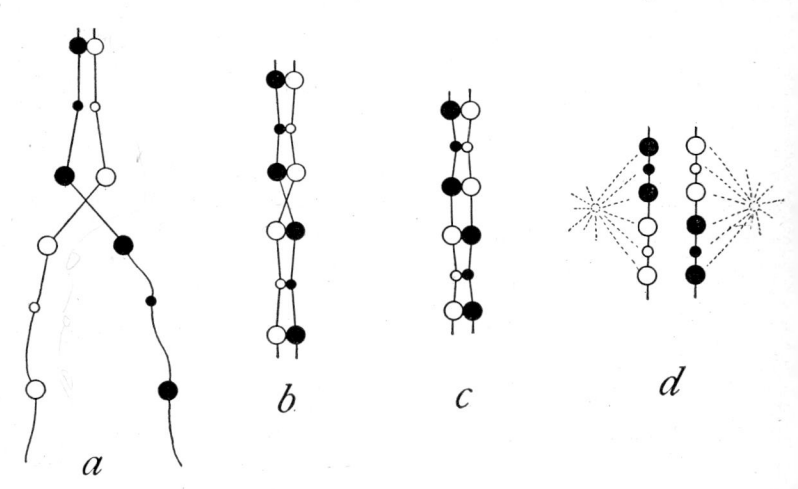

Fig. 68. Diagram Showing the Probable Chromosomal Mechanism by which "Crossing Over" is Caused. Pairs of chromosomes, one from the father, the other from the mother, are shown in synapsis (*a, b, c*) and in the reduction division (*d*). Homologous chromomeres (or allelomorphic genes) are represented by the black and white circles at the same level. In *a* and *b* the chromosomes are shown "crossing over"; in *c* they have broken at the point of crossing and half of each chromosome is joined to half of the other one; in *d* the "crossed over" chromosomes are separating in the reduction division (from Morgan).

THE CELLULAR BASIS

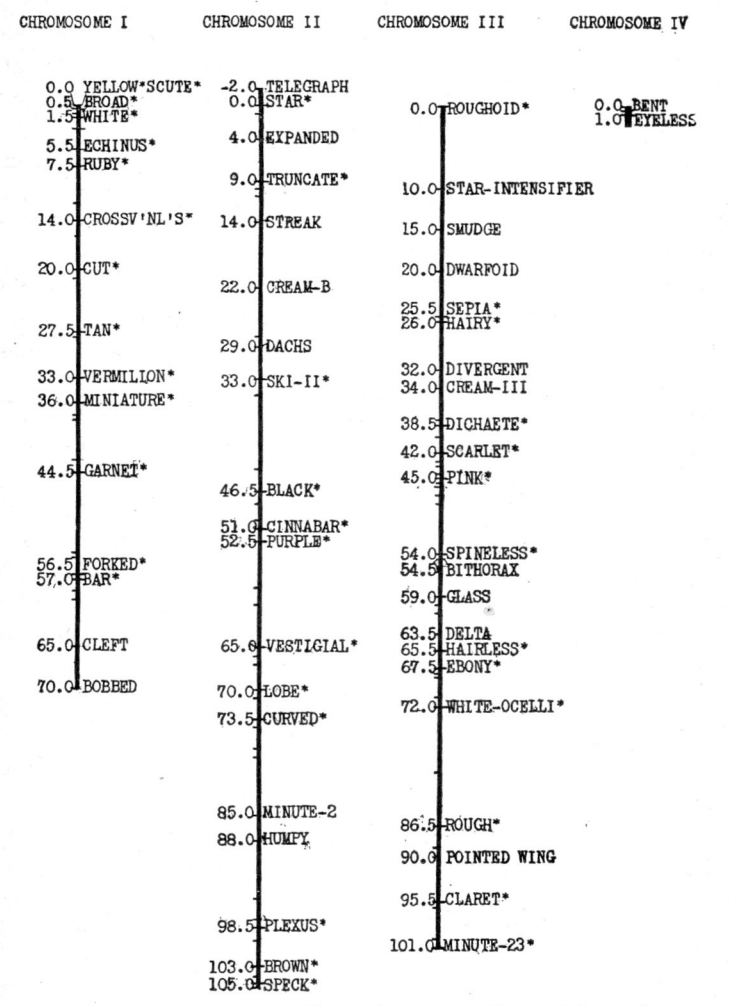

CHROMOSOME I	CHROMOSOME II	CHROMOSOME III	CHROMOSOME IV
0.0 YELLOW*SCUTE*	-2.0 TELEGRAPH	0.0 ROUGHOID*	0.0 BENT
0.5 BROAD*	0.0 STAR*		1.0 EYELESS
1.5 WHITE*			
5.5 ECHINUS*	4.0 EXPANDED		
7.5 RUBY*			
	9.0 TRUNCATE*	10.0 STAR-INTENSIFIER	
14.0 CROSSV'NL'S*	14.0 STREAK	15.0 SMUDGE	
20.0 CUT*		20.0 DWARFOID	
	22.0 CREAM-B	25.5 SEPIA*	
27.5 TAN*		26.0 HAIRY*	
	29.0 DACHS		
33.0 VERMILION*	33.0 SKI-II*	32.0 DIVERGENT	
36.0 MINIATURE*		34.0 CREAM-III	
		38.5 DICHAETE*	
		42.0 SCARLET*	
44.5 GARNET*		45.0 PINK*	
	46.5 BLACK*		
	51.0 CINNABAR*		
	52.5 PURPLE*	54.0 SPINELESS*	
56.5 FORKED*		54.5 BITHORAX	
57.0 BAR*		59.0 GLASS	
		63.5 DELTA	
65.0 CLEFT	65.0 VESTIGIAL*	65.5 HAIRLESS*	
		67.5 EBONY*	
70.0 BOBBED	70.0 LOBE*	72.0 WHITE-OCELLI*	
	73.5 CURVED*		
	85.0 MINUTE-2		
	88.0 HUMPY	86.5 ROUGH*	
		90.0 POINTED WING	
		95.5 CLARET*	
	98.5 PLEXUS*	101.0 MINUTE-23*	
	103.0 BROWN*		
	105.0 SPECK*		

FIG. 69. MAPS OF THE FOUR CHROMOSOMES OF *Drosophila melanogaster* howing positions of a few of the nearly 400 mutant genes; those that have een most accurately located are marked with a star. (From information urnished by Morgan and Bridges.) Figures of several of these Drosohila Mutants will be found on pp. 284-286.

have been determined most accurately (Fig. 69). Thus not only do they locate particular genes in particular chromosomes but they are able to locate the relative positions of these genes in each chromosome. This is in all respects the most remarkable work which has ever been done in this field; for the first time it gives us a detailed picture of what Weismann called the "architecture of the germplasm,"—for the first time it assigns to different genes "a local habitation and a name."

3. CYTOPLASMIC INHERITANCE.—The most direct and the earliest recognized correlations between the oosperm and the developed animal are found in the polarity and symmetry of the egg cytoplasm and of the animal to which it gives rise.

(a) *Polarity.*—In all eggs there is polar differentiation, one pole, at which the maturation divisions take place, being known as the animal pole, and the opposite one being known as the vegetative pole. The substance of the egg in the vicinity of the animal pole usually gives rise to the ectoderm, or outer cell layer of the embryo; the portion of the egg surrounding the vegetative pole usually becomes the endoderm or inner cell layer. The axis which connects these poles, the chief axis of the egg, becomes the gastrular axis of the embryo and in every great group of animals it bears a constant relationship to the chief axis of the adult animal. The polarity of the developed animal is thus directly connected with the polarity of the egg from which it came (Figs. 42-48, 70, 71).

(b) *Symmetry.*—In many cases the symmetry of the developed animal is foreshadowed in the cytoplasm of the egg. The eggs of cephalopods (Fig. 70) and of insects (Fig. 71) are bilaterally symmetrical while they are still in the ovary; in other cases, such as ascidians, *Amphioxus* and the frog, bilateral symmetry appears immediately after fertilization (Figs. 9, 10, 46, 47), though in some of these cases there is reason to believe that the eggs are bilateral even before fertilization; in still other cases bilaterality does not become visible until later in development and we do not

THE CELLULAR BASIS 197

now know whether it is present in earlier stages or not; but wherever it can be recognized in the earlier stages it is probable that the bilateral symmetry of the egg becomes the bilateral symmetry of the developed animal.

(c) *Inverse Symmetry.*—In most animals bilateral symmetry is not perfect, certain organs being found on one side of the midline and not on the other, or being larger or differently located on

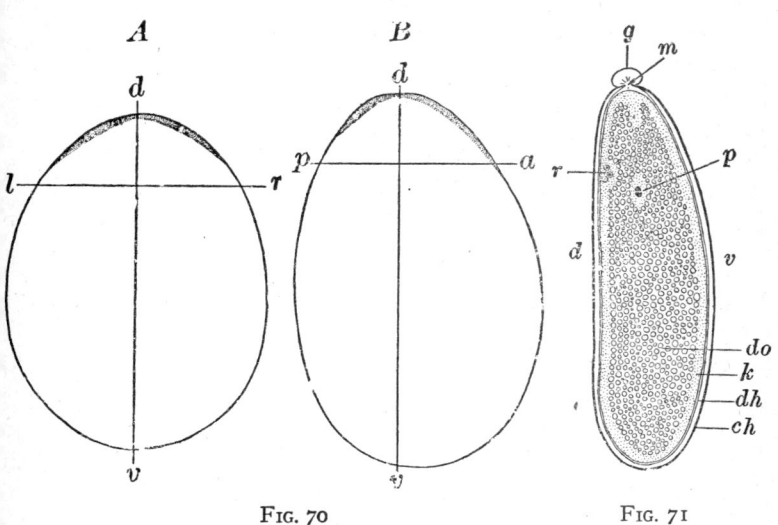

FIG. 70   FIG. 71

FIG. 70. OUTLINES OF THE UNFERTILIZED EGG OF A SQUID, *Loligo*, showing the polarity and symmetry of the egg with reference to the axes of the developed animal; *d*, dorsal; *v*, ventral; *l*, left; *r*, right; *a*, anterior; *p*, posterior. (After Watase.)

FIG. 71. MEDIAN SECTION THROUGH EGG OF A FLY, *Musca*, just after fertilization, showing the relations of the polarity and symmetry of the egg to the axes of the developed animal; the long axis of the egg corresponds to the antero-posterior axis of the animal; *d*, dorsal; *v*, ventral; *m*, micropyle through which sperm enters the egg; *g*, glutinous cap over the micropyle; *r*, polar bodies; *p*, egg and sperm nuclei; *do*, yolk; *k*, peripheral layer of protoplasm; *dh*, vitelline membrane of egg; *ch*, chorion. (After Korschelt and Heider.)

one side as compared with the other; among all such animals variations occasionally occur which show a complete reversal of these asymmetrical organs, *i.e.,* in man the heart and arch of the aorta may occur on the right side instead of the left, the pyloris and chief portion of the liver on the left instead of the right, etc. Among certain snails this inversion of symmetry may occur regularly in certain species and not in others, the inverse form being known as sinistral and the ordinary form as dextral (Fig. 74). In these sinistral snails, and probably in other animals showing inverse symmetry, the embryo is inversely symmetrical and every cleavage of the egg from the first to the last is the inverse of that which occurs in dextral snails (Figs. 72-74). There is good reason to believe that in such cases the unsegmented egg is also inversely symmetrical as compared with the more usual type (Fig. 72). In all of these cases there is a direct correspondence between the polarity and symmetry of the oosperm and the polarity and symmetry of the developed animal (Figs. 70-74).

(d) *Localization Pattern.*—In many animals the ectoderm and mesoderm may be traced back to areas of peculiar protoplasm in the oosperm, but, in addition to this, one can recognize in the ascidian egg areas of peculiar protoplasm which will give rise to mesenchyme, muscles, nervous system and notochord, and these substances are present in the oosperm in the approximate positions and proportions which they will have in the embryo and larva (Figs. 10, 11, 46-48).

Indeed there are types of localization of these cytoplasmic materials in the egg which are characteristic of certain phyla; thus there are the ctenophore, the flat-worm, the echinoderm, the annelid-mollusk and the chordate types of cytoplasmic localization (Fig. 75). The polarity symmetry and pattern of a jellyfish, starfish, worm, mollusk, insect or vertebrate are foreshadowed by the characteristic polarity, symmetry and pattern of the cytoplasm of the egg either before or immediately after fertilization. In all of these phyla, eggs may develop without fertilization, either by natural or by artificial parthenogenesis, and in such cases the characteristic polarity, symmetry and pattern of the adult are found in the cytoplasm of the egg just as if the latter had been fertilized. The conclusion seems to be justified that these earliest and most fundamental differentiations which distinguished the eggs of various phyla are not dependent upon the entering spermatozoon.

Share of Egg and Sperm in Heredity.—All of these correspondences between the polarity, symmetry and pattern of the egg and of the developed animal are found in the cytoplasm. No doubt the differentiations of the cytoplasm of the egg as well as the peculiar form and structure of the spermatozoon have arisen, during the genesis of these cells, under the influence of patenal and maternal chromosomes as well as of the environment, just as in the differentiation of any tissue cell; but in the case of the spermatozoon these cytoplasmic differentiations are lost when it enters the egg, whereas those of the egg persist. In short ontogeny begins in the egg before fertilization whereas the sperm can influence ontogeny only after, and usually a considerable time after, it enters the egg.

The fact remains that *at the time of fertilization the potencies of the two germ cells are not equal, the polarity, symmetry, type of cleavage, and the pattern, or relative positions and proportions of future organs, being foreshadowed in the cytoplasm of the egg cell, while only the differentiations of later development are influenced by the sperm. In short the egg cytoplasm determines the early development and the sperm and egg nuclei control only later differentiations.*

We are vertebrates because our mothers were vertebrates and produced eggs of the vertebrate pattern; but the color of our skin and hair and eyes, our sex, stature, and mental peculiarities were determined by the sperm as well as by the egg from which we came. The chromosomes of the egg and sperm are the seat of the differential factors or determiners for Mendelian characters,

200 HEREDITY AND ENVIRONMENT

Figs. 72, 73, 74. The Cause of Inverse Symmetry in Snails. In each case the right-hand column represents dextral forms, the left-hand column sinistral ones.

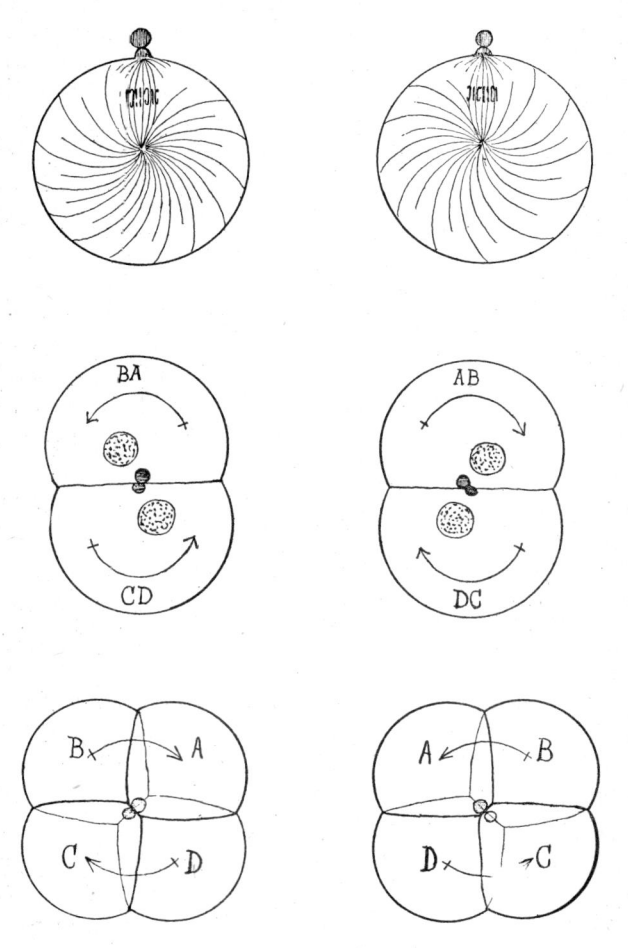

Fig. 72. Normal and Inverse Symmetry in the Unsegmented Egg and in the First and Second Cleavages.

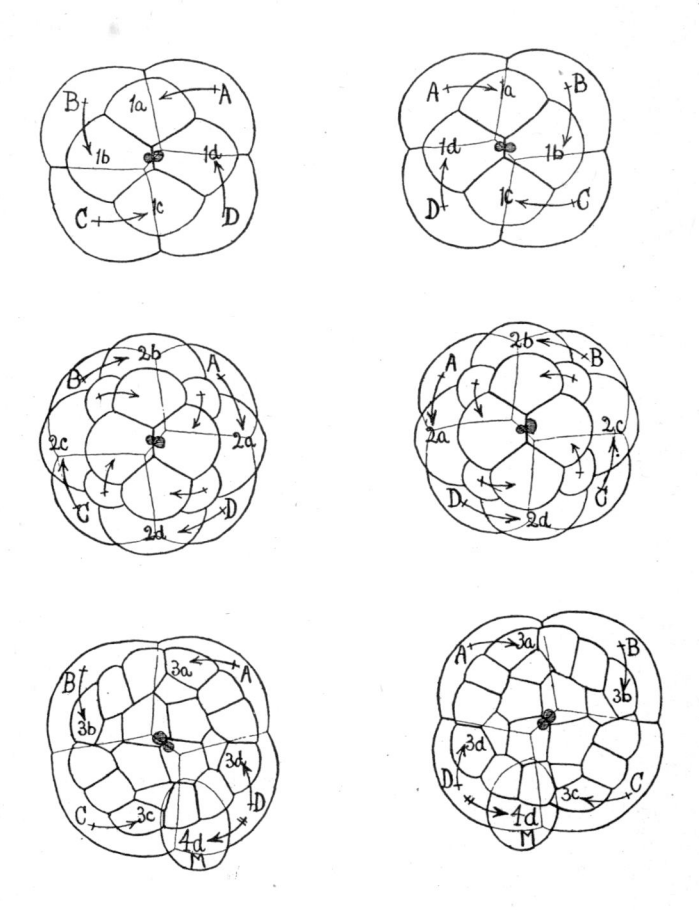

FIG. 73. NORMAL AND INVERSE SYMMETRY OF THE 3D, 4TH, 5TH AND 6TH CLEAVAGES. The cells 1a-1d, 2a-2d and 3a-3d give rise to all the ectoderm; 4d or M gives rise to mesoderm; A, B, C, D to endoderm.

202   HEREDITY AND ENVIRONMENT

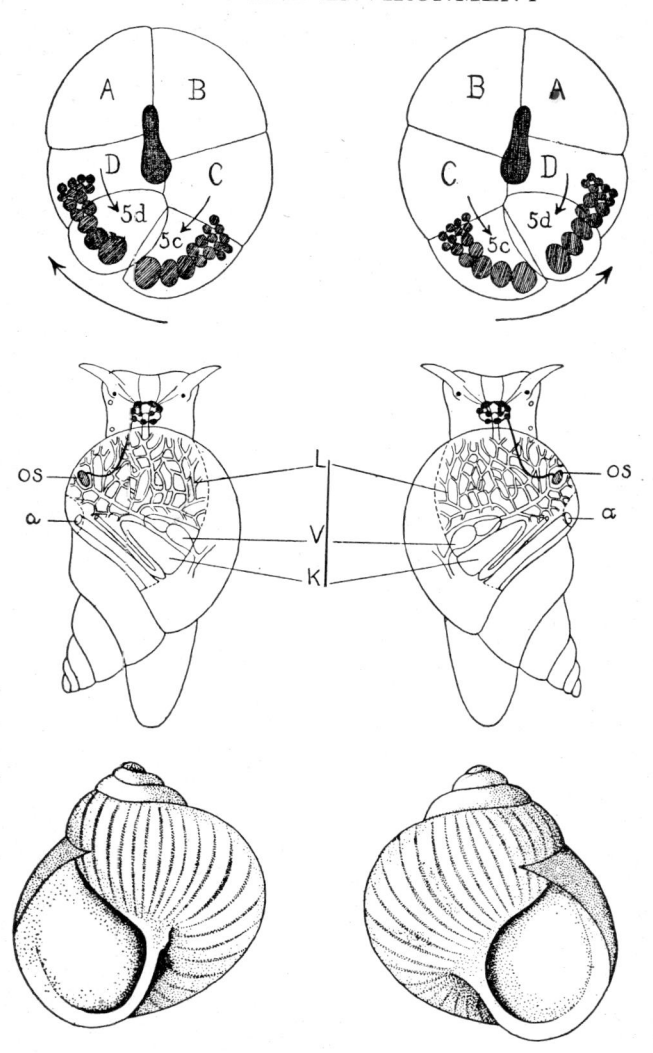

Fig. 74. Normal and Inverse Symmetry in Late Embryos and Adult Stages. In 1, black area is blastopore; cells shaded by lines, mesoderm; other cells, endoderm; the spiral twist of the snail begins in opposite directions in the two embryos. In 2, the adult organization is shown with all organs inversely symmetrical; *os*, olfactory organ, *a*, anus; *L*, lung; *V*, ventricle; *K*, kidney. In 3, sinistral and dextral shells of adult snails are shown.

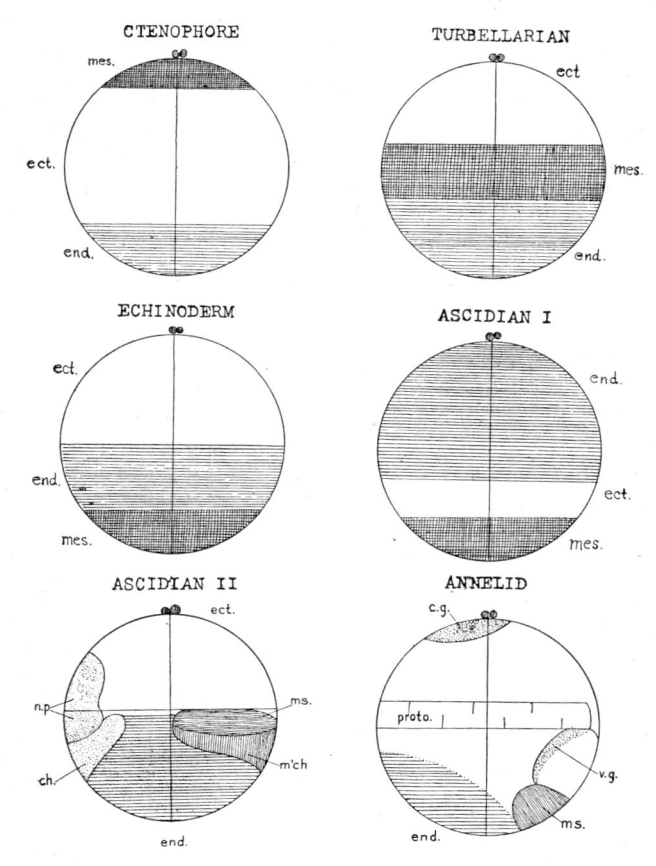

FIG. 75. TYPES OF EGG ORGANIZATION IN DIFFERENT PHYLA; cross-hatched area, mesoderm or mesenchyme (*mes*); horizontal lines, endoderm (*end*); clear area, ectoderm (*ect*). In the first four figures the pattern of localization is that which is found at the close of the first cleavage; in Ascidian II the pattern is that which is found at the close of the second cleavage; in the annelid egg the localization of later stages is projected upon the egg; *n.p.*, neural plate; *ch.*, chorda; *m'ch.*, mesenchyme; *ms.*, muscle; *c.g.*, cerebral ganglion; *v.g.*, ventral ganglion; *proto.*, protoderm.

but the general polarity, symmetry and pattern of the embryo are egg characters which were determined before fertilization. But these egg characters, like any other character of the female, were probably determined by chromosomes derived from her father and mother; if so they are Mendelian characters inherited by the mother through her chromosomes and carried over to the first filial generation, not as factors but as developed characters. Such cases may be called "maternal inheritance" since the characters come only from the egg, or "preinheritance" since these egg characters are developed before fertilization. (See pp. 113, 114.)

Share of Chromosomes and Cytoplasm in Heredity and Development.—It will be observed that the correlation between chromosomes and adult characters is different in kind from that between the cytoplasm of the egg and adult characters; in the latter case polarity, symmetry and pattern of localization are characters of the same kind in the egg and in the adult, and the correspondence comparatively close; in the former there is no correspondence in *kind* between the chromosomal peculiarities and the peculiarities of the adult. This fact suggests that the chromosomal organization is more fundamental than that of the cytoplasm; the chromosomes contain the germplasm, the cytoplasm is the somatoplasm; the chromosomes are chiefly concerned in heredity, the cytoplasm in development.

E. THE MECHANISM OF DEVELOPMENT

Development consists in the transformation of the oosperm into the adult. What is the mechanism by which this transformation is effected? There is progressive differentiation of the germ into the developed organism but by what process is this differentiation accomplished?

Many different processes are concerned in embryonic differentiation. From the standpoint of the cell the most important of these are (1) the formation of different kinds of substances in cells, (2) the localization and isolation of these substances, (3) the transformation of these substances into various struc-

THE CELLULAR BASIS

tures which are characteristic of the different kinds of tissue cells. We shall here describe only the first and second of these processes which are of more general interest than the last.

1. *The Formation of Different Substances in Cells.*—Embryonic differentiation consists primarily in the formation of different kinds of protoplasm out of the protoplasm of the germ cells. It is plain that different kinds of protoplasm are present in the two germ cells before they unite in fertilization, but in the course of development the number of substances present and the degree of difference among them greatly increase.

Actual observation shows that by the interaction with one another of substances or parts originally present and by their reactions to external stimuli new substances and parts appear which had no previous existence just as new substances result from chemical reactions. This is "creative synthesis" in general science, epigenesis in development. Differentiations appear chiefly in the cytoplasm but only as the result of interaction between cytoplasm and nucleus. Similarly, it may be argued, smaller units of organization such as chromosomes or chromomeres do not in themselves give rise to any adult parts, but only as they interact upon other units are new parts formed.

In many cases the first formation of such new substances appears in the immediate vicinity of the nucleus and, like assimilation itself, is evidently brought about by the interaction of nucleus and cytoplasm. In certain cases it can be seen that the chromatin and oxychromatin which escape from the nucleus during cell division take part in the formation of new substances in the cell body, and since the oxychromatin is derived from the chromosomes of the previous cell division, it is probable that the chromosomes are indirectly a factor in this process.

Weismann maintained that the chromosomes and the inheritance units contained in them undergo differentiation by a process of disintegration and that these disintegrated units escape into the cell body and there produce different kinds of cytoplasm in different cells. A somewhat similar view was advanced by deVries in his theory of "intra-cellular pangenesis." However, as we have seen already, there is good evidence that the chromosomes do not undergo progressive differentiation in the course of development; they always divide with exact equality, and even in highly differentiated tissue cells their number and form usually remain as in embryonic cells.

On the other hand the cytoplasm undergoes progressive differentiation, and when by pressure or centrifugal force it is brought into relations with other nuclei the differentiations of the cytoplasm are not altered thereby, thus showing that the different nuclei are essentially alike and that differentiations are mainly limited to the cytoplasm. Thus the differentiations of cells are not due to the differentiations of their nuclei, but rather the reverse is true, such differentiations of nuclei as occur are due to differentiations of the cytoplasm in which they lie. Nevertheless differentiations do not take place in the absence of nuclear material, and it seems probable that the interaction of nucleus and cytoplasm is necessary to the formation of the new cytoplasmic substances which appear in the course of development.

2. *Segregation and Isolation of Different Substances in Cells* —But differentiation consists not only in the formation of different kinds of substances in cells but also in the separation of these substances from one another. This separation is brought about to a great extent by flowing movements within cells which are associated especially with cell division.

In all these processes of heredity and development cell division plays a particularly important part. If cell divisions were always exactly alike there could be no initial difference between the daughter cells, and unless acted upon by different stimuli all cells would remain exactly alike. But there is much evidence that daughter cells are often unlike from the time of their formation, and that different stimuli act upon them to increase still further this initial difference.

THE CELLULAR BASIS

(a) *Differential and Non-differential Cell Division.*—When each half of any dividing unit is like the other half the division is non-differential. So far as we know the divisions of all the smallest elements of the cell are of this sort; there is no good evidence that the plastosomes, the chromomeres, or the chromosomes ever divide into unlike halves, though in the maturation divisions the separation of whole chromosomes leads to the appearance of a differential division of the chromosomes. But while all of the cell elements may be supposed to grow and divide into equivalent halves there may be an unequal distribution of these elements in cell division, so that the two daughter cells are unlike. This is what is known as differential cell division and it plays a most important part in differentiation. While the chromatin is equally distributed to the daughter cells, except in the case of the maturation divisions, the achromatin and the oxychromatin of the nucleus are not always distributed equally and this is probably an important factor in development. The divisions of the cytoplasm of the egg are frequently differential and such divisions are known to play a great part in embryonic differentiation.

(b) *Isolation of Cytoplasmic Substances by Division Walls.*—In the differential divisions of the cytoplasm unlike substances become localized in certain parts of the cell body, chiefly by means of definite flowing movements of the cytoplasm, and when cell division occurs these substances become permanently separated by partition walls. In this way irreversible differentiations are formed. If the formation of partition walls is prevented the different substances within the cell body may freely commingle, especially during nuclear division when the cytoplasmic movements are especially active; in such cases differentiation may be arrested even though nuclear division continues. In the developing eggs of most animals partition walls between daughter cells are necessary to prevent the commingling of different kinds of substances, which are sorted by the movements within the cell and are isolated by the partition walls. In some cases, as for example in certain protozoa, the commingling of different kinds of protoplasm within a cell may be prevented by the viscosity of portions of the protoplasm, or by the formation of intracellular membranes, or by a reduction to a minimum of the mitotic movement within the cell by the persistence of the nuclear membrane during division. In general the degree of differentiation may be measured by the degree of unlikeness between different cells, and by the completeness with which the protoplasm of different cell is kept from intermingling.

3. *The Chromosome Theory of Heredity Applied to Embryonic Differentiation.*—According to the chromosome theory of heredity the inheritance factors are located in the chromosomes and the cytological evidence shows that chromosomes always divide equally and presumably every cell of an individual contain the same kinds of chromosomes and the same kinds of inheritance factors. How then is it possible to explain embryonic differentiation? How can identical factors give rise to different product in different cells?

This is evidently due to the fact that while the division of chromosomes is non-differential, that of the cell body is often differential and the same chromosomes and genes acting upon different kinds of cytoplasm will produce different results. But differential cell-division is the result of definite movements in the cytoplasm, of definite orientations of spindles and cleavage planes and ultimately of a definite polarity and symmetry of the cytoplasm. There is abundant evidence that these cytoplasmic orientations are not the immediate results of chromosomal activity and even if some of them may be the remote results of such activity it is logically impossible to place all the differential factors of development in non-differentiating genes.

On the other hand if embryonic differentiations are produced by the interaction of chromatin and cytoplasm, and if the chromatin does not undergo differentiation, it follows that some of the

THE CELLULAR BASIS

differential factors of heredity and development must be located in the cytoplasm. Such factors would probably not be genes and would not be transmitted in Mendelian fashion, but they would need to be present in the cytoplasm from the very beginnings of ontogeny. They need not be numerous—in fact they are probably few in number—but they are absolutely indispensable to development. If a few orientating differentiations such as polarity and symmetry are present in the cytoplasm at the beginning of ontogeny all other differentiations of development can be explained as due to the interaction of non-differentiating genes on different parts of this cytoplasm, but there is no mechanism by which embryonic differentiations could come from the action of non-differentiating genes on a homogeneous cytoplasm. The genes or Mendelian factors are undoubtedly located in the chromosomes and they are sometimes regarded as the only differential factors of development, but if this were true these genes would of necessity have to undergo differential division and distribution to the cleavage cells; since this is not true it must be that some of the differential factors of development lie outside of the nucleus and if they are inherited, as most of these early orientations are, they must lie in the cytoplasm.

Summary

All the phenomena of life, including heredity and development, are cellular phenomena in that they include only the activities of cells or of cell aggregates. The cell is the ultimate independent unit of organic structure and function. In sexual reproduction the only living bond between one generation and the next is found in the sex cells and all inheritance must take place through these cells. Inherited traits are not transmitted from parents to offspring but the germinal factors or causes are transmitted, and under proper conditions of environment these give rise to developed characters. Every oosperm as well as every developed organism differs more or less from every other one and this remarkable condition is brought about by extremely numerous permutations in the distribution of the chromosomes of the sex cells in maturation and fertilization. Sex is an inherited character dependent, primarily, upon an alternative distribution of certain chromosomes to the germ cells. There is much evidence that the factors for all sorts of Mendelian characters are associated with the chromosomes. The differentiation of the oosperm into the developed organism is accomplished in part by the interaction of chromosomes and cytoplasm which leads to the formation of new materials, and in part by the segregation and localization of these materials in definite cells.

Germ cells and probably all other kinds of cells are almost incredibly complex. We know that former students of the cell greatly underestimated this complexity and there is no reason to suppose that we have fully comprehended it. What Darwin said of the entire organism may now be said of every cell: It "is a microcosm—a little universe, formed of a host of self-propagating organisms, inconceivably minute and numerous as the stars in heaven."

CHAPTER IV

INFLUENCE OF ENVIRONMENT

CHAPTER IV

INFLUENCE OF ENVIRONMENT

The development of an individual or the evolution of a race is dependent upon the interaction of two sets of factors or causes, the intrinsic and the extrinsic. The former are represented by the organization of the germinal protoplasm, the latter by all other conditions; the former are known as heredity or constitution, the latter as environment or education; or in the words of Galton, these two sets of factors may be called "nature" and "nurture." The great problem of development is the unraveling of these two factors, the assignment of its true value to each, and the ultimate control of development so far as this may be possible through the knowledge thus gained.

A. RELATIVE IMPORTANCE OF HEREDITY AND ENVIRONMENT

The distinction between these two factors of development is generally recognized and the question of the relative importance of the two has been discussed for ages. Which is the more important, constitution or environment? What characteristics are due to nature and what to nurture? To what extent is man the creature of heredity, to what extent the product of education? The old question "Which of you by taking thought can add one cubit to his stature," is a vital question to-day. To what extent may nature be modified by nurture? To what extent may education make up for deficiencies of birth?

1. *Former Emphasis on Environment.*—Formerly very great emphasis was placed upon influences of environment in phylogeny

and ontogeny. From the earliest times it has been believed that species might be transmuted by environmental changes and that even life itself might arise from lifeless matter through the influence of favorable extrinsic conditions. If environment could exert so great an influence on the origin of species or even of life itself much more could it affect the process of development of the individual. It is still popularly supposed that complexion is dependent upon the intensity of light, and stature upon the quantity and quality of food, that sex is determined by food or temperature, mentality by education, and that in general individual peculiarities are due to environmental differences.

Many philosophers of the seventeenth and eighteenth centuries taught that man was the product of environment and education and that all men were born equal and later became unequal through unequal opportunities. Descartes begins his famous "Discourse on Method" with these words:

"Good sense is, of all things among men, the most equally distributed . . . The diversity of our opinions does not arise from some being endowed with a larger share of Reason than others, but solely from this, that we conduct our thoughts along different ways, and do not fix our attention on the same objects."

Similar views were expressed by Rousseau and Diderot, and especially by John Locke and Adam Smith. The Declaration of Independence merely reflected the spirit of the age in which it was written when it held this truth to be self evident, "that all men are created equal." The equality of man has always been one of the foundation stones of democracy. Upon this belief in the natural equality of all men were founded systems of theology, education and government which hold the field to this day. Upon the belief that men are made by their environment and training rather than by heredity are founded most of our social institutions with their commands and prohibitions, their rewards and punishments, their charities and corrections, their care for the education and environment of the individual and

INFLUENCE OF ENVIRONMENT

their disregard of the inheritance of the race. To a large extent civilization itself means good environmental conditions, and the advance of civilization means improvement of environment.

2. *Present Emphasis on Heredity.*—On the other hand modern studies in genetics are emphasizing the immense, the overwhelming importance of heredity, in both phylogeny and ontogeny. No one now takes seriously the assertion that life can be experimentally produced at the present time from non-living matter. It is evident that the artificial production of life is a much more difficult problem than was once supposed, and it may be an insoluble problem. The first flush of enthusiasm over experimental methods in biology led to the expectation that we would soon be making species by the process of experimental evolution, but the results of two or three decades of such experimental work have been somewhat disappointing. Inherited variations do appear, incipient species arise, but there is very little evidence to show that they appear in response to environmental changes and at present we have no means of controlling such variations. Belief in the omnipotence of environment for the evolution of species has steadily waned in recent years, while a belief in the intrinsic causes of evolution, such as the mutation theory and orthogenesis, has increased.

In ontogeny also the environmental or extrinsic factors of development have been relegated to a subordinate place, while the intrinsic or hereditary factors appear more important than ever. The old view that men are chiefly the product of environment and training is completely reversed by recent studies of heredity. The modifications which may be produced by environment and education are small and temporary as compared with those which are determined by heredity.

3. *Both Indispensable.*—These conclusions are, in the main, well founded. The evidence of the tremendous importance of heredity is so complete that we may rest assured that thinking men will never again return to the position which prevailed until

a few years ago regarding the all-importance of environment. And yet there is danger of going too far in the opposite direction. Neither environment nor heredity is all-important, but both are necessary to development. The germ cells with all their inherent possibilities would forever remain germ cells were it not for environmental stimuli. The realization of germinal possibilities is dependent upon the responses of the germ to environmental stimuli, and although heredity is a relatively constant factor while environment is a more variable one, nevertheless the two are indispensable to development. Only by experiment can the relative importance of heredity and environment in development be determined. Extensive experiments have been made within recent years on developing animals and plants in order to discover the factors involved in development, and the modifications which may thus be produced are very striking.

B. EXPERIMENTAL MODIFICATIONS OF DEVELOPMENT

The study of development under experimental conditions has given rise to a new branch of biology, viz., experimental embryology or the physiology of development. By changes in environmental conditions notable modifications may be produced in adult organisms, but these modifications are much greater when the changed environment acts on the organism during the period of its development.

1. Developmental Stimuli

It is by no means easy to define such general terms as "environment," "stimulus," and "response." In its common use "environment" means all that lies outside the individual, if it is defined from the standpoint of the entire organism. But from the standpoint of an organ or cell it is the surrounding organs, cells or fluids of the body; the latter may be defined as "internal environment." If developmental stimuli arise outside the organism they are plainly extrinsic or environmental, but if they arise

INFLUENCE OF ENVIRONMENT

within the organism they are said to be intrinsic though they may be due to changes in the "internal environment."

Stimuli are chiefly energy changes of a physical or chemical nature. A stimulus to which an adult organism responds by movements or other activities may call forth or inhibit developmental responses when applied to germ cells or embryos.

These developmental stimuli may be classed as:

1. *Physical stimuli* including the following, (a) mechanical, (b) thermal, (c) electrical, (d) radiant, (e) light, (f) density of medium, (g) gravity and centrifugal force, etc.

2. *Chemical stimuli* include the action of (a) substances found in normal development, such as oxygen, carbonic acid, water, food, secretions of ductless glands, etc. and (b) substances not found in normal development, such as various salts, acids, alkalis, alcohol, ether, tobacco, etc.

3. *General vs. Specific Stimuli.*—In general the action of these stimuli during development does not call forth a perfectly specific and definite response of the organism; various stimuli may produce the same result. Thus artificial parthenogenesis has been produced by almost every stimulus named, and weakened or retarded development is produced by many different stimuli.

By the elimination of certain of these stimuli which are normally present, or by introducing stimuli which are not usually present, very important and even profound changes in development may be produced. In this way animals have been formed which are turned inside out, or side for side, or in which heads or nervous systems or muscles or backbones are lacking, or in which the various organs are not found in normal positions. In this way dwarfs and giants and one-eyed monsters as well as all sorts of double and partial embryos have been formed. In general monstrous and defective forms of development are due to alterations of the normal environment rather than to defective heredity.

II. DEVELOPMENTAL RESPONSES

The character of developmental responses to stimuli depends primarily upon (a) the nature of the organism and (b) the stage of development at which the stimulus acts. Modifications are more easily produced and are more profound during cell division than during intervening periods and at early stages of development than at later ones; indeed conditions which have no serious effect on an adult organism may greatly modify the development of an embryo or germ cell.

1. *Modifications of Germ Cells before Fertilization.*—It has been found by many investigators that development may be profoundly changed by influences acting upon the germ cells before fertilization. In general environmental changes acting during the growth of eggs or spermatozoa and especially during their maturation may produce marked changes in development though rarely if ever in heredity. Tower maintained that unusual conditions of temperature and humidity during the later stages of oogenesis and spermatogenesis may lead to the production of new races in the case of the potato beetle (Fig. 102) and MacDougal's experiments on plants led him to the conclusion that chemical substances may influence the ovules so as to change the hereditary character of the plant. Other workers have failed to confirm these results and it is doubtful whether these changes in hereditary constitution were caused by the changed environment. Bardeen and the Hertwigs have shown that great monstrosities may be produced if X-rays, radium or various chemical substances are allowed to act on spermatozoa before fertilization, but there is no evidence that these changes are inherited.

Effects of Alcohol.—Stockard subjected adult male and female guinea-pigs to the fumes of alcohol for some time before breeding them and then studied the effects of this drug on their offspring. He found that the influence of alcohol on the spermatozoa is as deleterious as when acting on the ova and that it produces sterility, or greatly reduced fertility, a great excess of still births, and weak and sickly offspring (Fig. 76). Later he and Papanicolau found that the deleterious effects disappear and after the fourth filial generation the offspring of alcoholized parents are stronger and live longer than normal controls. They attribute this to the elimination of weak germ cells, embryos and developed individuals that were most injured by the alcohol; consequently the final survivors are stronger and more vigorous than the controls in which the weak are preserved along with the strong.

Pearl found that the offspring of alcoholized chickens were on the whole stronger than those from normal animals and he attributes this to the elimination of weaker germ cells and embryos, so that only the most sturdy survive. The work of both Stockard and Pearl leaves no grounds for doubting that alcohol kills or injures many germ cells and Stockard has demonstrated that such injured cells may give rise to defective individuals and that this injury may persist through two or three generations. The fact that spermatozoa are affected even more than ova and that the injury persists to the third filial generation indicates that the chromatin of these cells is injured. Undoubtedly chromatin as well as cytoplasm may be injured by various unfavorable conditions, and if the injury is not too great it may persist through several generations and may cause defective development; but this is probably a different thing from the "inheritance of an acquired character"; its effects are seen not in particular characters but in a general weakening of development; not in mutative changes in genes, but in their temporary injury.

In venturing to apply Stockard's discoveries to human beings it should not be forgotten that his guinea-pigs were alcoholized to a degree far greater than ever occurs in man. Some of them that were five years old had been kept intoxicated for more than four years. It is probable that the use of alcoholic beverages never produces such serious effects on human germ cells as in the case of these guinea-pigs which were compelled to inhale the fumes of strong alcohol throughout the greater part of life. Elderton and

220   HEREDITY AND ENVIRONMENT

Pearson made a mathematical study of children, approximately nine years old, from temperate and from intemperate parents and they concluded that parental alcoholism had practically no effect upon them. However, the more serious the injury to germ cells the sooner they die and it may be that children that survive for nine years come from germ cells that were least injured, while those that were more seriously injured produced individuals that died before or shortly after birth.

Hoppe believes that a single drunken debauch may so injure

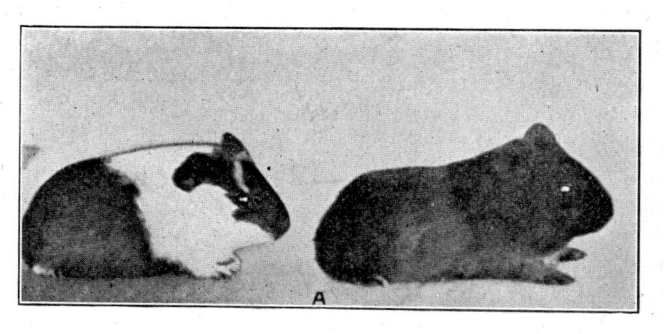

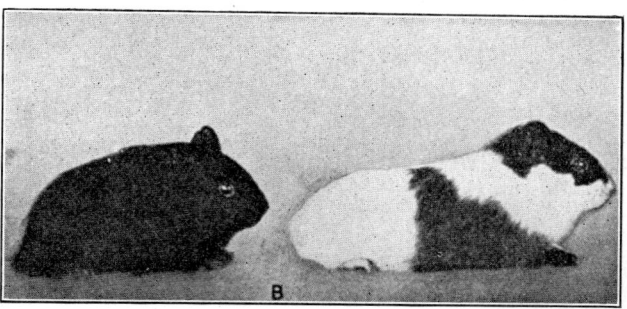

FIG. 76. DWARFED GUINA-PIGS ON THE LEFT AND NORMAL ONES ON THE RIGHT. All are of approximately the same age though the normal ones are nearly twice the weight of the dwarfs. The normals came from normal parents, the dwarfs from a normal mother and an alcoholic father; the dwarfing has therefore been produced by the influence of alcohol on the spermatozoa. (From Stockard.)

the germ cells of man as to produce abnormal and defective offspring. Hertwig concludes that the great prevalence of the drug habit may seriously affect the germ cells and their subsequent development; Forel has for many years maintained that one of the most serious causes of human malformations and degenerations is to be found in the effect of alcohol on the germ cells, especially at the time of conception. However these conclusions are by no means certain.

2. *Modifications During Fertilization Stages.*—Environmental changes acting during fertilization may cause more than one spermatozoon to enter the egg or may injure the egg or sperm; in either case the resulting development is abnormal. Where two or more spermatozoa enter the egg the nuclear divisions are usually abnormal, as Boveri has shown in the case of the sea-urchin; the distribution of chromosomes to different cleavage cells is unequal and such cells do not undergo typical development, while the embryo or larva produced is not capable of continued life. In cases where an egg is fertilized by a spermatozoon belonging to a different phylum or class (heterogeneous fertilization) the foreign sperm, after stimulating the egg to begin development, may itself die or remain inactive, in which case the hereditary traits which develop are those of the mother only. In many animals unfertilized eggs may be stimulated to begin development by a great variety of changes in the medium, all such cases being known as "artificial parthenogenesis."

3. *Modifications of Development after Fertilization.*—Environmental changes, acting upon the oosperm after fertilization, or upon the embryo, may produce an almost infinite variety of abnormal types of development, but so far as known none of these modifications becomes hereditary. It seems probable that changes in hereditary constitution take place in the main before fertilization and especially during the maturation divisions and all changes that affect germ cells must of course occur in the "germ track."

Isolation of Cleavage Cells.—If the cleavage cells are separated from one another in the 2-cell stage each of them may give rise to an entire animal (Fig. 77); in this way two complete animals may be derived from a single egg of a star-fish or sea-urchin, of an amphioxus, and of several other animal types. If the frog's egg is turned upside down in the 2-cell stage, double-headed or double-bodied embryos may result (Fig. 78). In such cases each cleavage cell is said to be totipotent, that is, it is capable of giving rise to an entire animal. Recent work shows that after the 2-cell stage individual blastomeres of these eggs are not entirely totipotent.

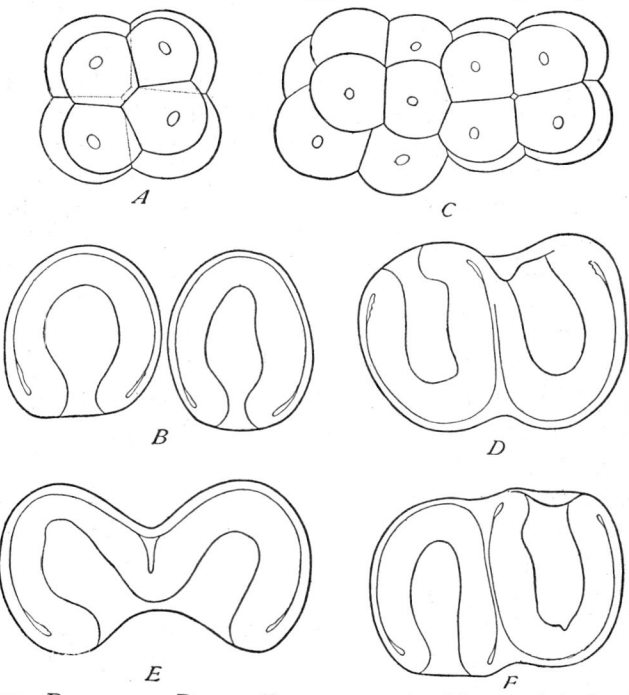

FIG. 77. DWARF AND DOUBLE EMBRYOS OF *Amphioxus*. *A*, isolated blastomere of the 2-cell stage segmenting like an entire egg. *B*, twin gastrulæ from a single egg. *C*, double cleavage resulting from partial separation of the first two cleavage cells. *D, E, F*, double gastrulæ arising from such forms as *C*. (From Wilson.)

On the other hand in certain animal phyla such as the ctenophores, mollusks, annelids and ascidians isolated cleavage cells give rise only to part of an animal; in this way one may get a right or left half of an animal (Fig. 79) from right or left cleavage cells; an anterior half (Fig. 80), or a posterior half (Fig. 81) from anterior or posterior cleavage cells; or any one of the cells of the 4-cell stage may produce the corresponding quarter of an entire animal. Such cases are known as "mosaic development."

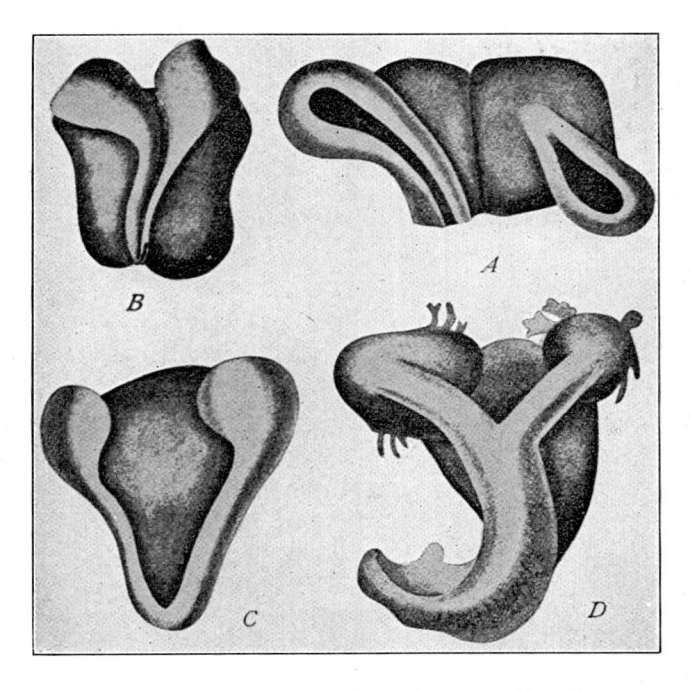

Fig. 78. Double Embryos of Frog Developed from Eggs Inverted when in the 2-cell Stage. *A*, twins with heads turned in opposite directions. *B*, twins united back to back. *C*, twins united by their ventral sides. *D*, double headed tadpole. (From Wilson after O. Schultze.)

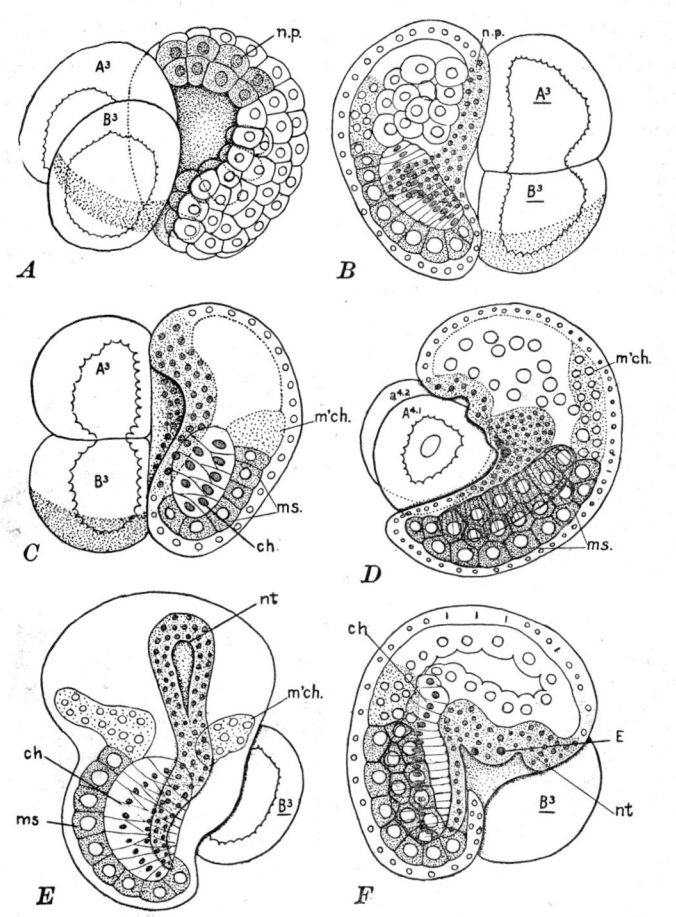

FIG. 79. HALF AND THREE-QUARTER EMBRYOS OF *Styela*. *np*, nerve plate; *nt*, nerve tube; *E*, eye; *mch*, mesenchyme; *ms*, muscle; *ch*, notochord. *A*, right half-blastula which developed after the left half of the egg, A_3B_3, had been killed. *B*, left half larva from the two left cells of the 4-cell stage, the right cells, A_3B_3, having been killed. *C*, similar right half-larva. The muscle cells (stippled) occur only on one side of the notochord. *D*, three-quarter larva, the left anterior cells been killed. *E, F,* three-quarter larvæ, the right posterior cell B_3 having been killed.

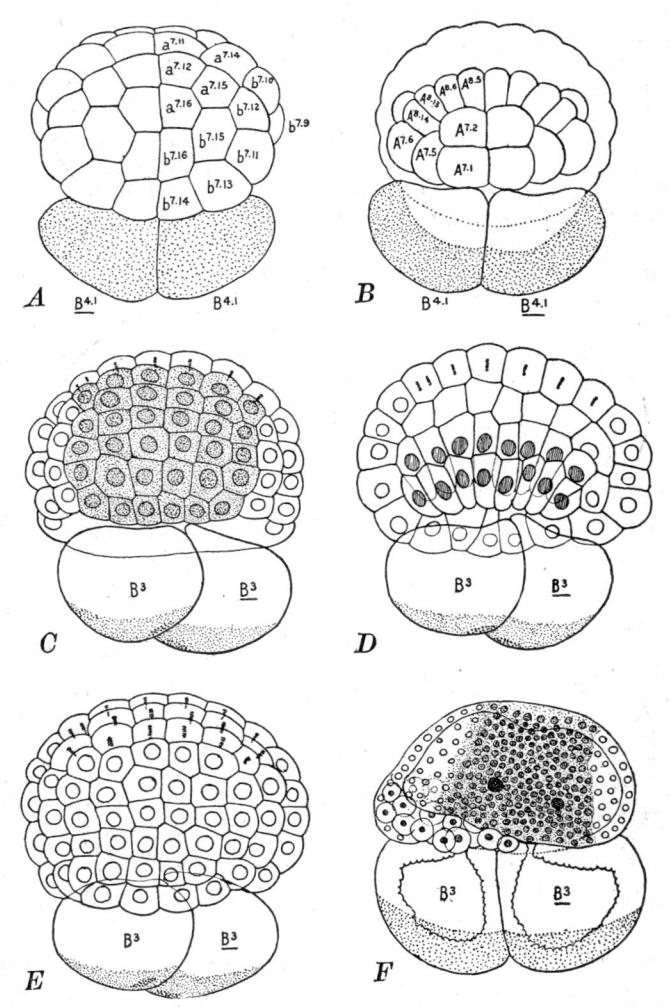

FIG. 80. ANTERIOR HALF-EMBRYOS OF *Styela*, the posterior cells having been killed in the 4-cell stage. Neural plate, eye-spots and chorda cells are present but no muscle cells or tail, which develop from the posterior cells only.

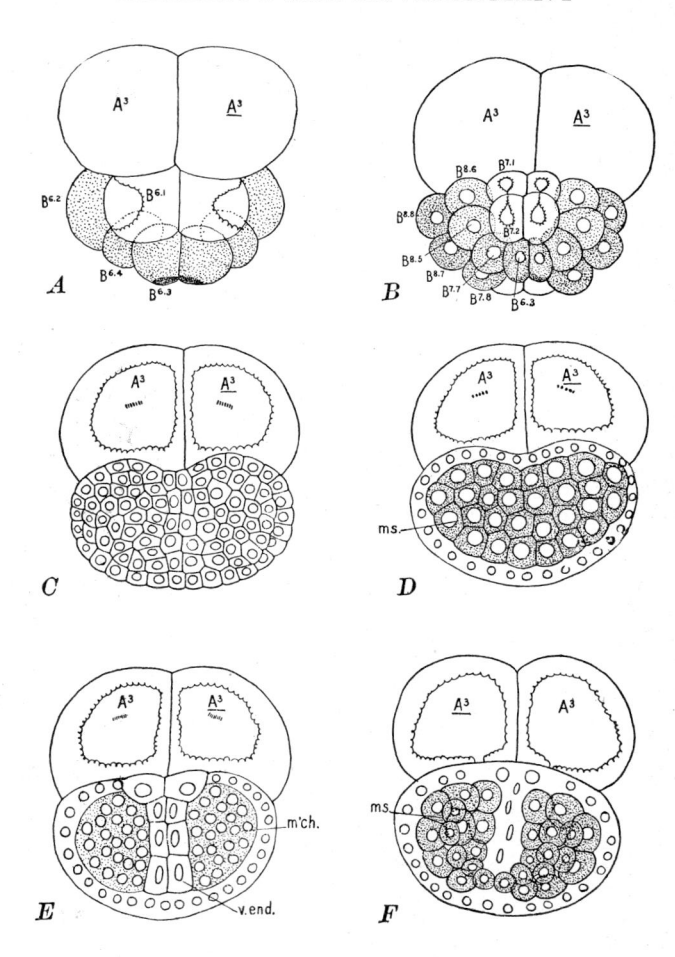

FIG. 81. POSTERIOR HALF-EMBRYOS OF *Styela*, the anterior cells having been killed in the 4-cell stage. Muscle cells and intestinal cells are present but no portion of neural plate or chorda, since these come only from the anterior cells.

There has been much discussion among biologists as to the meaning of these results. On the one hand it has been said that the totipotence of either of the first two cleavage cells proves that both of these cells are alike and that they have not yet begun to differentiate. On the other hand it is said that a part of an egg may give rise to a whole animal for the same reason that parts of certain adult animals may do the same thing, viz., because they have the power of regeneration. However there are many animals which are incapable of regenerating lost parts of their bodies, and similarly there are cases in which part of an egg cannot give rise to a whole animal. The evidence available at present favors the view that in cases where one of the cleavage cells is capable of giving rise to a whole animal there is a greater capacity of regeneration or regulation, and possibly also a lower degree of initial differentiation, than in those cases in which part of an egg is capable of producing only part of an animal.

Effects of Centrifugal Force.—If the fertilized egg is whirled rapidly on a centrifugal machine it may be subjected to a pressure several thousand times that of gravity. Under such conditions the heavier particles are thrown to one side of the egg and the entire substance of the egg becomes stratified into layers or zones. In the ascidian egg, where the different kinds of protoplasm give rise to different tissues and organs, this rearrangement of the egg substances may lead to a marked dislocation of organs; the animal may be turned inside out, having the endoderm on the outside and its ectoderm or skin on the inside, etc. (Fig. 82). On the other hand in some mollusks and echinoderms the development of centrifuged eggs is practically normal. In the former case the formative substances were dislocated; in the latter they were probably not.

Double Monsters and Identical Twins.—If the cleavage cells are only partially separated they may produce animals which are partially separated, such as Siamese twins, two-headed forms, etc. (Figs. 77, 78). Or these double monsters may be produced by division or budding of the embryo at a later stage of development. In the human species, no less than in other animals, all sorts of double monsters may be formed in this way by the partial division of a single egg or embryo (Fig. 83). If the division is slight the developed individual may show only the beginnings of a division into two, as in two-headed forms; if the division of the egg or embryo is complete two separate and perfect individuals may be formed from an originally single oosperm. When two individuals are formed from a single egg they have exactly the same heredity and accordingly they are always of the same sex and are so similar in appearance that they are known as "identical" or "duplicate" twins (Fig. 83, right end). On the other hand twins which develop from different eggs do not have the same heredity and may differ in sex as well as in other features; they are known as "fraternal" twins.

Other Monstrous Forms.—If the temperature or density of the surrounding medium is altered during the gastrula stages the endoderm may be caused to turn out instead of in (exogastrula), thus producing an animal which is turned inside out (Fig. 84). In other cases (vertebrates) the gastrula mouth may fail to close,

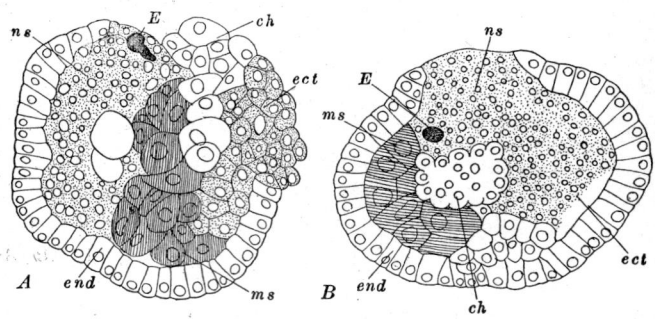

FIG. 82. TWO LARVAE OF *Styela* which were centrifuged in the 4-cell stage thereby changing the position of various organ-forming substances. Nervous system (*ns*), eyes (*E*), notochord (*ch*) and muscles (*ms*) have been displaced, and the larva has been turned inside out, the endoderm (*end*) being outside and the ectoderm (*ect*) inside.

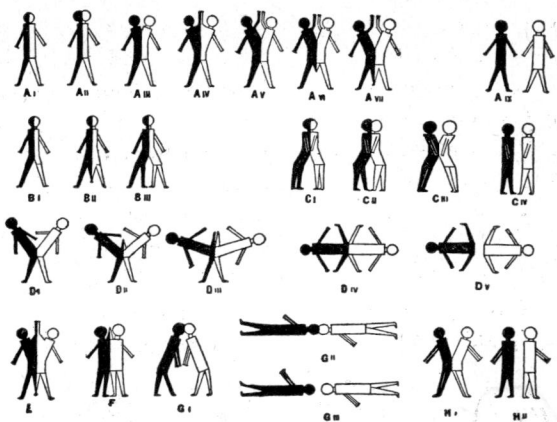

FIG. 83. DIAGRAM SHOWING THE DIFFERENT TYPES OF UNION OF DOUBLE HUMAN MONSTERS, each being produced by a partial division of a single egg or embryo. If the division is a complete one, duplicate twins are formed, as shown by the figures at the right end of each line. (From Wilder.)

thus producing animals in which the spinal cord and vertebral column are split in two (spina bifida); or the brain may be forced outside of the head or may be lacking altogether (anencephaly).

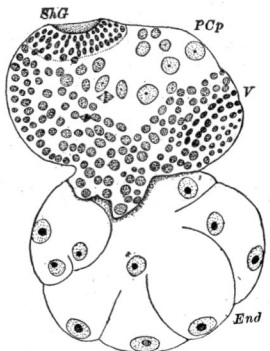

FIG. 84. EXOGASTRULA OF *Crepidula*. The endoderm (*End*) has been turned out instead of in, thus leaving the digestive layer of cells on the outside of the body; *Shg*, shell gland; *V*, velum.

In some cases eyes are wholly lacking, in others the two eyes fuse together into a single one as in the fabled Cyclops (Fig. 85). Practically all such cases of monstrous development are due to abnormal environmental conditions in early stages of ontogeny.

Effects of Food.—In addition to such monsters, which are incapable of long life, many peculiar if not abnormal types of animals are produced by the action of unusual environmental stimuli during later stages of development. Gudernatsch found that if tadpoles of the frog were fed on the thyroid gland they transformed into minute frogs, scarcely larger than flies, but if fed on thymus gland they grew to be large, dark-colored tadpoles

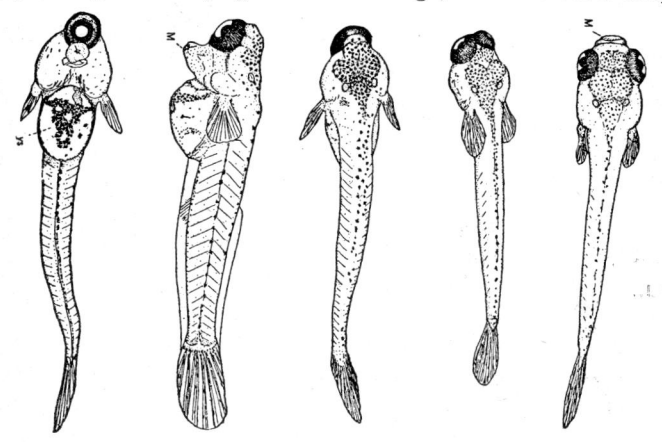

FIG. 85. YOUNG FISH. On the right a normal individual with two eyes; on the left cyclopean monsters with one eye; produced by treatment with magnesium solutions. (From Stockard.)

but did not change into frogs; if fed on the adrenal gland they produced extremely light-colored forms. If canary birds are fed on sweet red pepper they become red in color. If the larvæ of bees are fed on "royal jelly," which is a bee food rich in fats, they become fertile females or queens; if fed on ordinary "bee bread" they become infertile females or workers (Fig. 86). There

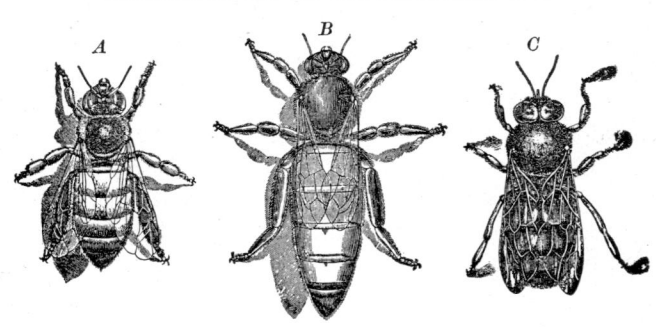

FIG. 86. THE THREE CASTES OF THE HONEY BEE. *A*, worker or imperfect female; *B*, queen or perfect female; *C*, drone or male. The differences between workers and queens are produced by the type of food supplied to the larvæ.

are marked structural differences between the workers and the queens but the differences in their habits and instincts are even more striking; all of these differences whether in bodily structure or in instincts are determined by the character of the food and not by heredity. Innumerable cases of a similar sort could be named which show the great effect of environmental stimuli on development but not upon heredity.

FUNCTIONAL ACTIVITY AS A FACTOR OF DEVELOPMENT

Another factor of development which is partly intrinsic and partly extrinsic is functional activity or use. Functional activity is response to stimuli which may be external or internal in origin. The entire process of development may be regarded as a series of such responses on the part of the organism, whether germ cell, embryo or adult. The nature of the response is determined by the nature and state of the organism and by the character of the stimulus. By the normal, or usual, series of stimuli certain parts are kept active while other parts are kept inactive or are inhibited.

Developmental Movements.—Normal development is dependent on the correlated activity of many parts of the organism. If in any part stimuli and responses are lacking the development of that part is arrested or inhibited. For example in the cleavage stages different substances are sorted and localized by protoplasmic movements within cells and these substances are then isolated by cell divisions and by the formation of partition walls between cells; these protoplasmic movements occur in response to stimuli and if these movements are stopped cleavage and differentiation are arrested. In later stages the infolding of the gastrula, or neural tube, or alimentary canal, and the foldings of layers in general, which play so important a part in development are due to the movements of substances within cells and to the movements of cells in the layers in which they lie, and if these movements are inhibited normal development is prevented.

Nutrition and Development.—Another type of functional activity which is a potent factor in development is found in the trophic or nutritive relations which exist between different parts of an organism. Organs long unused undergo regressive changes and may become rudimentary, for example the muscles of a limb, which has been paralyzed or placed in a cast, shrivel; on the other hand use increases the size and strength of any organ. Inactivity or atrophy of one part usually leads to imperfect nourishment and development of related parts; for example, the optic nerve atrophies when the eye is lost, and muscles atrophy when the nerves leading to them are destroyed or paralyzed. In general the normal development of any part is dependent upon its proper nutrition and this is dependent upon the functional activity of this and other related parts.

Internal Secretions; Hormones.—Still another phase of functional activity is found in the effects of certain secretions and chemical substances which are formed by different glands and poured into the blood. In many cases the secondary sexual characters which distinguish the male or the female are due to chemical substances from the interstitial cells of the gonads (testes or ovaries), which stimulate or inhibit the formation of these characters. If the ovary is removed from a young hen she de-

INFLUENCE OF ENVIRONMENT

velops the larger size, more brilliant plumage and the peculiar comb, wattles and spurs of the cock. These secondary sexual characters of the male are potential in the female but are kept from developing or are inhibited by the activity of the ovary. On the other hand the castration of the young cock does not prevent the development of most of the secondary sexual characters of the male. In the case of mammals removal of the ovaries of a young female or of the testes of a young male does not lead to the development of the secondary sexual characters of the other sex, but both sexes remain in a sexually undeveloped or infantile condition, that is, the presence of ovaries or testes serves as stimulus to call forth the development of the secondary sexual characters in mammals, and not as inhibitors to prevent the development of the secondary sexual characters of the opposite sex, as in the female fowl. If bits of the ovary of a guinea-pig are inserted under the skin of a young male which has been previously castrated, the latter develops mammary glands similar to those of a normal female; in short he is, to this extent, "feminized" by the stimulus of substances from the ovary.

Another gland whose secretions exercise a profound influence on development is the thyroid, which is found in the neck near the "Adam's apple." If the secretion of this gland is over-abundant it causes rapid heart-beat, higher temperature, increased activity, and in general a higher rate of metabolism, and if the gland becomes much enlarged it gives rise in addition to goitre and protruding eyeballs. Excess of the thyroid secretion hastens certain processes of development; young tadpoles that have been fed on thyroid gland or its extract quickly transform into frogs which are sometimes "no larger than flies" (Gudernatsch). On the other hand if the thyroid secretion is deficient the rate of metabolism is decreased, body temperature is lowered, muscular movements become slow and awkward, the skin may become swollen with mucous (myxedema) and the mind becomes dull or stupid. Thyroid deficiency in a young child prevents the normal development of body and mind and in extreme cases causes that peculiar type of idiotic dwarf known as "cretin." If thyroid extract is administered early enough in such cases the most wonderful improvement is brought about; the body frequently assumes normal proportions and features and the mind becomes bright and active.

Another gland of internal secretion which plays a great part in the development of body and mind is the pituitary, or hypophysis, which lies on the ventral side of the brain; it consists of two parts, an anterior portion derived embryologically from the roof of the mouth, and a posterior portion which came from the floor of the brain. Excessive secretion of the anterior portion stimulates the growth of the bones and if this occurs in early life it leads to a great lengthening of the bones, especially of the arms and legs, and in later life the bones of the hands, feet and face becomes enlarged so that the subject becomes a giant, though often with weakened body and mind; on the other hand if there is deficiency of secretion from the anterior pituitary there is a general dwarfing of the body. The secretion of the posterior pituitary acts as a stimulant to nerve cells, involuntary muscles and to the sex glands; deficiency of this secretion causes persistent "infantilism" or sexual immaturity.

Another pair of glands of internal secretion which exercise a profound influence on both the structures and functions of the body are the adrenals, or suprarenals, located just above the kidneys. These glands also are composed of two parts which produce secretions differing in their physiological action,—namely, a central part or *medulla* and a superficial part or *cortex*. The secretion of the medulla is known as adrenin or epinephrin and when introduced into the blood it strengthens the heart beat and causes the blood to be driven from the abdominal viscera to the skeletal muscles, heart, lungs and brain. It also causes the liver to set free, into the blood, sugar, which is "the optimum source of muscular energy"; it quickly restores fatigued muscle and increases the coagulability of the blood. Cannon has shown that in emotional excitement, such as pain, hunger, fear, or rage, there is increased secretion of adrenin into the blood, and he has called attention to the remarkably adaptive character of all these reactions. Probably the ability to "nerve oneself" to meet emergencies successfully depends upon this secretion. In cases of adrenin-insufficiency there is a lowering of blood pressure, lack of muscular tone and "loss of nerve," such as occur in neurasthenia, "shell shock," etc. On the other hand the secretion of the cortex is said to stimulate the development of the sex glands, and to hasten sexual maturity. Deficiency of this secretion, as in Addison's disease, causes a bronzing of the skin and it has been suggested by Keith that the skin colors of the darker races of mankind are due to a relative deficiency of this secretion, while "we Europeans owe the fairness of our skins to some particular virtue resident in the adrenal bodies." Keith has further suggested that many other racial characteristics such as shape and size of head, face, nose, eyes, teeth and lips, length of arms and legs, character and abundance of hair on various parts of the body, etc., may be correlated with the relative activity of the several glands of internal secretion. He believes that in general the white race possesses more of the internal secretions of gonad, thyroid, pituitary and adrenal than do the other races. While these suggestions are highly speculative, they do serve to emphasize the great importance of these secretions on the development of body, mind, and personality.

Since racial characters are inherited it is necessary to assume that in some way the chromosomes of the oosperm influence the development of the glands of internal secretion, and through these, various bodily and mental characteristics. The influence of the internal secretions on development does not disprove the importance of heredity, but rather it points out a mechanism through which heredity controls development.

Correlative-Differentiation and Self-Differentiation.—Many cases are known in which the development of one part is dependent upon the presence of another part; this is technically known as "correlative differentiation." Thus it has been found that the lens of the eye will develop from any portion of the ectoderm, or outer layer of the skin, if only the primitive retina, or optic cup, is brought near to this layer; if the optic cup is transplanted from the head to the thorax or abdomen a lens will form wherever the cup comes in contact with the ectoderm. If an embryonic limb is transplanted from its normal position to the middle of the back or belly, it will develop, and nerves and blood vessels will grow into it which would have had very different positions and distributions if the limb had not been there. If one of the first four cleavage cells is separated from the others it may develop into an entire animal though it would have formed only a quarter of an animal if it had remained in contact with the other three-quarters of the egg. All such cases are known as "correlative differentiation," implying that differentiation is dependent upon the stimuli which come from surrounding parts. On the other hand if the differentiation has already begun before the relation of a part to surrounding parts has been changed, it may continue to differentiate as if no change of position or relation had taken place. Thus if a right limb is transplanted to the left side of the body after it has begun to differentiate it remains a right limb and is not modified by its new relations (Harrison); if the cleavage cells are already differentiated in the four-celled stage, each cell when separated from the others will give rise to only one-quarter of an animal. In short the organ or cell is already set, or fixed, or differentiated to such an extent that it can not change its fate even though its environment should change. Such cases are known as "self-differentiation."

Many students of the physiology of development have been led to the view that the fundamental causes of development are to be found not in the egg cell itself but in environmental stimuli and in the interaction of the various parts. Driesch in particular regards the egg, or any cleavage cell, as an "harmonic equipotential system," that is, any part is capable of any fate and its actual fate is determined by its relation to other parts; in the striking phrase of Driesch, "The fate of a part is a function of its position." We now know that this expresses only a fraction of the truth. The fate of a part is primarily determined by its protoplasmic organization and only secondarily by its position.

These are only a few illustrations of the many kinds of abnormal development which may be caused by changed environment or by unusual functional activities. At all stages of ontogeny the course of development may be altered by extrinsic stimuli but earlier stages may be more profoundly influenced than later ones.

D. INHERITANCE OR NON-INHERITANCE OF ACQUIRED CHARACTERS

Few questions in biology have been discussed so fully and so fruitlessly as this. It is a problem of the greatest interest not only to students of biology but also to sociologists, educators and philanthropists and yet it is still to a certain extent an unsolved problem.

Opinions of Lamarck and Darwin.—It is well known that Lamarck taught that characters due to desire or need, use or disuse, and to changed environment or conditions of life were inherited and thus brought about progressive evolution. Long ago desire or need was repudiated as a factor of evolution. Lowell satirized it in his Biglow Papers in these words:

"Some filosifers think that a fakkilty's granted
 The minnit it's felt to be thoroughly wanted,
        *   *   *   *   *
 That the fears of a monkey whose holt chanced to fail
 Drawed the vertibry out to a prehensile tail."

Darwin wrote to Hooker, "Heaven forfend me from Lamarck's nonsense of adaptation from the slow willing of animals"; but although he repudiated this feature of Lamarckism he held that characters due to use or disuse and to changed conditions of life might be inherited and he proposed his hypothesis of pangenesis in order to explain the process of the transmission of such characters to the germ cells.

Weismann's Theories.—Weismann introduced a new era in biology by denying the inheritance of all kinds of acquired characters, and by challenging the world to produce evidence that would stand a rigorous analysis. But Weismann's greatest service lay in his constructive theories rather than in destructive criticism; he forever disposed of theories of pangenesis and the like by showing that the germ cells are not built up by contributions from the body and that characters are not transmitted from generation to generation; but on the other hand that there is transmitted a germplasm which is relatively independent of the body and which is relatively very stable in organization. This epoch-making theory of Weismann's has naturally undergone some changes, as the result of new discoveries. It is no longer believed that the germplasm is really independent of the body, nor that it is absolutely stable, as Weismann at one time held. There is no doubt that the germ cells and the germplasm are physiologically related to other cells and to other plasms, and similarly there is no doubt that the germplasm although very stable can and does change its constitution under some rare conditions. But in the main the germplasm theory is accepted by the great majority of biologists to-day, and recent work in genetics and cytology has brought many confirmations of this theory.

Distinction between Hereditary and Acquired Characters.—As long as it was believed that the developed characters of an organism could be transmitted as such to its descendants it was customary to speak of developed characters as hereditary or acquired and to talk of the inheritance or non-inheritance of acquired characters. This distinction is not a logical one for all developed characters are invariably the result of the responses of the germinal organization to environmental stimuli; and of course no developed character can be purely hereditary or purely environmental. But when a given character arises in many individuals of the same biotype under different environmental conditions it is probable that heredity, which is the constant factor in this case, is also the determining factor for that character. On the other hand if a character develops in response to peculiar stimuli and does not appear in other individuals of the same biotype in which such stimuli are lacking it is said to be an environmental or acquired character. In fine, inherited characters are those whose distinctive or differential causes are in the germ cells, while acquired characters are those whose differential causes are environmental.

Statement of Problem.—Briefly stated the question of the inheritance of acquired characters is this: Can the differential cause of a character be shifted from the environment to the germplasm? Can peculiarities of the environment which influence the development of somatic characters so affect the germ cells that they will produce these somatic characters in the absence of the peculiar environment? Can the characteristics of a developed organism enter into its germ cells and be born again in the next generation? Considering the fact that germ cells are cells and contain no adult characteristics, it seems very improbable that any peculiarity of environment whether of nutrition, use, disuse or injury, which brings about certain peculiarities of developed characters in the adult, could so change the structure of the germ cells as to cause them to produce this same character in subsequent generations in the absence of its extrinsic cause. How, for example, could defective nutrition, which leads to the production of rickets,[1] affect the germ cells, which contain no bones so as to produce rickets in subsequent generations, although well nourished? Or how can over-exertion, leading to hypertrophy of the heart, so affect the germ cells that they, in turn, would produce

[1] It has recently been shown that rickets may also be produced by absence of ultraviolet or other radiant energy. It may be prevented by direct exposure to sunlight, if taken in time.

hypertrophied hearts in the absence of over-exertion, seeing that germ cells have no hearts? Or how could the loss or injury of eyes or teeth or legs lead to the absence or weakened development of these organs in future generations, seeing that inheritance must be through germ cells which possess none of these structures?

Lack of Evidence for Inheritance of Acquired Characters.— But, apart from these general objections to the doctrine of the inheritance of acquired characters, there are many special difficulties. There is little or no conclusive and satisfactory evidence in favor of such inheritance. Almost all the evidence adduced serves to show only that characters are acquired, not that they are inherited.

It is a matter of common observation that mutilations are not inherited; (wooden legs do not run in families, although wooden heads do.) The evidence for the inheritance of peculiarities due to use or disuse is wholly inconclusive; for example, did the giraffe get his long neck because he browsed on trees, or does he browse on trees because he has by inheritance a long neck? Did attempts to fly lead to the development of wings in birds, or do birds fly because heredity has given them wings? Did life in caves make cave animals blind, or did blind animals resort to caves because the struggle for existence there was less severe for them? The evidence is in favor of the second of each of these alternatives rather than of the first.

There still remains the question of the inheritance of certain characters due to environment, though here also the most clear cut evidence is against this proposition. That unusual conditions of food, temperature, moisture, etc., may affect the germ cells so as to produce general and indefinite variations in offspring is probable, but this is a very different thing from the inheritance of acquired characters. The germ cells being a part of the parental organism may be modified by such changes in the environment as affect the body as a whole, they may be well nourished or starved, they may be modified by changed conditions of gravity

INFLUENCE OF ENVIRONMENT

salinity, pressure, temperature, etc., and these modifications of the germ cells probably lead to certain general modifications of the adult, which may be larger or smaller, stronger or weaker, according as the germ is well or poorly nourished, but it is incredible that the environment which produces rickets, or hypertrophied heart, or loss of sight in one generation should modify the germ cells in such a peculiar and definite way that they should give rise in the next generation to these particular peculiarities, in the absence of the extrinsic cause which first produced them. The inheritance of acquired characters is incredible, because the egg is a cell and not an adult organism; and in this case there is no suffi-

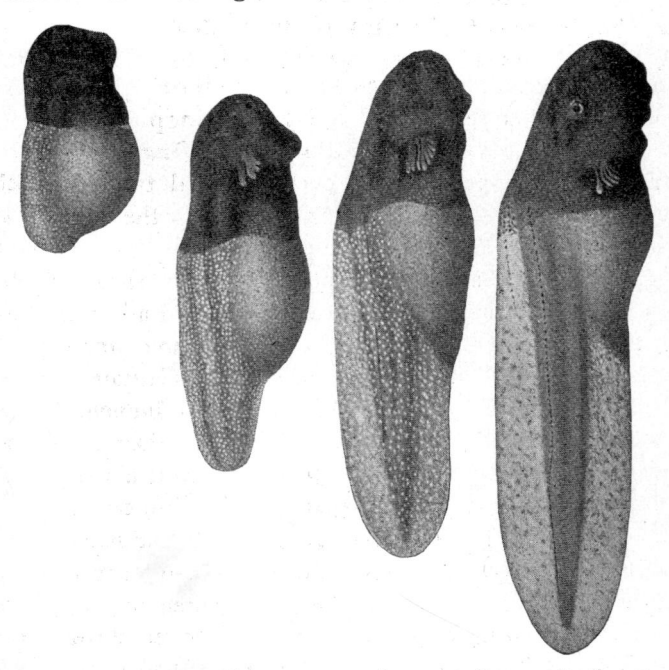

FIG. 87. GRAFTED FROG EMBRYOS, anterior part, *Rana sylvatica*, posterior part, *R. palustris*. In later stages, and even in the adult condition, the two parts preserve their peculiarities. (From Harrison.)

cient evidence that the thing which is incredible really does happen.

No Inherited Influence of Stock on Graft.—If specific changes of environment produced specific changes in heredity we should expect to find that where different plants or animals are grafted together each would modify more or less the hereditary constitution of the other. But this does not occur. Everybody knows that when a branch of a particular kind of fruit tree is grafted upon a tree of a different variety the quality of the fruit borne by that branch is not altered by its close union with the new stock. The same is true of all forms of animal grafts. Harrison cut in two young tadpoles of two species of frog, *Rana sylvatica* and *Rana palustris,* and spliced the anterior half of one to the posterior half of the other. These frogs and their tadpoles differ in color as well as in other respects, *R. sylvatica* being more deeply pigmented than *R. palustris*. In the grafted tadpoles each half preserved its own peculiarities even up to the adult condition (Fig. 87).

A still more striking case of the persistence of heredity in spite of environmental changes is found in experiments in which the ovaries are removed from one variety of animal and transplanted to another variety. Guthrie made such transplantations in the case of fowls and concluded that there was some influence of the foster mother upon the transplanted ovary, but Davenport, who repeated his experiments, was unable to confirm his results; on the contrary he frequently found that the engrafted ovary degenerated and that the excised ovary regenerated. Finally Castle and Phillips furnished the most conclusive demonstration that the hereditary characteristics of the transplanted ova are in no wise changed by the foster mother. They removed the ovary from a pure black guinea-pig and put it in the place of the ovary of a pure white animal. After recovery from the operation this white female with the "black" ovary was bred to a pure white male (Fig. 88). Three litters of offspring from these parents

INFLUENCE OF ENVIRONMENT

were all black as shown in Figure 89. Although both parents were pure white all the offspring of the F_1 generation were black because they came from "black" eggs and black is dominant over white. The fact that these "black" eggs developed in the body of a white female did not in the least change their hereditary constitution.

Dominants and Recessives Remain Pure.—A still more intimate union takes place when the dominant and recessive characters come together in any zygote. These characters, or rather the factors which determine them, may be intimately associated in every cell of the organism throughout an entire generation and yet we get a clean separation of these characters in the next generation; neither the dominant nor the recessive character has been at all modified by its most intimate association with the other.

Climatic Effects Not Inherited.—A striking instance of the purely temporary effect of the environment and of the long persistence of hereditary constitution amidst new environmental conditions, which have greatly changed the appearance of the developed organisms, is found in the case of alpine plants. Nägeli says that such plants, which have preserved the characters of high mountain plants since the ice age, lose these characters perfectly during their first summer in the lowlands.

Summary.—If acquired characters were really inherited we should expect to find many positive evidences of this instead of a few sporadic and doubtful cases. In particular why do we not find in plant or animal grafting that the influence of the stock changes the hereditary potencies of the graft? Why do we not find that transplanted ovaries show the influence of the foster mother as Guthrie supposed—a thing which has been disproved by Davenport and Castle (Figs. 88 and 89)? Why do dominant and recessive characters remain pure, even after their intimate union in a hybrid, so that pure dominants and pure recessives may be obtained in subsequent generations from this mixture? Why does every child have to learn anew what his parents learned so

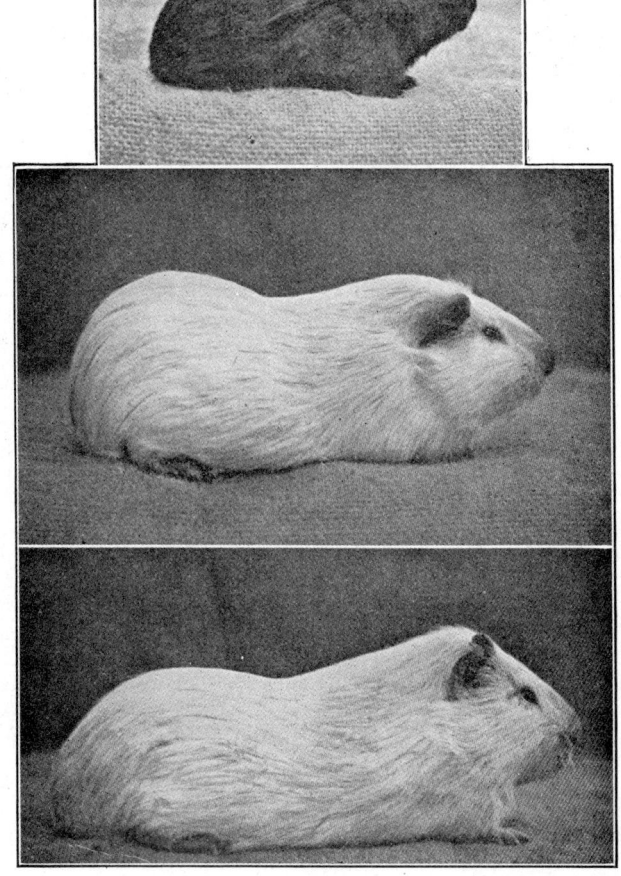

FIG. 88. EFFECT OF TRANSPLANTING OVARIES IN GUINEA-PIGS. Above, young black female; in the middle, mature white female; below, mature white male. The white female's ovary was removed and in its place was put the ovary from the black female. The white female (with "black" ovary) was then bred to the white male. (From Castle.)

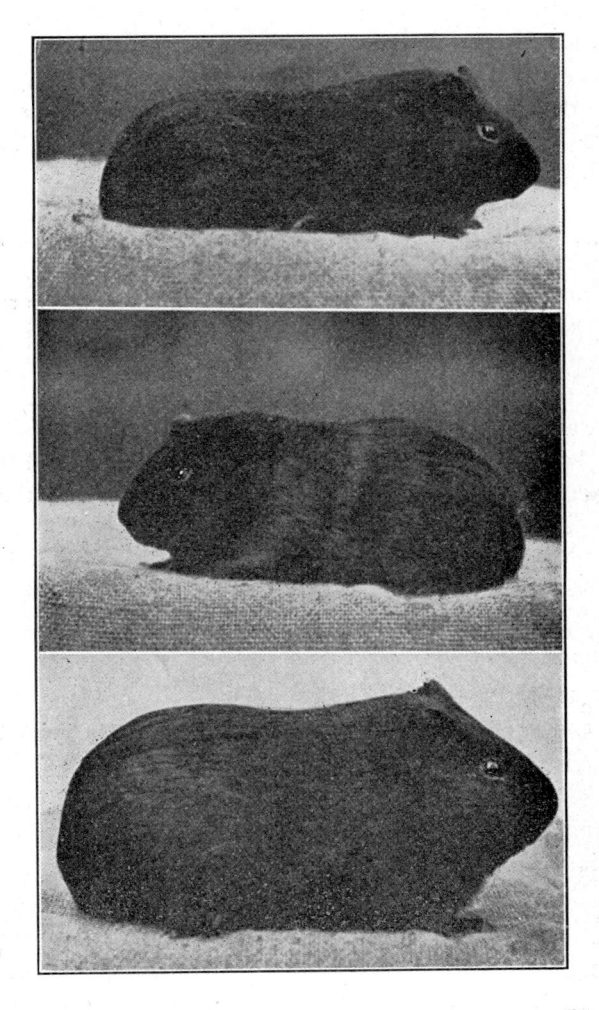

FIG. 89. RESULTS OF CROSS DESCRIBED IN THE PRECEDING FIGURE. All the offspring are black, though both parents are white, because the white female contains only "black" eggs and black is dominant over white.

laboriously before him? Even the strongest defenders of the inheritance of acquired characters are constrained to admit that it occurs only sporadically and exceptionally.

Neo-Lamarckism.—Many modifications of the Lamarckian hypothesis of the inheritance of acquired characters have been proposed in recent years. Foremost among these are the "mneme" theory of Semon and the "centro-epigenesis" theory of Rignano. To Semon as to many other biologists the apparent resemblance between memory and heredity has seemed significant, and this furnishes the basis of his theory. Semon holds that every condition of life, every functional activity of an organism leaves a permanent record of itself in what he calls an "engramme." If these conditions or activities are long continued their engrammes are heaped up and affect heredity. Semon does not ask if "acquired characters" are inherited, but rather "Are the hereditary potencies of the germ cells altered by stimuli acting on the parental body?" This is a very different thing from the inheritance of a particular acquired character, and there is some evidence that such stimuli may in rare instances produce changes in the hereditary constitution of the germplasm though these evidences are by no means conclusive.

Temporary Effects of Environment; "Induction."—On the other hand certain changes may be produced in germ cells or embryos which last for only a generation or two and then disappear. It is well known that plants grown in poor soil are smaller and generally produce smaller seeds than those grown in good soil, and deVries, Baur and Harris find that such seeds produce smaller plants having smaller seeds than do seeds of normal size. This is an after-effect of poor nutrition which changes the amount of food material in the seeds and through this the size of the plant which develops from the seed, but it does not change the hereditary constitution. Woltereck found that in *Daphnia* there is an after-effect of cold lasting for one or two generations, and this he calls "induction" when the effect lasts for one generation, or "pre-induction" when it lasts for two or three generations. Whitney found that rotifers poisoned with alcohol were weaker in resistance to copper salts and were less fertile than others, and when brought back to normal conditions the first generation was weak but the second was normal. On the other hand Stockard finds that the injurious effects of alcohol on guinea-pigs persist through two or more generations. In man alcohol may have an "induction" effect on offspring, but fortunately it does not seem to alter hereditary constitution. Probably of a similar character are Sumner's results; he found that mice raised in the cold have shorter tails than those raised at higher temperatures and this modified character appears in the next generation. If this is an after-effect or "induction" it should disappear in the following generations.

Kammerer found that salamanders with black and yellow spots when reared on yellow soil gradually lose their black color becoming more yellow, and their young continue to grow more yellow until finally almost all black may disappear. The offspring of such salamanders are said to be more yellow than normal; but this work has been called in question and needs confirmation. He also maintains that change of habitat of the "obstetrical" toad causes inherited changes in structure and function, and that repeated regeneration of the siphons of *Ciona* causes an inherited lengthening of the siphons. All of this has been denied or otherwise explained by other investigators.

Probably such cases are not instances of true inheritance; they do not signify a change in the hereditary constitution but an influence on the germ cells of a nutritive or chemical sort comparable with what takes place when fat stains are fed to animals; the eggs of such animals are stained and the young which develop from such eggs are also stained, though the germinal constitution remains unchanged. The very fact that the changed condition is reversible and that it disappears within a short time is evidence that it is not really inherited.

One of the most interesting and convincing cases of the inheritance of an experimentally induced character has been reported by Guyer and Smith with respect to certain eye defects in rabbits. They injected the pulped lenses of rabbits into fowls and, after the fowls had become sensitized to this foreign protein by the formation of anti-lens substances, their serum was injected into pregnant female rabbits. The effects on the injected rabbits were severe and many of them died, but there was no evidence that their eyes or lenses suffered injury; furthermore there was no evidence of any specific injury to their ovarian eggs, since in subsequent breeding they produced no young with eye defects. On the other hand, some of the embryo *in utero* did suffer *specific* injury; some were born with opaque lenses; sometimes their lenses were reduced in size, and when the lens was small the whole eye was usually small; sometimes the eyeball had collapsed leaving no trace of pupil or iris; finally these degenerative changes frequently increased and progressed after birth.

All of these defects might be explained by the direct action of the anti-lens substances of the fowl's serum upon the developing eyes of the embryos, which would be merely another case of "induction"; but the thing which cannot be explained so easily is the fact that these eye defects are inherited for at least five generations and that they do not gradually disappear but become more pronounced in successive generations. Furthermore it is not possible to assume that the lens anti-bodies are transmitted only through the cytoplasm of the egg, as plastids are in plants or fat-stains in animals, for the induced eye defects are inherited through the male as well as through the female line. The authors suggest the possibility "that the degenerating eyes are themselves directly or indirectly originating anti-bodies or other chemical substances in the blood-serum of their bearers which in turn affect the germ cells."

Other investigators (Stockard, Hansen, *et al.*) have found that similar eye defects may be produced by alcohol, X-rays, or other agents and it is a question whether the results obtained by Guyer and Smith are the *specific* results of anti-lens serum. But, however these results may be interpreted, it is evident that the old doctrine of the inheritance of acquired characters due to use, disuse, or external environment and the crude mechanisms proposed for such inheritance are not confirmed by these experiments. It remains to be seen whether the more subtle influences of the internal environment, such as hormones, anti-bodies and enzymes may affect the germplasm in a specific manner, and thus modify inheritance.

In conclusion: (1) Developed characters, whether "acquired" or not, are never transmitted by heredity, and the hereditary constitution of the germ is not changed by changes in such characters. (2) Possibly environmental stimuli acting upon germ cells at an early stage in their development may rarely cause changes in their hereditary constitution, but changes produced in somatic cells do not usually, if ever, cause corresponding changes in the hereditary constitution of the germ cells. (3) Germ cells like somatic cells may undergo modifications which are not hereditary; if starved they may produce stunted individuals and this effect may last for two or three generations; they may be stained with fat stains and the generation to which they give rise be similarly stained; they may be poisoned with alcohol or modified by unusual temperature and such influence may be carried over to the next generation without becoming permanent. All such cases are known as "induction" and many instances of the supposed inheritance of acquired characters come under this category. (4) Environment may profoundly modify individual development but it does not generally modify heredity.

3. *APPLICATIONS TO HUMAN DEVELOPMENT: EUTHENICS*

Man's Larger Environment.—Man's environment is more extensive than that of any other animal, and its influence on his development is correspondingly greater. In addition to chemical and physical stimuli which are potent factors of development in the case of all organisms, man lives in a world of psychical, social and moral stimuli which exert a profound influence on him. He is stimulated not merely by present environment but also by memories of past experiences and anticipations of future ones. Through intelligence and social cooperation he is able to control environment for particular ends, in a manner quite impossible to other organisms. On the other hand heredity as a factor of development is no more powerful in the case of man than in any other organism. Consequently the relative importance of heredity and environment is not the same in the development of an intelligent and social being, like man of the present age, as it is in lower organisms. For man and for every other living creature heredity fixes the possibilities of development, it "sets bounds about us which we cannot pass"; but the more complex those possibilities become the more complex must be the environment which calls them forth and the more varied become the results of development under altered conditions of life.

Capacity for Training and Education.—Functional activity also plays a larger part in man's development than in that of any other animal, owing to the longer period of his development and to the more extensive and varied training which he is capable of undergoing. It is a notable fact that the period of immaturity in man is longer than in any other animal, and it is during this formative period that environment and education have their greatest influence. Other animals develop much more rapidly than man but that development sooner comes to an end. The children of lower races of mankind develop more rapidly than those of higher races but in such cases they also cease to develop at an earlier age. The prolongation of the period of infancy and of immaturity in the human race greatly increases the importance of environment and training as factors of development.

The possible training of human faculties is also more varied

INFLUENCE OF ENVIRONMENT

and extensive than in other animals, not only because those faculties are more numerous but also because they are more plastic and are capable of higher development, that is, are more educable. Human faculties are functions and methods of reaction, which are dependent in part upon the bodily mechanism and in part upon environment and training. Habits are the usual methods of responding to stimuli, and they may be classified as inherent or acquired. The former are instincts or reflexes and are expressions of hereditary constitution; the latter are in a sense forced upon organisms by environmental conditions. All education is habit formation, and good education like good environment is such experience as leads to the formation of good bodily, intellectual, social and moral habits; it consists in placing the individual in such an environment and bringing such stimuli to bear upon him as to call forth desirable responses and to suppress undesirable ones.

Good and Bad Environment.—Only that environment and training are good which lead to the development of good habits and traits and to the suppression of bad ones. What we commonly call "good environment" is frequently the worst possible, what is often called a bad environment may be the best possible. We are all strangely blind with regard to these matters. We know of many cases in which men began their careers on a farm, in the backwoods, on a flat-boat, amidst hardships and discomforts of every sort and yet who achieved great distinction. And we speak of such men as winning in spite of disadvantages, forgetting that often these very disadvantages, hardships, discomforts, have been stimuli which have given them sturdy bodies, good judgments, good morals, and have called forth all their best qualities. On the other hand under different circumstances or with different men such conditions may prove to be too hard, too severe, and the result be disastrous. But environment may be too good as well as too hard. Food may be too rich and too abundant for good health, life may be too easy and luxurious for the development of character. Luxury, easy lives, refined surroundings have less of educational value than we commonly suppose and they may be a positive menace. Any environment is bad, however cultured, refined or pleasant it may be, which leads to the development of bad traits of body or mind. In general the best environment is one which avoids extremes, one which is neither too easy nor too hard, one which calls for sustained effort and produces maximum efficiency of body and of mind.

In education also we are strangely blind as to proper aims and methods. Any education is bad which leads to the formation of habits of idleness, carelessness and failure, instead of habits of industry, thoroughness and success. Any religious or social institution is bad which leads to habits of pious make-believe, insincerity, slavish regard for authority and disregard for evidence, instead of habits of sincerity, open-mindedness and independence.

Frequently the training of the human being, like the training of a star-fish, consists in limiting his activities to particular lines. Some physical defect which prevented a child from engaging in the usual activities of children has often turned his attention to scholarship. Galton says that great divines have usually had very poor health. Genius is frequently associated with physical defects. Great specialization is associated with corresponding limitations in other directions. Society needs the genius and the specialist, but for the general good of mankind the generalized type is needed even more than the specialized.

No given environment or training can be good for every individual, nor for the same individual at every stage of development. Every individual is unique and if the best results are to be had he must have unique environment and training, which must be supplied by omniscient intelligence. Such an ideal may not be practicable but the impossibility of securing the absolutely best conditions of development need not prevent society from securing better conditions than those which now prevail.

Relative Importance of Heredity, Environment, Education.—It

INFLUENCE OF ENVIRONMENT

is plain that environment and education play a greater part in the development of man than in that of other animals, whereas heredity plays a similar part; but it is difficult if not impossible to determine the relative importance of these three factors. In the field of intellect and morals most persons are inclined to place greater weight upon the extrinsic than upon the intrinsic factors, but this opinion is not based upon demonstrable evidence. So far as organisms below man are concerned there is general agreement that heredity is the most important factor, and this opinion is held also for man by those who have made a thorough study of heredity. Galton has made the best scientific study of this subject in the case of identical twins, in which as we know heredity is the same in the two, both individuals having come from the same oosperm (Fig. 83). In bodily and mental characters such twins are remarkably alike; the differences which exist are slight and may usually be traced to different environmental and educational influences, and particularly to different illnesses. Galton sums

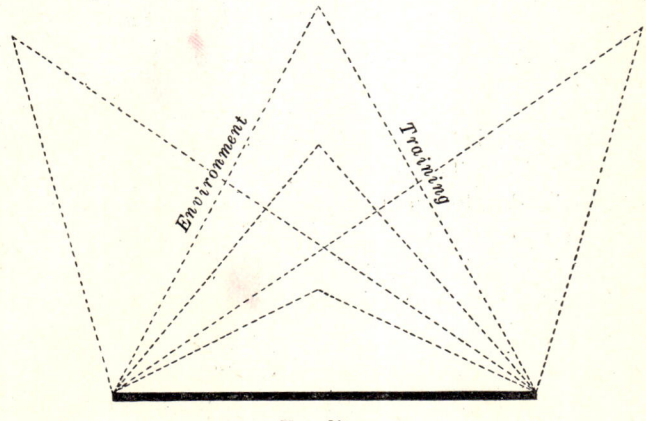

Heredity

FIG. 90. DIAGRAM TO SHOW THE INFLUENCE OF HEREDITY, ENVIRONMENT AND TRAINING IN THE DEVELOPMENT OF AN INDIVIDUAL. Various types of individuals (represented by the triangles) may be produced from the same germ cells (heredity) if the environment and training are variable.

up his study with these words: "There is no escape from the conclusion that nature prevails enormously over nurture when the differences of nurture do not exceed what is commonly to be found among persons of the same rank of society and in the same country."

The part played by these different factors of development may be graphically illustrated by the accompanying diagram (Fig. 90), in which the base line represents heredity and the other lines represent the extrinsic factors of environment and education. For each individual heredity is a constant factor but environment and training are variables. With a given heredity the characteristics of the developed organism may vary enormously depending upon the extrinsic factors. Hereditary possibilities are not changed by accidents of environment but development is so changed. After the fertilization of the egg the hereditary potencies of every organism are unalterably fixed but the extrinsic factors remain variable and may be controlled.

All of our social and ethical institutions such as government, education and religion deal only with extrinsic factors of development and of life. Nevertheless there is no evidence that such extrinsic influences ever modify heredity, no evidence that the effects of good environment or good training ever change the germinal constitution. The influences of environment and education affect only the development of the individual and not the constitution of the race, and consequently such influences are temporary in effect and must be repeated generation after generation.

Social versus Germinal Inheritance.—But though the effects of environment and training are not inherited, the environment and training and experience of former generations are handed down to later generations through custom, tradition, history. We do not inherit through the germ cells the effects on our ancestors of their training and environment, but we do inherit, in the property sense of that word, their environment, customs, institutions. In short the experiences and accomplishments of past generations

INFLUENCE OF ENVIRONMENT

are not inherited through the germ cells but are inherited through society. In this sense "we are the heirs of all the ages."

Man alone of all animals can profit largely by the experiences of others and especially by the experiences of former generations. In the human species only are successive generations born into new environments, created by the activities of preceding generations. To a certain extent the young of higher animals learn useful lessons from their parents, and in social animals the environment has to a certain extent been made by preceding generations, but such social inheritance is founded largely on instinct and it changes almost, perhaps quite, as slowly as does the germ-plasm. However, when social inheritance is founded largely on intelligence it may change and progress very rapidly; intelligence is a great time-saver as compared with instinct, and owing to this rapid change in social inheritance every human generation is born into a new environment, different from that of any previous generation. On the other hand, even the most intelligent wild animals such as monkeys, wolves, foxes and elephants, do not by their own activities change their natural and social environments from generation to generation.

Because of our social inheritance the extrinsic conditions of life continue to grow more complex age after age, while our inherited natures remain relatively unchanged. All moralists, all religions, have recognized the very general experience among men of a sense of imperfection and of disharmony with social and ethical standards. Huxley held that the spirit of ethics was opposed to the spirit of evolution. Metchnikoff finds these disharmonies due to the survival of bestial instincts in man. Galton finds the sense of sin to be due to the fact that the development of our inherited nature has not kept pace with the development of our moral civilization. Our psychical, social and moral environment has come to us from the past with ever-increasing increments, every age standing on the shoulders of the preceding one. The aspirations, impulses, responsibilities of modern life have become enormous and our inherited natures and abilities have not essentially improved. Social heredity has outrun germinal heredity and the intellectual social and moral responsibilities of our times are too great for many men. Civilization is a strenuous affair, with impulses and compulsions which are difficult for the primitive man to fulfil, and many of us are hereditarily primitive men. The frequent result is disharmony, poor adjustment, a struggle between primitive instincts and high ideals, with a resulting sense of discouragement and defeat which often ends in abnormal states of mind. The prevalence of crime, alcoholism, depravity and insanity is an ever-increasing protest and menace of weak men against high civilization. We are approaching the time when one or the other must give way, either the responsibilities of life must be reduced and the march of civilization stayed, or a better race of men, with greater hereditary abilities, must be bred.

Wars and revolutions shake off the burdens of social inheritance but they destroy the good along with the bad and afford only local and temporary relief. Mankind cannot return permanently to barbarism or savagery; civilization must be and will be preserved; but if society is really to advance from age to age the natures of men must improve as well as their environment.

CHAPTER V

CONTROL OF HEREDITY: EUGENICS

CHAPTER V

CONTROL OF HEREDITY: EUGENICS

It is the aim of science to interpret phenomena and as far as possible to control them. To what extent is it possible to control heredity and thus to improve the race, as well as the individual?

A. DOMESTIC ANIMALS AND CULTIVATED PLANTS

The history of domesticated animals and cultivated plants shows that it is possible to control or rather guide phenomena of heredity and evolution. Very many species of wild animals have been tamed by man but only about 40 species may be classed as domesticated. DeCandolle recognized 247 species of cultivated food plants, 193 of which still exist in the wild state.[1] In a number of instances the wild stocks from which these domestic forms came are known and it is possible to compare them with their modified descendants and thus to determine the degree of change which has been brought about under human guidance. In other cases where the original wild species are unknown it is possible to determine the amount of modification which has taken place within recent times.

The degree of change which has taken place under human guidance is very remarkable. In some cases dozens and even hundreds of races have been formed, showing the most remarkable differences in size, structure and proportion of parts, as well as in functions, instincts and behavior. The extent to which hered-

[1] Bailey says that more than 20,000 kinds of plants in cultivation are described in the *Standard Cyclopedia of Horticulture,* although not nearly all of these species are domesticated.

260   HEREDITY AND ENVIRONMENT

ity may be guided by man is forcibly illustrated by our present races of domesticated pigeons which Darwin said would be

Fig. 91. Races of Domestic Pigeons, the wild rock pigeon being shown in the center. (From Romanes, "Darwin and After Darwin.")

CONTROL OF HEREDITY: EUGENICS

classed by any naturalist who did not know their origin in not less than twenty different species and three different genera, though all of them have descended from the wild rock pigeon (Figs.

FIG. 92. RACES OF DOMESTIC PIGEONS, continued. (From Romanes.)

262   HEREDITY AND ENVIRONMENT

91, 92); or by the numerous races of dogs varying in size from a toy dog to a Great Dane or St. Bernard and showing almost unbelievable differences in structure and disposition; or by the

Fig. 93. Races of Fowls. (From Romanes.)

CONTROL OF HEREDITY: EUGENICS

great variety of domestic fowls, all of which have probably descended from the wild jungle-fowl; or by modern races of horses, cattle, sheep, or swine (Figs. 93-96). There are only a few of the

FIG. 94. RACES OF FOWLS, continued. (From Romanes.)

264   HEREDITY AND ENVIRONMENT

many illustrations which could be cited of the practical control of heredity and evolution for human purposes. How have the present races of domesticated animals and cultivated plants been produced?

FIG. 95. DIFFERENT BREEDS OF BRITISH CATTLE. (From Romanes.)

CONTROL OF HEREDITY: EUGENICS

Fig. 96. Wild Boar Contrasted with Modern Domestic Pig. (From Romanes.)

1. The Influence of Environment in Producing New Races

There is a popular belief that the improvement of cultivated races is due to good environment, good food, good soil, protection from enemies, etc., and that if turned out to shift for themselves such races revert at once to the original wild stock. This is not strictly true but if it were there are two ways in which it could be explained. It is conceivable that new races might be produced:

1. By the direct inheritance of somatic or personal characters acquired under the stimulus of the environment. In spite of popular opinion in favor of this view there is no evidence that this occurs. There is no doubt that environment has much to do with individual development, but it does not usually modify the hereditary constitution of the race.

2. It is possible that environmental changes acting upon germ

cells at a sensitive period of their development may produce germinal variations or mutations and thus give rise to new races. But while this is possible there is little evidence of a satisfactory nature that it actually occurs and there is no evidence that such changes in hereditary constitution are reversible and that the race reverts to its former state when the old environment is restored. Such reversible changes undoubtedly occur in somatic characters but they are not inherited; they are modifications of development, not of heredity; they are personal fluctuations, not racial mutations. Not infrequently reversion of cultivated races to wild stock is due to hybridization, as in the sweet peas described on pages 102-104, or in hybrids of some races of domestic pigeons (p. 82). Since wild types in general are more perfectly adapted to a natural environment than are domesticated forms they persist while the latter are eliminated.

II. Artificial Selection

Since the beginning of historic times, and probably through long prehistoric ages, breeders have followed the method of selecting desirable individuals for propagating their stock. There can be no doubt that almost all that man has done in the production of domestic animals and cultivated plants has been accomplished by this process of selection.

How Has Selection Acted? Darwin's Views.—Until very recently it was generally believed that continued selection of individuals which showed desirable characters gradually led to the improvement of those characters and thus to the production of new races; it was supposed that the character in question was "built up" by continued selection in one direction, and that the average development of the character in all the offspring was thus increased in successive generations to an indefinite extent. It was this view as to the supposed action of artificial selection which formed the basis of Darwin's theory of natural selection.

Non-effect of Selection on Pure Lines.—On the other hand it has been known for a long time that the limits of the possible improvement of any character by artificial selection are soon reached and that thereafter selection serves only to maintain the character at its high level but not to advance it. The probable explanation of this fact has been found only in recent years. The researches of deVries, Johannsen, Jennings, Tower, Pearl, Morgan and others have shown that in some cases at least selection merely isolates mutants or distinct hereditary lines which are already present in a mixed population but that it does not "build up" characters nor produce new mutations; in short it does not create the variations on which it acts.

Johannsen found that from a single species of beans he was able, by keeping the progeny of each individual bean separate from the others, to isolate 19 different "pure lines," each differing in certain respects from every other line. These lines were not created by selection but were merely sorted out of the general species where they existed already. He further found that when extremely large or small individuals from any pure line were selected and propagated, none of the progeny showed that character in a still more extreme degree but all merely fluctuated within the original extremes of that line. He concludes therefore that selection within a pure line is absolutely without effect in modifying any character in the offspring of that line.

Jennings found that different races of *Paramecium* differ in size, structure and rate of division, and that these differences are "as rigid as iron." With respect to average length of body he was able to isolate eight lines which constantly differed more or less from one another. Within each of these lines there was considerable fluctuation in size, but he was unable by selecting extremes to increase these fluctuations, the progeny of any line always fluctuating about the mean of that line (Fig. 22).

Similarly Tower found in his studies on the potato beetle that he was unable to shift the mean or the extremes of any character by selection of extreme forms of an inbred line.

Pearl also made an extensive study of the records of breeding experiments extending over many years in which the attempt was made to increase the egg-laying capacity of hens by selecting for breeding in each generation only those which had a high record for egg production. It was found that certain "blood lines" produced a larger number of eggs than other lines, but by no amount of selection was it possible to increase the egg production within any line.

On the other hand Castle strenuously defended the importance of the selection of fluctuations in "building up" a character. From among the descendants of piebald rats and rabbits he selected through many generations individuals showing the largest and others showing the smallest extent of color in the coat and he thus produced one line which was nearly all-black and another nearly all-white. He maintained that not only inherited characters, but also their factors are variable and that by means of selection of plus or minus variations the mode, or mean of a character may be shifted in one direction or the other. This is the old view which flourished before a distinction was made between fluctuations and mutations. Breeders have long been acquainted with similar results of selection from a mixed population containing different hereditary lines; however Castle was careful to employ as pure a race of rats and of rabbits as he could obtain, but it is not possible to get as pure a race of these animals or of any organisms in which cross fertilization occurs as in the case of self-fertilizing plants such as beans. Johannsen defines a "pure line" as "all individuals which are derived from a single, absolutely self-fertilizing, homozygous individual" and within such a pure line he maintains that selection is unable to change any character.

Adherents of the "pure line" hypothesis explain Castle's results in one of two or three ways: either his material may not have been genetically pure, or mutations may have occurred during the course of his experiments and his selection served merely to isolate distinct hereditary lines already present; or, more probably extent of color in his animals may depend upon multiple factors or modifying factors, which are more numerous in some individuals than in others, as is the case for example in "blending" inheritance (pp. 109-112), and selection merely sorted out some individuals with a larger or smaller number of factors, and consequently with a larger or smaller development of the character in question, but it did not in the least modify any individual factor. This view finds support in the work of McDowell, of Zeleny and Mattoon, and of Morgan and his associates on the effects of selection on certain characters of *Drosophila*. In a later paper Castle (1918) himself has abandoned his former position and has adopted this explanation of his results, and thus this controversy comes to an end.

The crux of this whole controversy lay in the question as to whether inheritance factors fluctuate or not; Castle maintained that they do, Johannsen that they do not. If inheritance factors fluctuate they may be changed gradually in one direction or another by selection; if they do not fluctuate, but mutate only, changing only rarely and not in all directions, selection can act only by sorting out mutations, but can do nothing to produce them. In general fluctuations are due to environment and are not inherited, therefore they concern development rather than heredity, developed characters rather than inheritance factors. There is much evidence that inheritance factors are relatively stable and that when they change, they undergo a complete change or mutation comparable to what occurs in a chemical reaction. As we have seen it is possible to explain almost all phenomena of inheritance on this basis even including Castle's results.

In another paper Jennings has shown that continued selection in one direction does, apparently, shift the mode of certain characters in a race of asexually reproducing *Difflugia,* and Middleton has found the same to be true of *Stylonychia.* These results differ totally from Jennings' earlier work on *Paramecium,* which has been repeated and confirmed by Ackert. It is a hard thing to believe that different organisms differ irreconcilably in so fundamental a matter and it seems much more probable that these discrepancies are due to an incomplete analysis

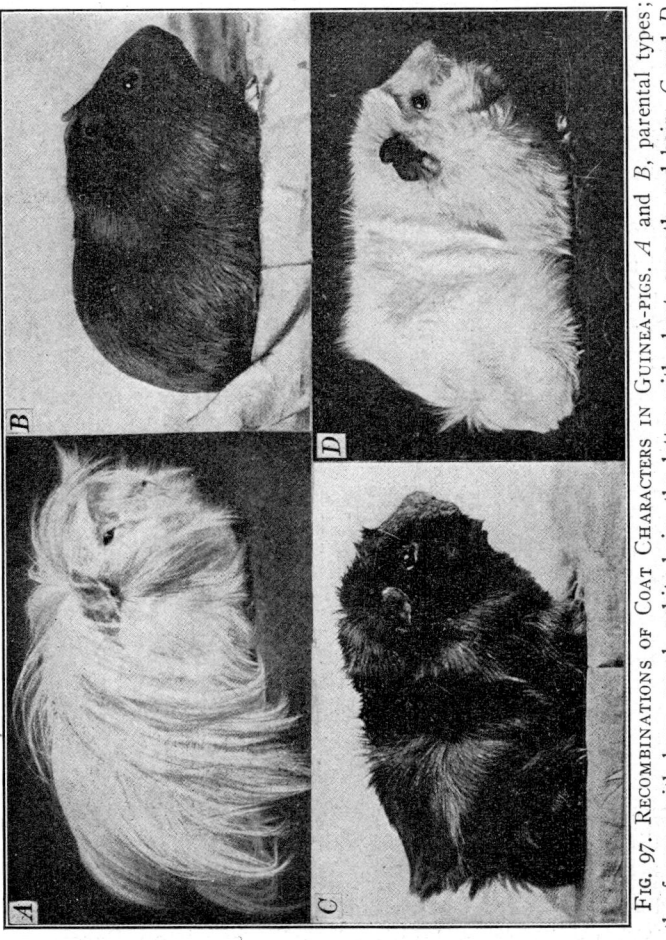

FIG. 97. RECOMBINATIONS OF COAT CHARACTERS IN GUINEA-PIGS. *A* and *B*, parental types; the former with long, rough, white hair; the latter with short, smooth, red hair. *C* and *D*, two of the types of the F₂ generation produced by crossing *A* and *B*. (From Castle.)

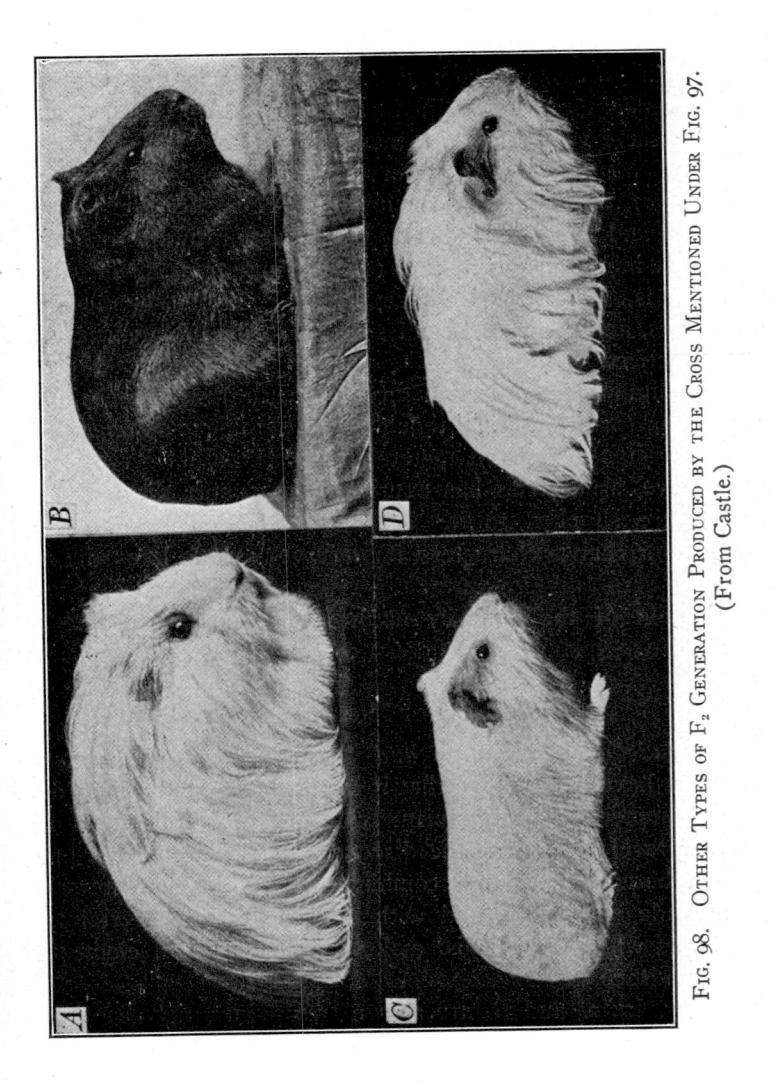

FIG. 98. OTHER TYPES OF F_2 GENERATION PRODUCED BY THE CROSS MENTIONED UNDER FIG. 97. (From Castle.)

of the phenomena in question. It is possible that the divisions of the cell body in these protozoans is not always into exactly equivalent halves in which case variations might take place in the descendants, which might then be heaped up by selection; or perhaps there are multiple or modifying factors in this case also, so that selection has acted as in Castle's rats.

Value of Selection.—In conclusion, the evidence which is most clear-cut and abundant indicates that selection by itself is unable to change inheritance factors or to cause mutations. Nevertheless selection is of great service in separating good lines or races from poor ones, and this is the chief significance of the artificial selection practiced by breeders.

The elimination of certain races by natural selection is an important factor in evolution though it has nothing to do directly with the formation of new characters or new races but serves merely as a sieve, as deVries has expressed it, to sort the individuals which are supplied to it. Although selection has no power to make or change characters, it preserves certain lines and eliminates others and thus fixes the type of a species. Finally the elimination of the unfit by natural selection is still the only mechanistic explanation of fitness, or adaptation, in organisms.

III. METHODS OF MODERN GENETICS

1. *Mendelian Association and Dissociation of Characters.*—Breeders have long known that it is possible to get certain desirable characters of an organism from one race and others from another race. But since the discovery of the Mendelian principles of heredity such new combinations of old characters have been made repeatedly, and with almost the same certainty of results as when the chemist makes combinations of elements or compounds.

In Fig. 97, A and B, are shown two guinea-pigs, one having long (L), rough and tumbled (T), white (W) hair, and the other having short (S), smooth (Sm), red (R) hair. When such animals are crossed one should get in the F_2 generation 27 genotypes and 8 phenotypes, one of each of the latter being homozygous and breeding true, as is shown in Fig. 32 for trihybrid peas. These 8 phenotypes of this cross are *STR, STW, SSmR, SSmW, LTR, LTW, LSmR, LSmW*. In Fig. 97, *C* and *D,* and Fig. 98, *A, B, C, D,* are shown 6 of these 8 phenotypes which were obtained by Castle from this cross. These figures well illustrate the new combinations of Mendelian characters which may be obtained by cross breeding.

Hybridization.—This is the chief method employed by Burbank in producing his really wonderful "new creations in plant life." By extensive hybridization he brings about many new combinations of old characters, a few of which may be commercially valuable, and sometimes actually new characters or mutations appear, possibly as a result of the interaction of old characters, or rather of their factors. Lotsy, for example, maintains that the sole source of variation is crossing, and Bateson says that the new breeds of domestic animals made in recent times are the carefully selected products of the recombination of pre-existing breeds, and that most of the new varieties of cultivated plants are the outcome of deliberate crossing.

One of the striking results of modern work in plant-breeding has been the discovery of the greatly increased vigor of certain hybrids as compared with either pure-bred parent. In general it is not possible to tell without previous experience what the character of the hybrid of two races or lines will be; sometimes it is more and sometimes less vigorous than either parent, but not infrequently it is more vigorous. East and Shull have shown that hybrids between two races of corn may be very much larger and more fertile than either parent. In some instances the yield of corn per acre has been increased from 20-30 bushels to 80-90 bushels, and in one case to more than 250 bushels per acre (Figs. 99, 100). Unfortunately such hybrid races of corn do not continue to breed true and the crossing must be made anew in each

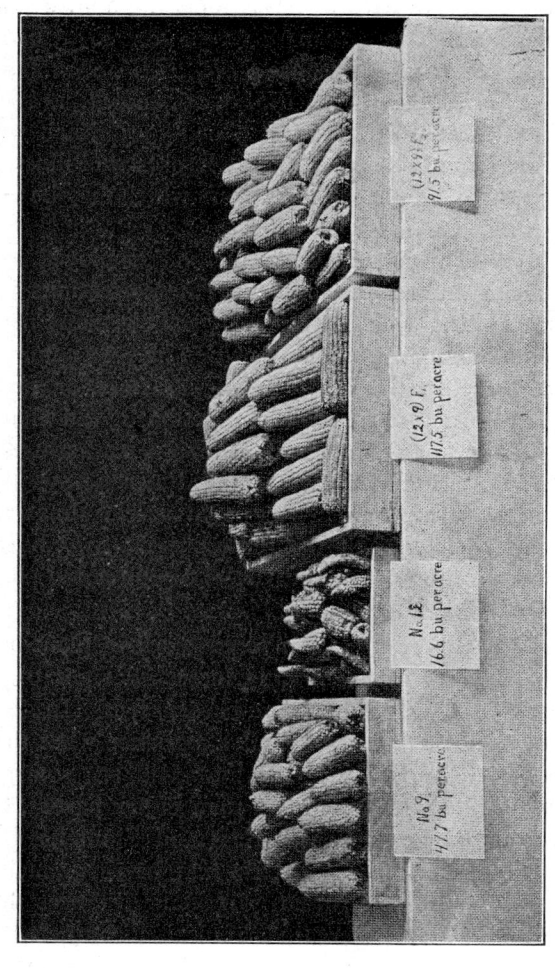

FIG. 99. INBRED STRAINS OF LEAMING DENT CORN (Nos. 9 and 12) compared with the F_1 and F_2 generation of a cross between these strains. (From East.)

CONTROL OF HEREDITY : EUGENICS

FIG. 100. GROWING CORN OF STRAINS SIMILAR TO THOSE SHOWN IN FIG. 99; two inbred strains on left, crosses between these on right. (From East.)

generation if maximum results are to be had. Nevertheless this stimulation by hybridization or "heterosis," as it has been called, offers an extremely important means of quickly producing very vigorous and fruitful individuals, but not lines or races which breed true.

2. *Mutations.*—Mendelian association and dissociation of characters produce new forms of adult animals and plants but not new hereditary characters. Permutations of Mendelian characters we may have almost without number, of new combinations of these there may be no end, but no new unit characters are formed by such temporary combinations, no new inheritance factors are created or evolved. New combinations of factors may be compared to new combinations of chemical elements; you can always get out of the combination what went into it, whereas new factors are comparable to the changes which take place in certain atoms, for example radium, by which the element itself is changed in an irreversible manner. The discoveries of Mendel show us how to follow inheritance factors through many combinations and through many generations, but they do not show us how new factors arise. These discoveries have given us an invaluable method of sorting and combining hereditary qualities, but Mendelian inheritance as such does not furnish the initial materials for evolution.

In 1901 Hugo deVries startled the scientific world by the publication of his great work on the "Mutation Theory" of evolution in which he proved that the evening primrose, *Oenothera lamarckiana* occasionally produced "sports" or "mutations" which differed so much from the parent form that they deserved to be called new species (Fig. 101). He discovered and studied a large number of these mutations in *Oenothera* as well as in some other plants and concluded that evolution takes place by steps or jumps rather than by "creeping on from point to point" as Darwin believed.

Several geneticists have expressed doubt as to whether there are any such things as mutations in the sense of deVries, maintaining that his results may be explained by assuming that *Oenothera* is a hybrid and that the various "mutations" which he has described are due to segregation and recombination of old factors rather than to the appearance of new ones. Indeed Davis has made up an *Oenothera* by hybridization that is similar in some respects to *O. lamarckiana*. There are many things which seem to indicate that this species and probably other species of *Oenothera* are not genetically pure and it is probable that some of deVries' results may be due to this fact.

However, if *O. lamarckiana* were an ordinary hybrid it should show phenomena of Mendelian splitting with the usual Mendelian ratios. The fact that the "mutations" observed by deVries occur only very rarely shows that they are not the ordinary emergence of Mendelian recessives. But the discovery of lethal factors and of "crossing over" in *Drosophila* has pointed out a possible method of accounting for many *Oenothera* "mutations." In *Drosophila* it is known that when homologous lethal factors are received from both parents the zygote is non-viable; however, if the lethal factor received from one parent is balanced by a normal one from the other parent the zygote is viable; in other words, all individuals that are homozygous for any lethal disappear, while only those that are heterozygous survive.

If recessive factors are linked with a lethal they can not come to expression, for recessives appear only when mated with other recessives. But if "crossing over" should take place in such a way as to break the linkage between the lethal and the recessive factors, the latter would, when homozygous, come to expression as ordinary Mendelian recessives (Cf. Fig. 68).

More than half of the mutants of *O. lamarckiana* which were first described by deVries are probably due to this cause. Shull's very extensive work on this species shows that it is a persistent heterozygote (hybrid) in which certain recessive characters do not appear as in ordinary Mendelian segregation, because the fac-

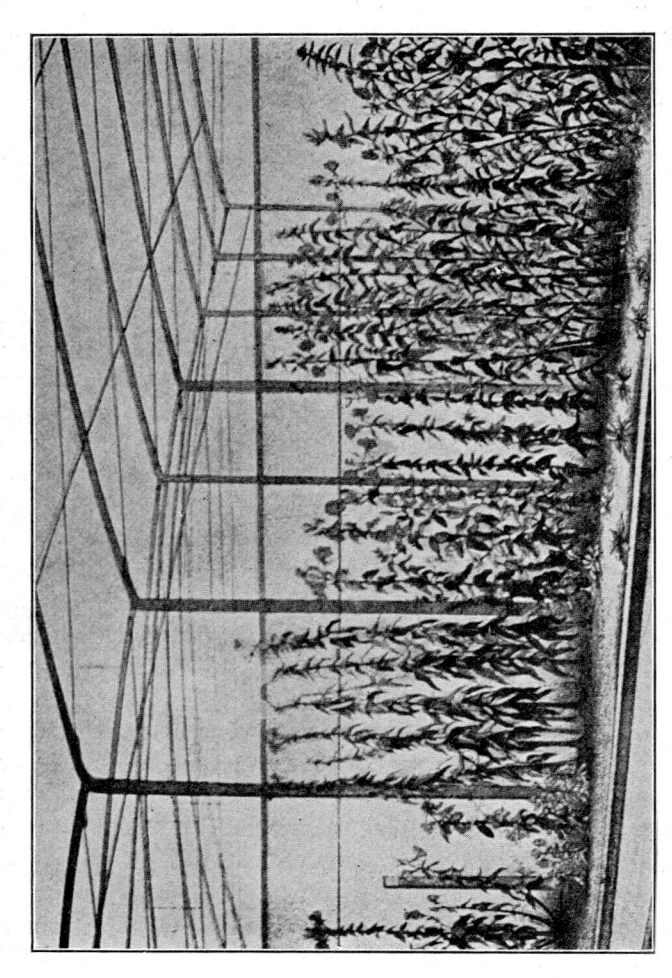

FIG. 101 LAMARCK'S EVENING PRIMROSE, *Oenothera Lamarckiana*, on the left and seven of its mutants. (From deVries.)

tors for these characters are linked with a lethal factor and it is only when this linkage is broken that they have a chance to develop.

However, it is certain that mutations do take place in species where there is no evidence of genetic impurity, as for example in *Drosophila melanogaster,* and it is an extraordinary circumstance that many of the mutations of *Oenothera* upon which deVries founded his great theory are probably not mutations at all but are, as Muller (1917) has said, "merely the emergence into a state of homozygosis, through crossing over, of recessive factors constantly present in the heterozygous stock." DeVries himself had previously suggested this explanation for his double reciprocal crosses and as Muller says, "it probably lies at the root of nearly all the unusual genetic phenomena of this genus." Shull has established the truth of this suggested explanation of the peculiar behavior of *Oenothera.* That there are lethal factors also in *Oenothera* which produce their effect upon the gametes rather than upon the zygotes is indicated by the partial or complete failure to form fertile pollen in certain forms of this genus.

It is probable that many natural or Linnean species, other than *O. lamarckiana,* are not pure and homozygous; within every such species there are usually found many "elementary species" and by intercrossing of these a mixture of many lines or strains results from which new forms may occasionally arise by segregation. Lotsy maintains that all mutations arise in this way. But such an explanation does not account for the existence of the original "elementary species" and if they be referred to still earlier crossings it is evident that we only put off the explanation to a more remote period. After all, the fundamental problem here concerns the *origin,* rather than the *segregation,* of dominants, recessives, lethals, etc. The exact and exhaustive work of Morgan and his associates on *Drosophila* has proved that the mutations in this species are not due to Mendelian segregation and it plainly indicates that they are caused by sudden transformations in the Mendelian factors themselves, comparable to changes in chemical composition.

3. *Causes of Mutations.* The causes of new hereditary characters, or rather of mutations in genes, are obscure. Practically all of the earlier workers and writers on evolution found the principal causes of transmutation in the action of extrinsic or environmental forces on the organism. As the result of years of labor on this subject Darwin concluded that "variability of every sort is due to changed conditions of life"; but in the light of modern genetics such a statement is too sweeping.

It is well known that environmental changes produce many kinds of modifications in organisms, and in general these modifications are the more profound the earlier they occur in ontogeny; it is known that slight alterations of the germ cells may produce great modifications of adult structure, and yet one of the most striking results of recent work is to show the small effect of environmental changes on hereditary characters. Marked individual modifications may be produced which do not become racial. Usually not one of thousands of variations which occur has any evolutionary value. These fluctuations come with changing environment and with changing environment they disappear. Very rarely a sudden variation or mutation of the germplasm arises which forms the basis of a new race (Fig. 101, 102, 103). In most cases such mutations manifest themselves in the failure of some old character to develop rather than in the addition of a new one, but at least they represent modifications of hereditary constitution and as such they furnish material for evolution. Whence and how they appear we do not know, for like the kingdom of heaven they come without observation. Their infrequency amidst the multitude of fluctuations indicates the wonderful stability of racial types and teaches respect for Weismann's doctrine of a germplasm relatively stable, independent and continuous.

This distinction between somatic and germinal variations, between those which concern only the individual and those which are inherited and furnish material for evolution, marks the great-

CONTROL OF HEREDITY : EUGENICS 281

est advance in the study of evolution since the work of Darwin. And just as these germinal variations are the only ones of importance in the process of evolution so the question of their origin is the greatest evolutionary problem of the present day. How are such germinal variations produced?

There is some evidence that environmental changes of the right sort acting upon germplasm at the right stage may lead to permanent modification of the hereditary organization. Extrinsic influences acting upon germ cells at the time of their maturation divisions may lead to new distributions of chromosomes or even to changes in the composition of individual chromosomes,

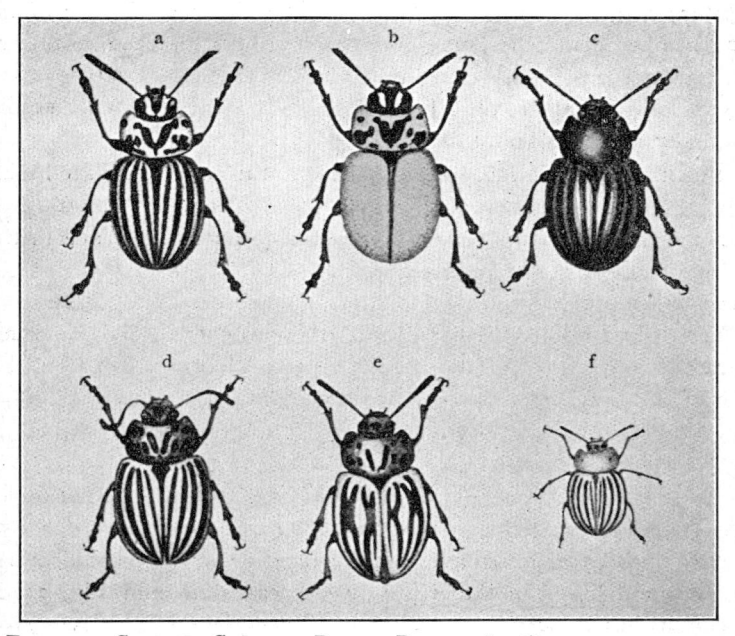

FIG. 102. COMMON COLORADO POTATO BEETLE, *Leptinotarsa*, AND SOME OF ITS MUTANTS. *a*, Normal *L. undecemlineata*, *b*, the mutant *augusto-vittata*, *c*, the mutant *melanothorax*, *d*, normal *L. decemlineata*, *e*, the mutant *tortuosa*, *f*, the mutant *defectopunctata*. (From Plate after Tower.)

thus producing new hereditary types. Certain mutants of *Oenothera* (Fig. 101) seem to be of this sort as Gates, Miss Lutz and Stomps especially have emphasized; for example *O. lamarckiana* has 14 chromosomes, *O. lata, O. albida* and *O. scintillans* 15 each, *O. semi-gigas* 21, *O. gigas* 28, and these variations in the number of chromosomes are probably due to abnormalities in their divisions, such as non-disjunction, irregular distribution, or chromosome division without cell division. It is significant that the mutants *lata* and *semi-gigas* have occurred several times, whereas *gigas* appeared but once; this may be explained by the fact that the chances of the doubling of chromosomes in both germ cells (*gigas*) are very few compared with the chances of their doubling in one germ cell (*semi-gigas*) or of their increase by one in a single germ cell (*lata*). Blakeslee has recently shown that many mutations of *Datura,* the jimson weed, are associated with abnormal numbers of chromosomes.

But it is probable that mutations are not usually associated with changes in the number of chromosomes. Where the number of chromosomes remains constant the change may take place in the composition of single chromosomes, as in "cross-overs" or in the number or composition of the chromomeres or units of the next lower order. But such changes as these concern only the emergence into visibility of more fundamental changes, for whatever the cellular changes may be which accompany mutations, it is certain that the initial changes take place in the inheritance factors themselves.

In addition to gene mutations, such as the transformation of the gene for the red eye of *Drosophila* into one for white eye, some genes appear to have become inactive or to have dropped out altogether, as in the white sweet peas shown in Fig. 33, in the wingless or eyeless mutants of *Drosophila* (Figs. 103-105), and in many other cultivated races of plants and animals, thus producing regressive mutations. Indeed most of our domestic animals and cultivated plants appear as if they had arisen by the omission of something which their wild ancestors had. It was

CONTROL OF HEREDITY : EUGENICS

this appearance of omission which led to the "presence-absence hypothesis" (p. 97) which has now been abandoned. Phenomena of "multiple allelomorphism" (p. 98) prove that recessive characters are not the result merely of the absence of dominant ones. On the other hand many things suggest that recessive genes may originate from dominant ones by a process of degradation. In many of the mutations studied by deVries, Bateson, Morgan and others some factor seems to have undergone degradation and Bateson suggests that at present all new forms arise only by the loss of factors or by the fractionation of factors and that new factors are not added from without. This leads him to inquire "whether the course of evolution can at all reasonably be represented as an unpacking of an original complex which contained within itself the whole range of diversity which living things present." This is as extreme preformation in the field of inheritance factors as was the old theory of "emboîtement" in the field of developed characters; it is devolution rather than evolution since it assumes that the earliest organisms were genetically the most complex.

But if, as is often said, evolution from amœba to man necessarily involves the addition of many new inheritance factors it does not involve additions *from without* as Bateson implies. New hereditary factors are to be thought of as we think of new chemical compounds, which are formed by new combinations of the same old elements, or as we think of new elements, such as helium and radium emanation, which are formed by dissociation of radium. As compared with chemical elements the factors of heredity are probably very complex things and the new factors which appear in the course of evolution probably arise as new combinations of factors or parts of factors previously present. Nowhere in the entire process of organic evolution is there any evidence that new factors are "extrinsic additions" or are created *de novo*. The whole process is one of evolution, that is of new combinations of existing units, having new qualities which are the results of these new combinations.

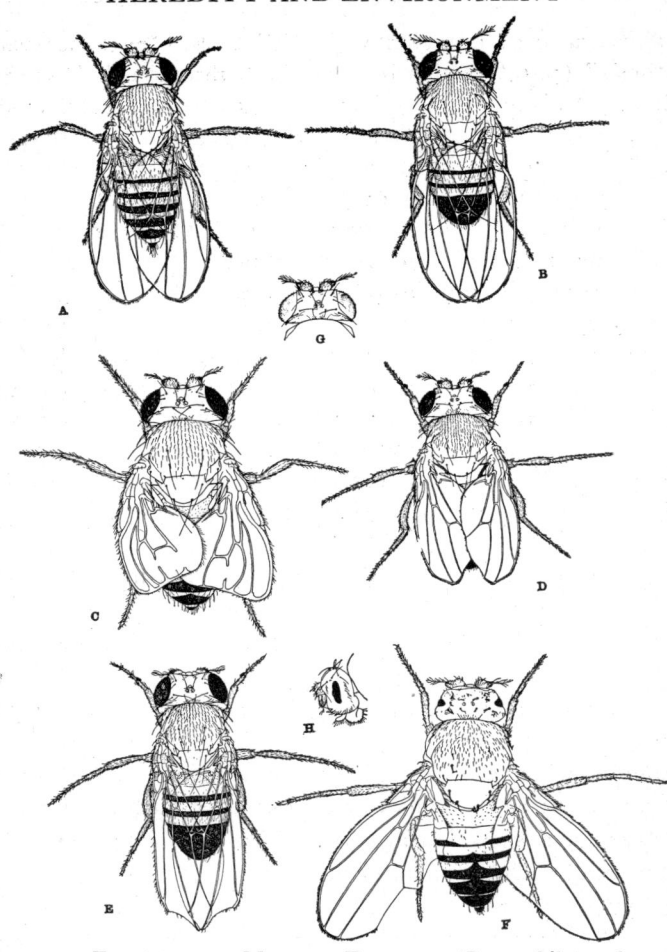

FIG. 103. TYPICAL AND MUTANT FORMS OF *Drosophila melanogaster*. *A*, Typical Female. *B*, Typical Male. *C* to *H*, Mutants resulting from mutations of genes in Chromosome Pair I (XX or XY) at "Loci" indicated by numbers in parenthesis: See also map of Chromosome I, Fig. 69, p. 195. *C*, "Rudimentary" Wing (54.5). *D*, "Miniature" Wing (36.0). *E*, "Cut" Wing (20.0). *F*, "Forked" Bristles (56.5) and "Bar" Eye (57.0). *G*, "White" Eyes (1.5). *H*, "Bar" Eye (57.0) seen from left side. (Figures loaned by Professor Morgan.)

CONTROL OF HEREDITY: EUGENICS 285

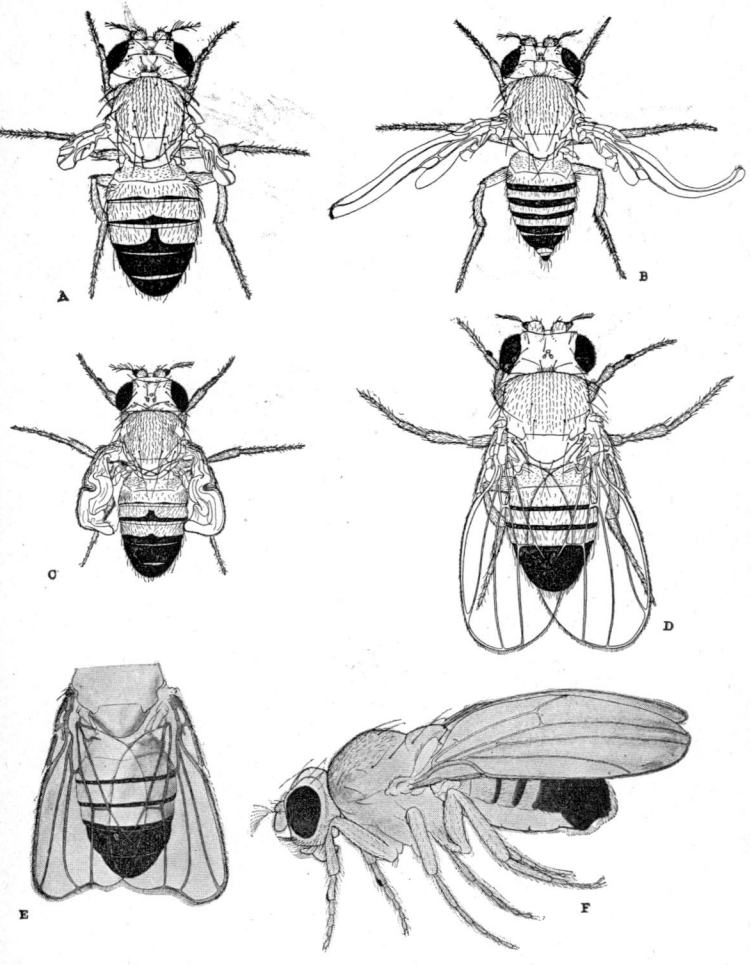

FIG. 104. MUTANTS OF *Drosophila melanogaster* resulting from mutations of genes in Chromosome Pair II at Loci indicated by numbers in parenthesis: See also map of Chromosome II, Fig. 69, p. 195. *A*, "Vestigial" Wing (65.0). *B*, "Strap" Wing (65.0, allelomorph of "Vestigial"). C_1 "Flipper" Wing (±28.0). *D*, "Squat" Body (±35.0). *E*, "Truncate" Wing (9.0). *F*, "Arc" Wing (97.5). (Figures loaned by Professor Morgan.)

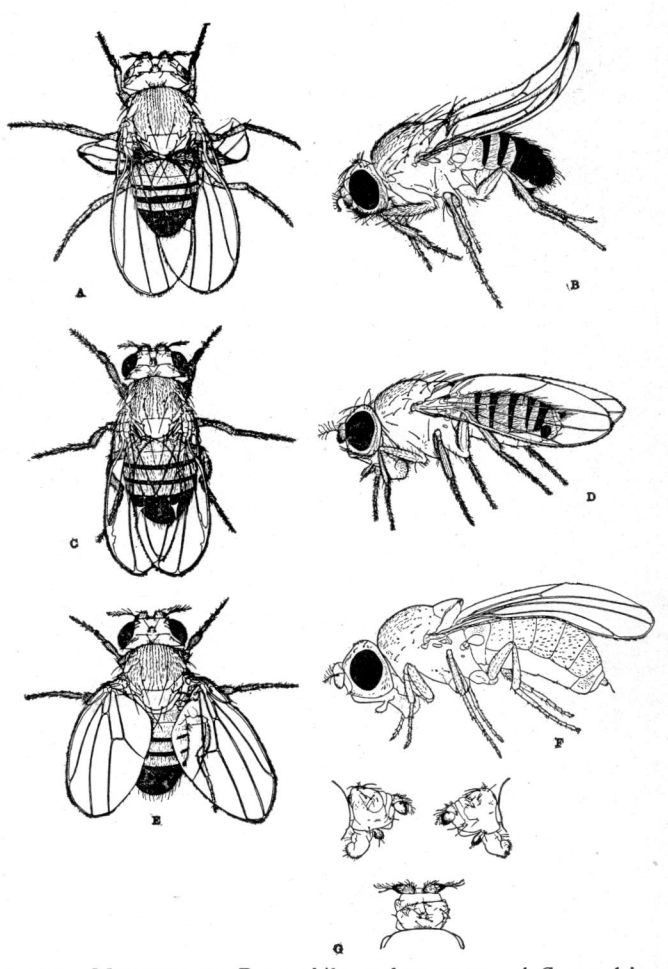

FIG. 105. MUTANTS OF *Drosophila melanogaster*. *A-C*, resulting from mutations of genes in Chromosome Pair III; *E-G*, due to mutant genes in Chromosome Pair IV at the Loci indicated by numbers in parenthesis. See also map of Chromosomes III and IV, Fig. 69, p. 195. *A*, "Bithorax" (III, 54.5). *B*, "Ski" Wings (III, 43.5). *C*, "Delta" veins in Wings (II, 63.5). *D*, "Roof" Wings (II, ±77.0). *E*, "Bent" Wings (IV, 0.0). *F*, "Shaven," few hairs (IV, ±0.5). *G*, "Eyeless" (IV, 1.0) left, right, and dorsal views. (Figures loaned by Professor Morgan.)

If these changes in the germplasm may be induced by extrinsic conditions, then a real experimental evolution will be possible; if they cannot be so induced but are like the changes taking place in the radium atom we can only look on while the evolutionary processes proceed, selecting here and there a product which nature gives us but being unable to initiate or control these processes.

B. CONTROL OF HUMAN HEREDITY: EUGENICS

1. Past Evolution of Man

There is every evidence that man also, no less than domesticated animals, has evolved from a natural or wild state. The most primitive types of men are known only from a few fossil remains, which indicate that these primitive men belonged to different species, and some of them even to different genera, from *Homo sapiens* (Fig. 106). Later stages in the evolution of man are known from many remains, implements and handiwork, as well as from certain primitive races or tribes which have persisted to the present time. The grades of culture represented by these extinct or persistent tribes and by modern men are usually classified as savagery, barbarism and civilization. There must have been much greater evolution of human types during prehistoric times than since the beginnings of civilization. The physical, mental and moral changes which took place in men from the earliest stages of savagery down to the beginnings of civilization were very great, but they were nevertheless slight compared with the tremendous changes which must have occurred in those long ages before the ancestors of man actually became men. Within the historic period the evolutionary changes in man have been very small. Minor changes have occurred and are still going on, as Osborn has shown in his "Cartwright Lectures on Contemporary Evolution in Man," but the species has remained relatively stable during the historic epoch as compared with the much longer prehistoric period.

288   HEREDITY AND ENVIRONMENT

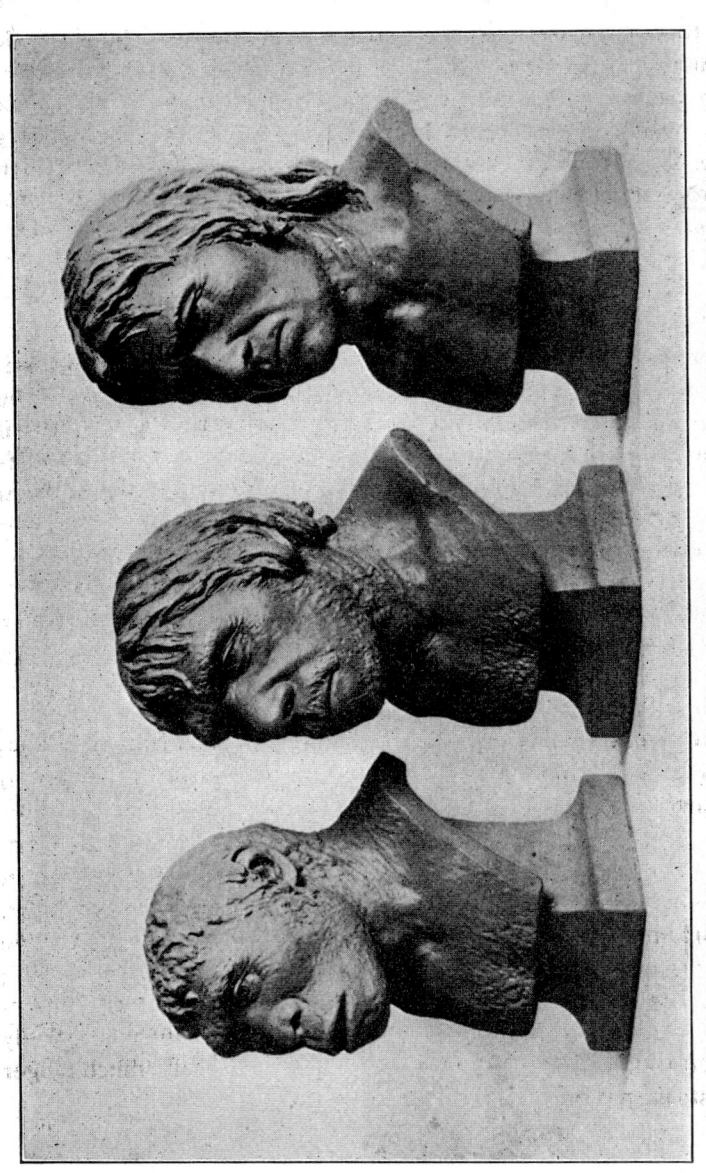

FIG. 106. TYPES OF PREHISTORIC MEN. Heads of *Pithecanthropus erectus* (Erect Ape Man); *Homo neanderthalensis* (Neanderthal Man); *Homo sapiens* (Cro-Magnon Man), modelled on restored skulls by Dr. J. H. McGregor, of Columbia University, and the American Museum of Natural History, New York City.

CONTROL OF HEREDITY : EUGENICS

The past history of man has been a long one, no one can say how long, but probably not less than half a million years have passed since the Hominidae appeared, and not less than fifty thousand years since the present species arose. There is every reason to believe that the future history of man will be even longer. Barring great secular changes, catastrophes or cataclysms, which cannot be foreseen nor provided against, man controls his own destiny on this planet.

It is a curious fact that in prescientific times the instability of nature especially appealed to men. How often in the past have men looked forward to a "speedy end of the world"! It may well have seemed to our ancestors a useless thing to take any thought for the morrow if very soon the heavens are to be rolled up as a parchment and the elements dissolved in fervent heat; it would be folly to plan for future ages if the time is at hand when the angel shall stand with one foot on the sea and the other on land and declare that time shall be no more. But science has taught us something of the wonderful stability of nature, something of the immensity of past time and of future ages, something of the eternity of natural processes. Compared with this infinite stability and eternity of nature what are our little systems and customs! Our years and centuries fall like grains of sand into this abyss of time. Our individual lives are like drops of water in this great ocean of life. What intellectual developments, what social institutions, what control of natural processes may come in the long ages of futurity it has not entered into the heart of man to conceive. And yet so far as we may judge by the small portion of the record of the past which we can read there has been no *necessary* progress. There has been "eternal process moving on," but not eternal progress. Stagnation, degeneration, elimination, as well as progression, have occurred all along the path of evolution. And yet on the whole evolution has been progressive and there is no reason to suppose that the elimination of the unfit and the preservation of the fit will cease to be the law of future evolution, as it has been that of the past.

Existing Human Types.—There are three principal types of the human species—white, yellow, and black—and many subtypes and races. These types and races differ in many regards in physical, mental and social characteristics, and their comparative value has frequently been discussed. It is difficult to take an impartial view of such a matter, though I suppose there would be little question on the part of any well informed person that the white and yellow types have contributed most to what we call civilization. Nevertheless every race probably has good qualities which could be made of service to society. The various races of cattle, horses, sheep, etc., are all useful to man, but in different ways and degrees, and the same is true of various races of men with respect to civilization. In general the dominant races are the most capable intellectually and socially, while those which have been left behind or have been eliminated have been the less capable ones. And yet some very good races, possibly with capacities for much social and intellectual development, have been completely exterminated, as for instance the Lucayan Indians of the West Indies, and the aborigines of Tasmania.

Race Extermination.—Few animals have suffered more wholesale destruction than have the more primitive races of men in different parts of the earth. Several species of man have become entirely extinct, leaving only, as is generally believed, a single existing species, *Homo sapiens.* Race extermination has been witnessed in relatively recent times and on a large scale in the West Indies, North and South America, Africa, Australia, New Zealand and the Islands of the Pacific. But in the disappearance of native races extermination is usually supplemented by amalgamation. After the most warlike members of a race have been destroyed the more peaceful remnants are generally incorporated in the conquering race. Thus the Maoris of New Zealand, the finest native race with which the English have come in contact in

CONTROL OF HEREDITY : EUGENICS

their colonies, were estimated to number more than a quarter of a million at the end of the eighteenth century. Owing to imported diseases and to destructive wars among the tribes and with the English there are not fifty thousand of them today, and these are gradually being absorbed into the white race.

Undoubtedly there has been a great growth of altruism in the modern world; there is a relatively new feeling among men that nothing so becomes a strong nation as the exercise of justice toward weaker ones, and many idealists maintain that every race and every people has the right to live its life in its own way. But however philanthropic they may be in theory, the practice of all nations demonstrates that weaker and inferior peoples are not permitted to stand in the way of dominant ones. When such peoples occupy territory which is desired by more powerful neighbors, they are either exterminated, expelled, exploited or amalgamated with the conquering race. In practice their rights are usually of small concern as compared with the desires of the invaders, and the inaccessible or undesirable parts of the earth, the deserts and mountains and regions of polar ice, become the refuge of the less capable races, just as "the conies, who are but a feeble folk, make their houses in the rocks." This is an illustration of the great law of evolution, the survival of the strong and capable and intelligent, and even though ideal justice be meted out to weaker peoples, dominant races will still dominate and possess the earth. The only recourse which the inferior peoples have, and it is a terrible revenge, is to amalgamate with the superior race and thus lower its hereditary qualities.

From the way in which primitive races have gone down before more cultured ones there is reason to believe that in general the principle of the elimination of the unfit and the survival of the fit has characterized human evolution no less than that of other organisms. Undoubtedly intelligence has played a great part in the evolution of man, as is at once apparent when we consider the infinitely varied experiments by which he has worked his way from savagery to civilization. And yet he has not consciously set before himself an evolutionary goal to be attained by intelligent attention to principles of good breeding.

II. Can Human Evolution Be Controlled?

Almost all that man now is he has come to be without conscious human guidance. If evolution has progressed from the amœba to man without human interference, if the great progress from ape-like men to the most highly civilized races has taken place without conscious human control, the question may well be asked, Is it possible to improve on the natural method of evolution? It may not be possible to improve on the method of evolution and yet by intelligent action it may be possible to facilitate that method. Man cannot change a single law of nature but he can put himself into such relations to natural laws that he can profit by them.

1. *Selective Breeding the only Method of Improving the Race.*—It is surely not possible to improve on nature's principle of eliminating the worst lines from reproduction. This has been the chief factor in the establishment of races of domesticated animals and cultivated plants, though as we have seen it has probably had nothing to do with the origin of mutations. The history of such races shows that evolution may be guided to human advantage by intelligent elimination and selection, and probably any hereditary improvement of the human race must be accomplished by this means, though of course such elimination and selection can apply only to the function of reproduction. The method of evolution by the elimination of persons, the destruction of the weak and cowardly and antisocial, which was the method practiced in ancient Sparta, is repugnant to the moral sense of enlightened men and cannot be allowed to act as in the past; but the worst types of mankind may be prevented from propagating, and the best types may be encouraged to increase and multiply. This is apparently the only way in which we may hope to improve permanently the human breed.

2. *No Improvement in Human Heredity within Historic Times.*—The improvement of environment and of opportunity for individual development enables men at the present day to get more out of their heredity than was possible in the past. Advance of civilization has meant only improvement of environment. But neither environment nor training has changed the hereditary capacities of man. There has been no perceptible improvement in human heredity within historic times, nothing comparable with the changes which have occurred in domesticated animals. Indeed no modern race of men is the equal of certain ancient ones. Galton has pointed out the fact that in the little country of Attica in the century between 530 and 430 B.C. there were produced fourteen illustrious men, one for every 4,300 of the free born, adult male population. In the two centuries from 500-300 B. C. this small, barren country with an area and total population about equal to that of the present State of Rhode Island but with less than one-fifth as many free persons produced at least twenty-five illustrious men. Among statesmen and commanders there were Miltiades, Themistocles, Aristides, Cimon, Pericles, Phocion; among poets Æschylus, Euripides, Sophocles, Aristophanes; among philosophers and men of science Socrates, Plato, Aristotle, Demetrius, Theophrastus; among architects and artists Ictinus, Phidias, Praxiteles, Polygnotus; among historians Thucydides and Xenophon; among orators Æschines, Demosthenes, Isocrates, Lysias. In this small country in the space of two centuries there appeared such a galaxy of illustrious men as has never been found on the whole earth in any two centuries since that time.

These illustrious men came from a remarkable race composed of individuals drawn together from all the shores of the Mediterranean by a process of unconscious but severe selection. Athens was the intellectual and social capital of the world and to it the most ambitious and most capable men were irresistibly drawn. It was good immigration as well as good native stock that made

Athens famous. Galton concludes that the average ability of the Athenian race of that period was, on the lowest estimate, as much greater than that of the English race of the present day as the latter is above that of the African negro.

But this marvellously gifted race declined, as all such races have in time declined:

> Social morality grew exceedingly lax, marriage became unfashionable and was avoided, many of the more ambitious and accomplished women were avowed courtesans and consequently infertile, and the mothers of the incoming population were of a heterogeneous class. . . . It can be therefore no surprise to us, though it has been a severe misfortune to humanity, that the high Athenian breed decayed and disappeared, for if it had maintained its excellence and had multiplied and spread over large countries, displacing inferior populations (which it well might have done, for it was naturally very prolific), it would assuredly have accomplished results advantageous to human civilization to a degree that transcends our powers of imagination. (Galton, "Hereditary Genius," page 331.)

Bateson suggests that the high intellectual qualities of the ancient Athenian race were due to the inbreeding of homogeneous and very superior phratries and gentes, but when foreign marriages were sanctioned, and aliens and manumitted slaves were admitted to citizenship by the "reforms" of Cleisthenes (507 B. C.) the population gradually became mongrelized and its intellectual superiority declined.

3. *Why the Race Has Not Improved.*—If mankind has made no progress in hereditary characteristics since the time of the Greeks the cause is not far to seek. There have been gifted races and families, doubtless many notable human mutations have occurred, but most of these have been diluted, squandered, lost. There has been persistent violation of all principles of good breeding among men. For example, there has been for ages a futile reliance upon good environment to improve heredity. Men do not so improve the races of animals and plants, and thousands

of years of human history show that this method is of no avail in improving the human breed.

But the case is far worse than this; such efforts though futile are at least well intentioned, but on the part of most men and governments there has been complete disregard of the entire question of the improvement of the human stock. Natural selection which has through countless ages eliminated the worst and conserved the best-fitted and thus has led on the whole to the survival of the fit is so far as possible nullified by civilized man; the worst are preserved along with the best and all are given the same chance of reproduction. The mistake has been not in nullifying natural selection by preserving the weak and incompetent, for civilized men could not well do otherwise, but in failing to substitute intelligent artificial selection for natural selection in the propagation of the race. Instead of this there has been perpetuation of the worst lines through sentimental regard for personal rights, even when opposed to the welfare of society; and both church and state have cheerfully given consent and blessing to the marriage and propagation of idiots and of diseased, defective, insane and vicious persons. Finally there has been extinction of the world's most gifted lines by enforced celibacy in many religious orders and societies of scholars; by almost continuous wars which have taken the very best blood that was left outside of the monastic orders; by luxury and voluntary sterility; by vice, disease and consequent infertility.

Is it any wonder that the inheritance of the human race has not improved within historic times? Is it not rather an evidence of the broadcast distribution of good and wholesome qualities in the race that in spite of such serious violations of the principles of good breeding mankind remains as good as we find it today?

III. Eugenics

If a superior power should deal with man as man deals with domestic animals no doubt great improvement could be effected in the human breed. Society is in some respects such a power and can do what the individual, because of self-interest, short life or lack of ability, cannot accomplish. In matters of public health and comfort, security of life and property society is superior in power to the individual; in matters of the perpetuation of the race the individual is still supreme. In animal societies the race, the breed, is to the swift and strong and fit, and the same was probably true of primitive men. But it is impossible to return to the conditions of primitive society in this respect, and the social body itself must in some way control the breeding of men.

There are millions of men in civilized countries whose mental equipment places them on a plane with barbarians or savages, and they have on the average more offspring than their civilized contemporaries. There are millions of others who are so seriously defective in body or mind, owing to hereditary causes, that they can never take care of themselves and must always be a charge upon the state, and yet in many civilized countries they are permitted to perpetuate their kind and produce a never-ending supply of mental and moral defectives, whose maintenance must seriously interfere with the proper education and development of the normal population and whose unrestrained existence constantly threatens to pollute purer streams of heredity. The practice of society regarding marriage and reproduction up to the present has been to allow all sorts, good, bad and indifferent, to propagate with the belief that good environment and training will make up for deficiencies of birth. But recently the conviction has been growing that good environment is far less important than good heredity and that in some way society must influence the race of men at its source. This is the doctrine of eugenics, which Galton defines as follows:

The science of improving stock, which is by no means confined to questions of judicious mating but which, especially in the case of man, takes cognizance of all influences that tend in however remote a degree to give to the more suitable races or strains of

CONTROL OF HEREDITY : EUGENICS

blood a better chance of prevailing speedily over the less suitable than they otherwise would have had. ("Inquiries into Human Faculty.")

Fortunately or unfortunately the methods which breeders use cannot be rigidly applied in the case of man. It is possible for breeders to eliminate from reproduction all except the very best stocks, and this is really essential if evolution is to be guided in a definite direction. If only the very worst are eliminated in each generation, the standard of a race is merely maintained, but the more severe the elimination is the more does it become a directing factor in evolution. In the case of man, however, even the most enthusiastic eugenicists have never proposed to cut off from the possibility of reproduction all human stocks except the very best, and if only the very worst stocks are thus eliminated, we must face the conclusion that no very great improvement can be effected. It is impossible, then, to apply rigidly to man the methods of animal and plant breeders. Society cannot be expected to eliminate from reproduction all but the very best lines. The great majority of mankind cannot be expected voluntarily to efface itself. The most that can be hoped for in this direction is that the great mediocre majority may eliminate from reproduction a very small minority of the worst individuals.

Furthermore, other and perhaps even more serious objections to the views of extreme eugenicists are to be found in human ideals of morality. Even for the laudable purpose of producing a race of supermen, mankind will probably never consent to be reduced to the morality of the breeding-pen with a total disregard of marriage and monogamy. The geneticist who has dealt only with chickens or rabbits or cattle is apt to overlook the vast difference between controlling reproduction in lower animals and in the case of man where restraints must be self-imposed.

Another fundamental difficulty in breeding a better race of men is to be found in a lack of uniform ideals. A breeder of domestic animals lives long enough to develop certain races and see them well established, but the devotee of eugenics cannot be sure that his or her ideals will be followed in succeeding generations. The father of Simon Newcomb is said to have walked through the length and breadth of Nova Scotia seeking for himself a suitable mate, but neither he nor any other eugenicist could be sure that his descendants would follow a similar course, and long continued selection along particular lines must be practiced if the race is to be permanently improved. Mankind is such a mongrel mixture, and it is so impracticable to exercise a strict control over the breeding of men, that it is hopeless to expect to get pure or homozygous stocks except with respect to a very few characters and then only after long selection.

But granting all these difficulties which confront the eugenicist, there is no doubt that something may be gained by eliminating merely the worst human kinds from the possibility of reproduction, even though no marvelous improvement in the human race can be expected as a result of such a feeble measure.

1. *Possible and Impossible Ideals. Supermen.*—What the future evolution of the human race may lead to is an interesting speculation, but it is and can be only a speculation. There is no present evidence that there will ever be a higher animal than man on the earth, and the only evidence that there may be a higher species than *Homo sapiens* is to be found in the fact that there have been lower species of men in the past and that evolution has been on the whole progressive. The idea that by the aid of that infant industry, eugenics, a new race of supermen is shortly to be produced is an iridescent dream, and the fantastic demand of some enthusiasts for changes in racial fashions has served to bring this whole subject of eugenics into disrepute among thoughtful men.

Hereditary Classes.—To a considerable extent ideals regarding individuals and society have differed among different races in the past, but with the closer communications which have been established between all parts of the earth in modern times there has developed a greater uniformity of ideal. In a complex society all types of service are needed and different types of individuals are socially useful. If the social good were the supreme end, as it is in a colony of ants or bees, the greatest differentiation of individuals for particular kinds of service would be desirable. There should be an hereditary class of laborers, of business men, of scholars, of artists, etc., and for the improvement of each class there should be inbreeding in that class. Such methods are now used by breeders of various races of domestic animals and cultivated plants with the best of results. No breeder would think of trying to improve draft horses by crossing with race horses, nor of improving milk cows by crossing with beef cattle. In other countries and ages the development of hereditary classes and castes in human society has been tried, and survivals of it persist to this day, but they are only vestigial remnants of an old order which is everywhere being replaced by a new ideal in which the good of the individual as well as that of society is the end desired.

The whole development of modern society is in the direction of racial solidarity and away from hereditary classes. Government, education and religion; socialism, syndicalism, bolshevism all reflect the movement for individual liberty, fraternity and equality. The modern ideal individual is not the highly specialized unit in the social organism, as in the case of social insects, but rather the most general all-round type of individual, the man who can when conditions demand it combine within himself the functions of the laborer, business man, soldier and scholar. For such a generalized type the methods of inbreeding or close breeding used by the breeder of thoroughbreds are wholly inappropriate. On the other hand such a generalized type must include the best qualities of many types and races and Mendelian inheritance shows how it is possible to sort out the best qualities from the worst.

Nowadays one hears a lot of high sounding talk about "human thoroughbreds," which usually means that those who use this phrase desire to see certain narrow and exclusive social classes perpetuated by close inbreeding; it usually has no reference to good hereditary traits wherever found, indeed such traits would not be recognized if they appeared outside of "the four hundred." Such talk probably does neither harm nor good; the "social thoroughbreds" are so few in number and so nearly sterile that the mass of the population is not affected by these exclusive classes.

Galton advocated the segregation and intermarriage of the most highly intellectual members of society, such as the prize scholars in the colleges and universities; but if the human ideal is the generalized rather than the specialized type it would be better if the prize scholars married the prize athletes. A race of highly specialized scholars or athletes is not so desirable as a race in which these and other excellences are well balanced. From this point of view the person who is voted the "best all-round man in his class" is nearer the eugenical ideal than the prize scholar.

No man can trace his lineage back through many generations without realizing that it includes many hereditary lines differing greatly in value. The significance of sexual reproduction lies in this very fact that it brings about the commingling of distinct lines and thereby makes every individual different from every other one. The entire history of past evolution testifies to the value of this process, although it causes the gardener, the breeder, the eugenicist serious trouble. But the gardener can propagate his choice fruits by budding and grafting, the breeder can for a time preserve his choice stock by close inbreeding, but the eugenicist cannot shut out the influence of foreign blood, and perhaps it is well that he cannot for if he could do so the progress of the race might soon come to an end.

Racial Amalgamations.—In the human species the only absolute barrier to the intermingling of races is geographical isolation. Every human race is fertile with every other one, and though races and nations and social groups may raise artificial barriers against interbreeding we know that these artificial restraints are frequently disregarded and that in the long run amalgamation does take place; and in general the further amalgamation progresses the faster it goes. In Australia and New Zealand, after little more than a century's contact with white races, there are about as many "half castes" as there are full blooded aborigines. In the United States one-quarter of all persons of African descent contain more or less white blood; there are about eight million full blooded negroes and two million mulattoes, and during the past twenty years the latter have increased at twice the rate of the former. In Jamaica, where there are about seven hundred thousand blacks and fifteen thousand whites, there are about fifty thousand mulattoes. A similar condition prevails wherever different races occupy the same country. Even the Jews, who were long supposed to be a peculiarly separate and distinct people, have received large admixtures of Gentile blood in every country in which they have lived.

Whether we want it or not hybridization of human races is going on and will increase. Partition walls between classes and races are being broken down; complete isolation is no longer possible, and a gradual intermixture of human races is inevitable. We are in the grip of a great world movement and we cannot reverse the current, but we may to a certain extent direct the current into the more desirable channels.

There is a popular belief that hybrid races are always inferior to pure bred ones, but this is by no means the case. Some hybrids are undoubtedly inferior to either of the parents but on the other hand some are vastly superior; only experience can determine whether a certain cross will yield inferior or superior types. Society may well attempt to prevent those crosses which produce inferior stock while encouraging those which produce superior types.

Immigration.—It is race mixture which makes the problem of immigration so serious. Generally immigration is regarded merely as an economic and political problem, but these aspects of it are temporary and insignificant as compared with its biological consequences. In welcoming the immigrant to our shores we not only share our country with him but we take him into our families and give to him our children or our children's children in marriage. Whatever the present antipathies may be to such racial mixtures we may rest assured that in a few hundred years these persons of foreign race and blood will be incorporated in our race and we in theirs. From the amalgamation of good types good results may be expected; but fusion with inferior stock, while it may help to raise the lower type, is very apt to pull the higher one down. How insignificant are considerations of cheap labor and rapid development of natural resources when compared with these biological consequences!

2. *Negative Eugenical Measures. Late and Early Marriages.*—Galton said nothing about sterilization or elimination from reproduction of less valuable lines in his "Inquiries into Human Faculty" which was first published in 1883. He proposed no radical policy but rather one which he thought would be practical and might meet with general favor. He suggested a social policy which would delay the age of marriage among the weak and hasten it among the vigorous, whereas present social agencies act in the opposite direction. He showed by statistics that, on the average, marriage at the age of 22 would produce at the end of one century four times as many offspring as marriage at 33 and at the end of two centuries ten times as many. He particularly emphasized the great harm which would be done by an application of the theory of Malthus among the better classes. For the prudent to put off marriage and to limit offspring while the imprudent continue to reproduce at the present rate would be to give the world to the imprudent within a few centuries at most.

Segregation and Sterilization.—His suggestions, which were at first received with indifference or ridicule, were much less radical than the legal requirements in many of our States today. Public sentiment has been greatly aroused on this question; the apparent increase in the number of defectives and criminals has seemed to call for radical action and a flood of hasty but well intentioned legislation has been the result. We may confidently expect that in a very short time the marriage of the feeble-minded, hopelessly insane or epileptic, the congenitally blind, deaf and dumb, and those suffering from many other inherited defects which unfit them for useful citizenship will be prohibited by law in all the States. Our immigration laws already exclude such aliens, and the number of persons of the types named who seek legal consent to marry is not large so that it need not be expected that such laws will quickly improve the general population. If in addition such persons are either segregated or sterilized the danger of their leaving illegitimate offspring will be removed; such precautions have been taken in certain of our States and will probably become general, though at present few of the laws on this subject are strictly enforced.

The study of heredity shows that the normal brothers and sisters, and even the more distant relatives, of affected persons may carry a defect as a recessive in their germ plasm and may transmit it to their descendants though not showing it themselves. It will be more difficult, perhaps an impossible thing, to apply rigidly the principles of good breeding to such persons and to exclude them from reproduction; but if in each generation those persons in whom this recessive trait appears are prevented from leaving offspring the number of persons affected will gradually grow less, other conditions being equal.

But while such negative eugenical measures are wholly commendable when applied to such defects as those named, which are certainly inherited and which render those affected unfit for citizenship, the wholesale sterilization of all sorts of criminals, alcoholics and undesirables without determining whether their defects are due to heredity or to conditions of development would be like burning down a house to get rid of the rats; and the only justification which could be offered for the general sterilization of the inmates of all public institutions, which is urged by some of our modern crusaders, would be the defense which some persons make for war, namely that there are too many people and that anything which will prevent the growth of population is to be welcomed.

Effects of War on Race.—Advocates of war never cease to point out its beneficial effects on the race,—how it makes men strong, courageous, unselfish, how it makes nations great, powerful, progressive. There is no doubt that war like any other great crisis discovers great men and furnishes opportunities for the development of great qualities that might otherwise remain undeveloped and unknown. But there is also no doubt that it takes the very best blood of the nations. Those who go to war are the young, the strong, the capable, while the weak, incompetent and degenerate are left behind as unfit for military service. If conditions could be reversed and the bungled and botched, the feeble-minded and insane, the degenerate and debauched could be put in the forefront of battle some benefit to the race might result, but no increase of national greatness can compensate for the awful waste of the best thing which any nation possesses—its best blood.

Realizing that progress in evolution has been won only through struggle and that the human race owes much to the fact that man is by nature and instinct a "fighting animal" many persons who have recognized the evil effects of war have endeavored to find some substitute for modern warfare, which is no longer the wager of personal combat, but a vast impersonal mechanism of destruction. In view of the fact that "intrepidity, contempt for softness, surrender of private interests, obedience to command, must still remain the rock upon which states are built," William James proposed, as a "moral equivalent for war," compulsory service in hard and difficult occupations where dangers and hardships would be incentives to effort and where struggle for success would "inflame the civic temper as past history has inflamed the military temper."

Professor Cannon, whose work has demonstrated that the adrenal glands are *par excellence* the glands of combat and virility, and who recognizes the importance to the human race of maintaining the functional activity of these glands, has proposed athletics and especially international athletic contests, such as the Olympic Games, as a "physical substitute for warfare."

The eugenical ideal is not a life of "peace, perfect peace," nor a millennium in which all struggle shall cease, but rather a life of adventure, conflict and hard-won success. Inaction and satiety end in degeneration, and progress can be purchased only by struggle. But it is not only unnecessary, it is positively irrational, to resort to war to secure these ends. As civilization advances more and more substitutes are found for war. Among these are not only athletics and sports but also struggles with natural difficulties and forces in the great warfare which is being waged for the conquest of nature. Even intellectual and political contests and competitions in skill and workmanship may to a great extent replace war as a field of adventure and emprise.

3. Positive Eugenical Measures.—Positive eugenical measures are much more difficult to apply and are of more doubtful value than are negative ones. Of course compulsory measures requiring the best types to intermarry and have children are out of the question and encouragement and advice alone are feasible. Giving advice regarding matrimony is proverbially a hazardous performance, and it is not much safer for the biologist than for others.

Eugenical Predictions Uncertain.—With much more complete knowledge regarding human inheritance than we now possess it may be possible to give eugenical advice wisely, especially with respect to physical characteristics which are hereditarily simple and generally of minor significance. But where the character is an extremely complex one such as intellectual ability, moral rectitude, judgment and poise, which are the chief characteristics which distinguish the great man from his fellows, it will probably never be possible to predict the result before the event.

He would be a bold prophet who would undertake to predict the type of personality which might be expected in the children of a given union. Some very unpromising stocks have brought forth wonderful products. Could anyone have predicted Abraham Lincoln from a study of his ancestry? Observe I say "predict," and not "explain" after his appearance. Can anyone now predict from what kind of ancestral combinations the great scholars, statesmen, men of affairs of the next generation will come? The time may come when it will be possible to predict what the chances are that the children of given parents will inherit more or less than average intellectual capacity, but since germinal potentiality is transformed into intellectual ability only as the result of development such a prediction could not be extended to the latter unless the environment as well as the heredity were known.

Mankind is such a mongrel race, good and bad qualities are so mixed in us, marriage is such a lottery, the distribution of the germinal units to the different germ cells and the union of particular germ cells in fertilization is so wholly a matter of chance, the influence of even bad hereditary units on one another is so unpredictably good or bad as is shown in many hybrids, even the minor influences of environment and education which escape attention are so potent in development, that the chances were infinity to one against any one of us, with all his individual characteristics, ever coming into existence. If the Greeks or Romans had known of the real infinity of chances through which every human being is brought to the light of day not only would they have deified Chance but they would have made her the mother of gods and men.

Selective Mating.—But granting the impossibility of predicting the character of children it may well be asked if good general advice may not be given regarding the choosing of a mate. Many

CONTROL OF HEREDITY : EUGENICS

people have thought so, and if all that has been said or written on this subject were to be gathered together I suppose that there would not or should not be room for it in all the libraries of the world. It is generally admitted that few lines are wholly free from hereditary defects and the question has often been asked what the eugenical practice should be in such cases. Of course people with really serious hereditary defects should not have children. If the defects are slight Davenport has suggested that they may either be disregarded or weakness in any character may be mated with strength in that character. That people with only slight hereditary defects should not marry at all is a counsel of perfection.

On the other hand it would be a dangerous rule to propose that persons having really serious hereditary defects should be mated with those who are strong in those characters on the ground that in general strength in a character is dominant over weakness. It has been suggested that a normal man who marries a feeble-minded woman would have only normal children, since both genius and feeble-mindedness seem to be recessive when mated with mediocrity or normality. But in all such cases the weakness is not neutralized or removed but merely concealed in the offspring and is therefore the more dangerous. If a man chooses to marry a feeble-minded woman he at least does so with his eyes open and he need not be deceived. But the normal and perhaps capable children of such a union carry the taint concealed in their germ-plasm and if they should be mated with other normal persons carrying a similar taint some of their children would be feeble-minded, and thus the sins of the parents in mating weakness with strength would be visited upon the children to the nth generation. Such a policy of concealing weakness by mating it with strength is wholly comparable with the custom once prevalent of concealing cases of contagious diseases, and may properly be characterized as the "ostrich policy."

After all in the choosing of mates a combination of instinct and intelligence is probably the safest guide. Our instincts, built up through long ages, are generally adaptive and useful, and if they be guided by reason the result is apt to be better than if either instinct or reason acts alone. More need not be said on this subject, since it is treated *ad infinitum* in works of fiction and in ladies' journals.

4. *Contributory Eugenical Measures. General Education.*—In addition to these negative and positive eugenical measures many conditions may be classed as contributory to eugenics. One of the most important of all contributory measures is the general education of the people regarding heredity. The widespread ignorance on this subject is profound and very many offenders against the principles of good breeding have sinned through ignorance. Any general reform must rest upon enlightened public opinion, and the schools, the churches and the press can do no more important work for mankind than to educate the people, after they have been educated themselves, on this important matter.

Society too may cultivate a proper pride in good inheritance. Much of value would be accomplished if the silly pride in ancestral wealth or position or environment which touched our forebears only superficially and never entered into their germplasm, or the still sillier claims of long descent, in which we are all equal, could be replaced by a proper pride in ancestral heredity, a pride in those inherited qualities of body, mind and character which have made some families illustrious. A proper pride in heredity would do much to insure the perpetuation of a line and to protect it from admixture with baser blood.

Coeducation versus Monasticism.—Among other contributory measures which serve to promote good breeding among men must be reckoned coeducation, as well as other means of promoting good and early marriages. The president of a large coeducational institution once said that if marriages were made in heaven he was sure that the Lord had a branch office in his university. I had occasion a few years ago to investigate the eugenical record of a coeducational institution, which is not unknown in the world of scholarship, and I found that about 33 per cent of the recent graduates had married fellow students, that there had been no divorces and that there were many children. There is no doubt that coeducation promotes good and early marriages and that it is not necessarily inimical to good scholarship even though it violates the spirit of mediæval monasticism. There was a time when it was supposed that a scholar must live the monkish life of seclusion and contemplation, but the monasteries are disappearing the world over, and it is time that the monastic spirit should go out of the colleges and universities.

On the other hand the colleges exclusively for men or women appear to have a bad influence on the marriage rate and birth rate of their graduates. Johnson has shown that 90 per cent of all the women of the United States marry before the age of 40, but that among college women only half that number have married at the same age. As a result of investigations at one of the leading women's colleges he finds that the marriage and birth rate of the most brilliant students, who have been elected members of Phi Beta Kappa, is lowest of all. Cattell says that a Harvard graduate has on the average three-fourths of a son, a Vassar graduate one-half of a daughter.

At present early and fruitful marriages among able and ambitious people are very unfashionable and are becoming increasingly impracticable. If society has any regard for its own welfare all this must be changed. As Galton has shown, the race that marries at 22 instead of 33 will possess the earth in two or three centuries.

The good of society demands that we reverse our methods of putting a premium upon celibacy among our most gifted and ambitious young men and women, and if monastic orders and institutions are to continue they should be open only to those eugenically unfit.

5. *The Declining Birth Rate. Stationary Population Normal.*
—Among animals and plants in a state of nature the number of individuals in each species remains fairly constant from year to year; that is, only enough young are born and survive to take the places of mature individuals that die. But when a species is placed in new and favorable conditions it may for a while increase at an amazing rate until the pressure of population becomes sufficient to reestablish an equilibrium between the birth rate and the death rate. Thus when the English sparrow was introduced into the United States it increased at a phenomenal rate for a number of years, but now the number of individuals in any given locality remains about the same from year to year, the birth rate merely compensating for the death rate. This equilibrium is brought about in the main by increased mortality, especially among the young, though decreasing fecundity may play a minor part.

Essentially the same principles apply to human populations. Up to two or three centuries ago the populations of the older countries of the world were practically stationary. Fecundity was relatively high but the death rate was also very high, the excess of population in each generation being carried off in large numbers by war, pestilence and famine. Then owing to the developments of science and industry and to the opening up of new countries a period of remarkable expansion of population began. The population of Europe, which was about 175 millions in 1800, increased to 420 millions in 1900, and this in spite of the fact that about 35 millions migrated from Europe to new countries during this period. This great increase in the population of Europe was due primarily to reduction of the death rate since the birth rate also declined slightly during this period, while in the newer countries there was both an increase in the birth rate and a decrease in the death rate.

It is perhaps an open question how long the advances of science in rendering available the natural resources of the earth may be

CONTROL OF HEREDITY : EUGENICS

able to keep pace with increasing population, but it is evidently impossible for this great increase in the population of the world to go on indefinitely; sooner or later it must come to an end and the population again become stationary. Already the birth rate is decreasing more rapidly than the death rate in all the western countries of Europe and this movement must ultimately extend to all parts of the world and lead to a checking of the great increase in population which has characterized the last two hundred years. This approach to a stationary population is both a normal and a desirable thing, for no one could wish to see the population increase more rapidly than the supply of food or other necessaries of life; and of the two possible methods of checking population few would hesitate to choose a decreasing birth rate as preferable to an increasing death rate.

It is not therefore the declining birth rate in the general population that should cause alarm but rather the declining birth rate in the best elements of a population, while it continues to increase or at the least remains stationary among the poorer elements, and there is abundant evidence that this is just what is taking place. The descendants of the Puritans and the Cavaliers who have raised the cry for "fewer and better children" are already disappearing and in a few centuries at most will have given place to more fertile races of mankind. Many of the old New England families are dying out and their places are being taken by recent immigrants. The few exceptions are merely eddies in the current that is bearing them to doom. In Massachusetts the birth rate of the foreign born is twice that of the native population while the death rate is about the same for both. The same is true of the older families in many parts of the world.

Cattell has made a statistical study of the families of 917 American men of science and he finds that the average size of family of the parents of these men was 4.66 children, whereas the average size of family of these men is 2.22 children. In one generation the fertility of these lines has been reduced by more than half. The causes of this decline are chiefly voluntary being assigned to health, expense and other causes.

Death of Families.—But the causes of sterility are not only social and voluntary ones, which could be changed by custom and public opinion; there are also involuntary and biological causes of a deep-seated nature. Fahlenbeck has made a study of 433 noble families of Sweden which have become extinct in the male line, and he shows that the last male died unmarried in 45 per cent of these families, and before the age of 21 in 39 per cent, while the line ended in infertile marriage in 11 per cent and in daughters only in 5 per cent.

The extinction of families, however, is often confused with the extinction of family names, which means only that the family has died out in the direct male line. Biological inheritance does not necessarily follow family names. Owing to the elimination of one-half of the chromosomes in the formation of the sex cells and the replacing of these in fertilization by chromosomes from another source it happens that many persons bear the name of some progenitor but do not have a single one of his chromosomes or inherited traits; on the other hand, many persons who do not bear his name may have some of his chromosomes and traits. Assuming that there are 48 chromosomes in the human species and that these never break up or lose their identity it is evident that no person can inherit from more than 48 contemporary ancestors though he may be descended from an innumerable number.

Much confusion is caused also by the expression "hereditary lines," as if each family were separate and distinct from all others. But this is, of course, never true. The only hereditary lines which exist are those of individual chromosomes or genes and these divide and diverge like the branches of a tree. An individual containing many chromosomes received from many sources belongs to no single hereditary line, but rather to a network of many lines.

It has been said that if the birth rate of the "Mayflower" families continues to decrease at the present rate for the next 300 years, all the survivors at that time could be sent back in the original "Mayflower." But there is no reason to suspect that the decreasing birth rate will go on indefinitely at a constant ratio, and to assume that it will do so is merely to look forward to the extinction of all families, classes, races and nations in which the birth rate has been decreasing; this includes practically the entire population of the United States and Western Europe and it is evident that such a result while theoretically possible is not at all probable. Considering the large number of collateral lines which have come from the "Mayflower" stock and the enormous number of individuals who think they can trace their ancestry back to the "Mayflower," it is incredible that all these should be reduced to a company no larger than that which came over on that famous ship.

Broman points out that most noble families of Europe die out (probably the direct male line only is meant) after 100 to 250 years and generally do not live beyond the third generation. The same is true of the families of great scholars, artists and statesmen. Possibly one cause of such declining fertility may be found in too great brain activity, but there is no doubt that in many instances it is due to luxurious living. On the other hand bodily fatigue and simple living favor fertility in both animals and men. Wild animals brought into captivity where they have comfortable quarters and an unwonted abundance of rich food are usually infertile; and the conditions of life of the upper classes of society are almost as unfavorable to fertility as is captivity with wild animals. It is evident that if we had fewer luxuries we could have, and could afford to have, more children.

But animals in captivity may gradually become adapted to their new conditions so as to become fertile, and there is evidence that man also may undergo a slow adaptation in this regard to conditions of high civilization. Some royal families of Europe go back six or eight hundred years, and in general if a family survives the new conditions of affluence and luxury for more than three generations it may become more or less adapted to the new conditions.

Birth Control.—No eugenical reform can fail to take account of the fact that the decreasing birth rate among intelligent people is a constant menace to the race. We need not "fewer and better children" but more children of the better sort and fewer of the worse variety. There is great enthusiasm today on the part of many childless reformers for negative eugenical measures; the race is to be regenerated through sterilization or birth control. But unfortunately this reform begins among those who because of good hereditary traits should not be infertile. Sterility is too easily acquired; what is not so easily brought about is the fertility of the better stocks. Galton was far wiser than most of his followers for he realized the necessity of increasing the families of the better types as well as of decreasing those of the worse.

What Bernard Shaw regards as the greatest discovery of the nineteenth century, *viz.*, the means of artificially limiting the size of families, may prove to be the greatest menace to the human race. If it were applied only to those who should not have children or to those who should for various reasons have only a few children it would be a blessing to mankind. But applied to those who could and should have many children it is no gift of the gods. No one denies that the chief motive for limiting the size of families is personal comfort and pleasure rather than the welfare of the race. The argument that people should have no more children than they can rear in comfort or luxury assumes that environment is more important than heredity, which is contrary to all the biological evidence. In the breeding of horses or cattle or men heredity is more potent than environment; and it is more important for the welfare of the race that children with good inheritance should be brought into the world than that parents should live easy lives and have no more children than they can conveniently rear amid all the comforts of a luxury-loving age.

CONTROL OF HEREDITY : EUGENICS

The method of evolution in the past has been the production of enormous numbers of individuals and the elimination of the least fit. The modern method of improving domestic races is to select for reproduction the best types from large numbers of individuals. Nature has provided an almost infinite wealth and variety of potential personalities in human germ cells but only an infinitesimal number ever come to development. If this number is still further reduced by artificial means, and without regard to fitness, the race will be made the poorer not merely in quantity but also in quality. The optimism of those who believe that supermen may be produced by artificially limiting the number of children is a foolish and fatal optimism.

Finally for those who are denied the privilege of parenthood and upon whom sterility is forced by whatever circumstances there is a lesson of value to be drawn from the social insects. The sterile members of a colony of ants or bees are forever denied the possibility of having offspring of their own, but they become foster-mothers to the offspring of the queen. They tenderly nurse, care for and rear the young of the colony. There are many children in the world who need foster mothers and fathers; there are many men and women in the world, both married and unmarried, who need adopted children. "Go to the ant, thou sluggard; consider her ways and be wise."

CHAPTER VI

GENETICS AND ETHICS

319

CHAPTER VI

GENETICS AND ETHICS[1]

Modern studies of development are profoundly changing the opinions of men with respect to human personality. Observations of the relentless laws of heredity, of the inevitable influences for good or bad of environmental conditions over which the individual has no control, undoubtedly tend to produce a sense of helplessness and hopelessness. What light is thrown upon the great problems of freedom and determinism, of responsibility and irresponsibility, of duty and necessity by modern studies of development? Such questions cannot be dealt with quantitatively or experimentally, and they lie outside the field of exact science, but they are involved in all inquiries which have to do with rational and social beings; they lie at the foundation of the application of science to human welfare; they occupy a large place in the thought and conduct of all men.

1. THE VOLUNTARISTIC CONCEPTION OF NATURE AND OF HUMAN RESPONSIBILITY

Primitive men regarded their own activities and all phenomena of nature as the expression of will, and a similar view has been maintained by certain philosophers and theologians even in modern times. Nature was regarded as the immediate expression of a vast will which creates, rules, builds and destroys as it sees fit. The lightning is hurled from the hand of Jove, the sea is dis-

[1] A portion of this chapter was given as the presidential address before the American Society of Naturalists in January, 1913, and was published in *Science* under the title "Heredity and Responsibility."

turbed by angry deities, the winds are let loose or stilled, the earth trembles, the hills smoke, the sun and moon and stars travel in their appointed courses as the gods will.

In this primitive view of nature even inanimate objects were supposed to be endowed with wills of their own, and many modern men are sufficiently primitive to kick the chair over which they stumble, or to swear that the devil has gotten into the automobile. Of course the actions of all animate things were held to be the result of choice; the fly that dances on your head or gets into the soup is doing it to annoy you; the cats that yowl, the dogs that howl, the maniacs that screech are possessed of devils, evil wills, and should be punished. All good is the result of good will, all evil of evil will. Some being, some volition, is responsible for everything that happens. All nature is the expression of big or little wills, of good or bad wills, and the good should be rewarded and the bad punished.

This conception of nature finds its counterpart and probably its origin in similar views concerning human conduct and responsibility. According to this belief every man is the architect of his own character; the will is absolutely free; no taint of heredity or necessity rests on the mind or soul; character is a *tabula rasa*, upon which the self writes its own record as it chooses, and is responsible for the result. Conduct whether good or bad, benevolent or criminal, rational or irrational rests upon voluntary choice, and for such choices men must be held responsible. To a great extent this view of freedom and responsibility is the basis of present systems of government, education, ethics and religion.

II. THE MECHANISTIC CONCEPTION OF NATURE AND OF PERSONALITY

As contrasted with this voluntaristic view of nature and of man consider the scientific conception of nature as a vast mechanism, an endless chain of causes and effects. Science deals with "the unfailing order of immortal nature," with the universality of cause and effect with the eternal stability and inevitability of natural processes. Natural phenomena are not the results of volitions big or little, good or bad, but of all the events which have gone before. To the man of science nature is not the mere caprice of god or devil, to be lightly altered for a child's whim; nature is, as Bishop Butler said, that which is "stated, fixed, settled," eternal process moving on, the same yesterday, to-day and forever.

From sands to stars, from the immensity of the universe to the minuteness of the electron, in living things no less than in lifeless ones, science recognizes everywhere the inevitable sequence of cause and effect, the universality of natural law. Man also is a part of nature, a part of the great mechanism of the universe, and all that he is and does is limited and prescribed by laws of nature. Every human being comes into existence by a process of development, every step of which is determined by antecedent causes.

1. *The Determinism of Heredity.*—There can be no doubt that the main characteristics of every living thing are unalterably fixed by heredity. Men differ from horses or turnips because of their inheritance. Our family traits were determined by the hereditary constitutions of our ancestors, our inherited personal traits by the hereditary constitutions of our fathers and mothers. By the shuffle and deal of the hereditary factors in the formation of the germ cells and by the chance union of two of these cells in fertilization our hereditary natures were forever sealed. Our anatomical, physiological, psychological possibilities were predetermined in the germ cells from which we came. All the main characteristics of our personalities were born with us and cannot be changed except within relatively narrow limits. "The leopard cannot change his spots nor the Ethiopian his skin," and "though thou shouldst bray a fool in a mortar with a pestle yet will not his foolishness depart from him." Race, sex, mental capacity are determined in the germ cells, perhaps in the chromosomes, and all the possibilities of our lives were there fixed, for who by taking thought can add one chromosome, or even one determiner to his organization?

The thought of this age has been profoundly influenced by such considerations. We formerly heard that "all men were created free and equal"; we now learn that "all men are created bound and unequal." We were once taught that acts, if oft repeated, become habits, and that habits determine character; hereditarians of the stricter sort now teach that acts, habits and character were foreordained from the foundation of the family. We once thought that men were free to do right or wrong, and that they were responsible for their deeds; now we learn that our reactions are predetermined by heredity, and that we can no more control them than we can control our heart beats. For ages men have believed in the influence of example, in the uplift of high ideals, in the power of an absorbing purpose; for ages men have lived and died for what they believed to be duty and truth, and have received the homage of mankind; or they have lived malevolent and criminal lives and have been despised by men and punished by society. But if our reactions, habits, characters are predetermined in the germplasm such men have deserved neither praise nor blame. If personality is determined by heredity alone all teaching, preaching, government is useless; freedom, responsibility, duty are delusions; whether men are useful or useless members of society depends upon their inheritance, and the only hope for the race is in eugenics—always supposing that enough freedom is left to the individual or to society to control the important function of choosing a mate.

Already a few enthusiastic persons have begun to apply these doctrines to practical affairs. We are told that children should never be admonished or punished, for they do only what their natures lead them to do; the nature of the child must be respected and must be allowed to manifest itself in its own way. Lying and stealing will cure themselves like the mumps, or they will remain incurable, in which case the germplasm is to blame and nothing could have been done anyway. Laziness is due to inheritance or to hookworms; the latter kind may be cured, but not the former. Thriftlessness, alcoholism and uncleanness run

GENETICS AND ETHICS

in families and can be cured only by extermination. Men who prey upon society were born with wolfish instincts, and cannot help but eat the lambs. Villains, lawbreakers, murderers should be pitied but not punished; if blame attaches to their deeds it falls upon the marriage bureau and the parents. The world needs hospitals and sanatoria and sterilization institutes for the criminal and the vicious, but not courts and prisons, and all punishments should be visited only upon the parents to the third and fourth generations.

Do our studies of heredity lead us to any such radical conclusions? If they do we must accept them like brave men. "Truth is truth if it sears our eyeballs." But when theories lead to such revolutionary results it behooves us to examine carefully those theories to see if there is not somewhere a fundamental flaw in them.

One of the most difficult things in the world is to recognize a great truth, to feel its significance and yet not to be carried away by it. Great scientific errors are frequently due not so much to faulty observations as to sweeping conclusions. In biology the search for universal laws is a peculiarly dangerous pursuit. In philosophy great errors are often due not so much to false premises as to supposed logical necessities. As a test of truth logic is inferior to experience; its faults are not so much in its methods as in its premises and applications. For this reason a logical chain has led many a man into the bondage of error. Truth is not usually found in extremes, in "carrying out a process to its logical conclusions," but rather in some middle course which is less striking but more judicious.

Having observed that the main characteristics of our minds as well as of our bodies are inherited, it is easy and natural to go further and to conclude not only that all the possibilities of our lives are marked out in the germ but that all that will actually develop from the germ is there determined and cannot be altered. There are many similarities between such an extreme view and the old doctrine of preformation, and it contains a like absurdity.

It practically denies development altogether. If the germ is a closed system and receives nothing from without, and if adult characteristics are predetermined in the germ, they are as irrevocably fixed as if they were predelineated.

At the opposite extreme is the old voluntaristic view of absolute freedom and absolute responsibility. This view, like the old epigenesis, virtually postulates a new creation for each individual. As far as the mind and soul are concerned there is no hereditary continuity with past generations and none with future ones. But while such a view may be logically complete and theologically satisfying, it is not scientific, for it also contradicts the evidence.

The truth then seems to lie somewhere between these two extremes. Our personalities were not absolutely predetermined in the germ cells from which we came, and yet they have arisen from those germ cells and have been conditioned by them. When it is said that any characteristic is predetermined in the germ cell, what does this mean? What but that the development of that characteristic is made possible? Adult characteristics are potential and not actual in the germ, and their actual appearance depends upon many complicated reactions of the germinal units with one another and with the environment. In short, our actual personalities are not predetermined in the germ cells, but our possible personalities are. There are many possible personalities in each of us, and what we actually are is only a fraction of what we might have been.

2. *The Determinism of Environment.*—This determinism of heredity is matched by a corresponding determinism of environment. Life is possible only within rather narrow limits of physical and chemical conditions and in the main these limits are fixed by the constitution of nature. But apart from these antecedent conditions of life in general there are many minor conditions of environment which exercise a profound influence upon organisms, especially in the course of their development. Very slight changes in food, temperature, moisture and atmospheric conditions may produce great changes in the developing organ-

GENETICS AND ETHICS

ism, and these conditions are for the most part entirely beyond the control of the individual affected.

In all organisms the potentialities of development are much greater than the actualities. In many animals a small part of the body is capable, when separated from the remainder, of producing a whole body, though this potency would never have become an actuality except under the stimulus of separation. In like manner a part of an egg may, when separated from the remainder, give rise to an entire animal. By modifying the conditions of development animals may be produced which have one eye, many eyes or no eyes; animals in which the body is turned inside out, or side for side; animals in which all sorts of dislocation of organs have taken place; and the earlier the environmental forces act the more profound are the modifications.

But leaving out of account all forms which are so monstrous that they are incapable of reaching maturity we find that there are left many variations in the size and vigor of the body as a whole, as well as of its parts; many variations in the more or less perfect correlation of these parts with one another, which were determined by the conditions of development rather than by heredity. In a given germ cell there is the potency of any kind of organism that could develop from that cell under any kind of conditions. The potencies of development are much greater than the actualities. Anything which could possibly appear in the course of development is potential in heredity and under given conditions of environment is predetermined. Since the environment cannot be all things at once many hereditary possibilities must remain latent or undeveloped. Consequently the results of development are not determined by heredity alone but also by extrinsic causes. Things cannot be predetermined in heredity which are not also predetermined in environment.

Of all animals I suppose that man enjoys the most extensive and the most varied environment, and its effect upon his personality is correspondingly great. Of all animals man has the longest period of immaturity and it is during this period that the play of environmental stimuli on the organism is effective in modifying development. In addition to the material environment he lives in the midst of intellectual, social and moral stimuli which are potent factors in his development. By means of his power to look before and after, he lives in the future and past as well as in the present; through tradition and history he becomes an heir of all the ages. The modifying influences of all these environmental conditions on personality are very great. Each of us may say with Ulysses: "I am a part of all that I have met." So great is the power of environment on the development of personality that it may outweigh inheritance; a relatively poor inheritance with excellent environmental conditions often produces better results than a good inheritance with poor conditions. Of course no sort of environment can do more than bring out the hereditary possibilities, but on the other hand those possibilities must remain latent and undeveloped unless they are stimulated into activity by the environment.

Functional activity or use is one of the most important factors of development. Functional activity is response to stimuli, which may be external or internal in origin. The entire process of development may be regarded as an almost endless series of such responses on the part of the organism, whether germ cell, embryo or adult, to external and internal stimuli. It is a truism that use strengthens a part and disuse weakens it; it is likewise a truism that responses which are oft repeated become more rapid and more perfect, and in this way habits are formed. Practically all education, whether of man or of lower animals, consists in habit formation, in establishing constant relations between certain external or internal stimuli and certain responses of the organism. At first these stimuli are largely of external origin; later the external stimuli may be replaced more and more by internal ones; but whatever the source of the stimulus the response of the organism to these stimuli is one of the most important factors of development, whether of the body or of the mind.

The influence of environment upon the minds and morals of men is especially great. To a large extent our habits, words, thoughts; our aspirations, ideals, satisfactions; our responsibility, morality, religion are the results of the environment and education of our early years. It cannot be doubted that if we had been born in other countries or ages we should have been different from our present selves in many important respects; if we had been born and reared in the slums of great cities we should have been other than we are; indeed if the little illnesses, accidents and contingencies of our lives had been different we should have been different in our bodies and minds, as identical twins come to differ from each other under such circumstances. The conditions of early life and education have a great influence in shaping personality and are almost as much beyond the control of the individual as is heredity.

If personality in all of its main features is fixed by heredity and environment over which the individual has little or no control, and this is certainly true, personality is as inevitably determined by its antecedents as is any other natural phenomenon. This is, I believe, a conclusion from which there is no escape. How then is it possible to believe in freedom and responsibility? Is there not justification for the view so often expressed of late that man is never free and that responsibility and duty are mere delusions?

III. Determinism and Responsibility

Many persons who have thought upon these subjects have felt, apparently, that there was no tenable middle ground between extreme voluntarism and extreme mechanism; man has been regarded as a "free agent" or a mere "automaton," absolutely free or absolutely bound, wholly indeterminate or wholly predetermined. But these extreme views are unreal, unscientific and unjustifiable, for they contradict the facts of experience. We have the assurance of experience that we are not absolutely free nor absolutely bound, but that we are partly free and partly bound; the alternatives are not merely freedom *or* determinism, but rather freedom *and* determinism.

1. *Determinism not Fatalism.*—Whatever the philosophical meaning of "determinism" may be, all that is meant by that term in science and in actual life is that every effect is the resultant of antecedent causes and that identical causes yield identical results. Determinism does not mean predeterminism: the one finds every effect to be due to a long chain of preceding causes, the other attributes every effect to a single original cause; the one is scientific naturalism, the other is fatalism.

Applying this to personality, actual experience teaches that constant conditions of heredity and environment give constant results in development and that different conditions give different results. Undoubtedly the entire personality, body and mind, undergoes development, and modifications of either heredity or environment modify personality. This is scientific determinism, but it is not fatalism and it is not incompatible with a certain amount of freedom and responsibility.

2. *Control of Phenomena and of Self.*—Even the most extreme mechanists, who maintain that we are mere automata and that we could never do otherwise than we do, admit the possibility of a certain amount of control over phenomena outside ourselves. They tell us that the aim of science is not merely to understand but also to control nature. But if man may to a limited extent control physical, chemical and biological processes in the world around him, if he may control to a limited extent the behavior of a star-fish or dog or child, on what ground is it possible to deny a similar control of his own behavior? Does it not come to this that all such control means intelligent action, or rather the introduction of intelligence as a factor in the chain of cause and effect? Before the appearance of intelligence, whether in ontogeny or in phylogeny, no such control of phenomena or of self is possible, but when intelligence becomes a factor in behavior a limited control of the world and of the self is made possible.

Of course man has no control over events which have already happened. Our heredity and early development are accomplished facts which nothing can change. Development is not a reversible

GENETICS AND ETHICS

process; a man cannot enter a second time into his mother's womb and be born again. Once the sex cells are formed their hereditary nature is determined; once the egg is fertilized the hereditary possibilities of the new individual are fixed; once any stage of development has passed, that page in the book of life is closed and sealed.

And yet at every step in this long process of development there were one or more alternatives which might have been taken instead of the one which was taken. There were innumerable possible alternatives in the matings of our ancestors, there were billions of possible alternatives in the union of the millions of types of germ cells which each of our parents produced; at every step in the development of the oosperm from which each of us came there were many possible alternative stimuli and responses. But in each case one of these innumerable alternatives was taken and the others left. In every instance there was some cause that determined which alternative was taken, but these causes are so local and individual that they cannot be generalized; one cause works in one instance, another in another, and so we say that chance determines which alternative is taken, meaning by chance only this that the causes involved cannot be generalized. At critical stages in this process of development the alternatives are so evenly balanced that minor considerations, which we call chance, determine which path shall be taken; but there are no backward steps in development and once a path has been taken that particular crisis or turning point does not occur again.

Thus each of us has wandered through the maze of life, chance usually determining which path shall be taken of the many which heredity and environment offer, until he has come to a stage where associative memory makes it possible to profit by experience and where intellect and will make possible intelligent choice. With the growth of intellect and will there comes to be a limited degree of freedom and responsibility, and with increasing complexity of organization the number of alternative paths is greatly increased. The possible reactions of germ cells are relatively few and fixed, the possible reactions of a complex animal are relatively many and behavior is more plastic; and thus this very complexity and plasticity allow adaptations to the minutest alterations of environment.

3. *Birth and Growth of Freedom.*—In animals below man and in the stages of human development one may trace the birth and growth of freedom. Even in some of the simplest organisms one can observe inhibitions of responses and modifications of behavior which seem to be due to conflicting stimuli or to changes in the physiological state. In higher organisms such inhibitions or modifications proceed particularly from internal stimuli, which in turn are probably conditioned by hereditary constitution and past experience. The factors which determine behavior are not merely the present stimulus and the hereditary constitution, but also the experiences through which the organism has passed and the habits which it has formed.

A moth cannot avoid the impulse to fly toward the light, and it does not learn by experience to avoid the flame. Its reactions are relatively fixed and machine-like. Many other animals learn by experience to inhibit responses to certain stimuli; a tame fish or frog will take food from your hand, but if it is repeatedly frightened when it attempts to take food it will not come near you though it is starving,—it inhibits the strong impulse of a hungry animal to take food by the counter impulse of unpleasant memories or of fear. Here we have the beginnings of what we call freedom, the immediate response to a stimulus is suppressed, internal stimuli are balanced against external ones and final action is determined largely by past experience. Owing to his vastly greater power of memory, reflection and inhibition man is much freer than any other animal. Animals which learn little from experience have little freedom and the more they learn the freer they become.

In both ontogeny and phylogeny there has been development of freedom. The reactions of germ cells and of the lowest organisms are relatively fixed. In more complex organisms re-

GENETICS AND ETHICS

actions become modifiable through conflicting stimuli, intelligence, inhibitions. *Freedom is the more or less limited capacity of the highest organisms to inhibit instinctive and non-rational acts by intellectual and rational stimuli and to regulate behavior in the light of past experience. Such freedom is not uncaused activity, but freedom from the mechanical responses to external or instinctive stimuli, through the intervention of internal stimuli due to experience and intelligence.* To the person accustomed to think of will and choice as absolutely free this may seem to be a sort of freedom so limited as to be scarcely worth the having; and yet "it is the dawning grace of a new dispensation," the beginnings of rational life, social obligations, moral responsibility.

The only control over natural phenomena that is possible is in choosing between alternatives that are offered; and the only control which one who has reached the age of intelligence can have over his own development consists in choosing between the alternatives that are open to him. He may not choose his heredity or early development for the alternative paths which were once offered here have long since been passed; but to a limited extent he may choose his present environment and training, he may choose a path which leads to discipline and increased powers of self-control or the reverse, and to this extent only is he responsible for what he may become.

4. *Responsibility and Will.*—All organisms are capable of responding to chemical and physical stimuli but in addition normal men have the capacity of responding to stimuli of a higher order. By responsibility I understand the ability on the part of the organism to respond to rational, social and ethical stimuli or impulses and to inhibit responses to stimuli of an opposite nature, and the corresponding expectation on the part of others that the individual will so respond. The psychical stimuli which influence our behavior are not merely remembered experiences but the words, suggestions, admonitions, ideas which come to us from others, as well as the almost endless permutations of such memories and suggestions in our thoughts. The social and

ethical stimuli are not merely such as arise from love of reward and fear of punishment or the desire for praise and the fear of blame but also from the deep-seated social instinct to do good, which may reach the highest levels of altruism and self-sacrifice.

The higher the type of organization the larger is the range of stimuli to which it will respond and the larger the number and kind of responses which may be called forth; and at the same time the larger becomes the power of inhibition of responses whether through the balancing of one stimulus against another or from whatever cause. Human responsibility varies with the complexity of the stimuli involved as well as with the capacity of individuals to respond to those stimuli. A man might be quite responsible in savage society who would be quite irresponsible in civilized communities. In an infant there is no capacity to respond to rational, social or ethical stimuli but with increasing capacity in this respect comes increasing responsibility. Mental and ethical imbeciles, insane and mentally defective persons have a low capacity for such responses and inhibitions and consequently less is expected of them. There are in different men all degrees of responsibility, as there are all degrees of capacity. In one and the same individual responsibility varies at different times and under different circumstances; it rises and falls, like the tides, in every life. Varying capacity to respond to rational, social and ethical stimuli and to inhibit responses of an opposite nature depends not merely upon inheritance but also upon training, habits, physiological states. The common opinion that all normal men are equally responsible is not correct; in the eyes of the law this may be true, because legal obligations are so far below the capacities of normal men that all may be held equally responsible before the law, though in reality their responsibilities are as varied as their inheritance or their training.

Conversely the responsibility of society to the individual is universally recognized. Irresponsible persons must be cared for by older or wiser persons who become responsible for them; and in general the responsibility rests upon society to provide as favorable environment as possible for all its members. Experienced persons can to a certain extent choose their own environment and thus indirectly control their responses and habits but young children are almost if not quite as incapable of choosing their environment as of choosing their heredity, and it becomes the duty of society to see to it that the environmental stimuli are such as to develop rational, social and ethical habits rather than the reverse.

We need not think of the will as a *deus ex machina*, nor even as "a little deity encapsuled in the brain," but rather as the sum of all those psychical processes, such as memory and reason, which regulate behavior. In this sense the will is as free as the mind, and no freer. Indeed the will is the mind acting as internal stimulus, inhibition, regulation; in this sense the existence and power of will is no more to be doubted than the existence of those other mental conditions which we call intellect or memory.

Just as intellect or memory may be trained to accomplish results which would have been impossible to the untrained mind, so will may be trained to initiate, inhibit or regulate behavior in a manner quite impossible to one who has not had this training. It is one of the most serious indictments against modern systems of education that they devote so much attention to training memory and intellect and so little attention to the training of will upon the proper development of which so much depends.

5. *Our Unused Talents.*—Will is indeed the supreme human faculty, the whole mind in action, the internal stimulus which may call forth all the capacities and powers. And yet the will does not directly create nor even discover these powers; they are produced by the factors of development, by heredity, environment and training; and they are usually discovered by accident or under the stress of necessity. How often have we surprised ourselves by doing some unusual or prodigious task! What we have once done we feel that we can do again. We realize more or less clearly, depending upon our experience, that what we habitually do is far less than we could do. It is this reserve, upon which we can draw on special occasions, that gives us the sense of freedom.

In his inspiring address on "The Energies of Men" William James showed that we have reservoirs of power which we rarely tap, great energies upon which we seldom draw, and that we habitually live upon a level which is far below that which we might occupy. Darwin held the opinion, as the result of a lifetime of observation, that men differ less in capacity than in zeal and determination to utilize the powers which they have. In playful comment on the variety and extent of his own life work he said in modest and homely phrase, "It's dogged as does it." It may be objected that the zeal and determination were inherited, but here also the hereditary possibilities become actualities only as the result of use, training, the formation of habits.

It is generally admitted that no constant distinction can be recognized between the brain of a philosopher and that of many a peasant. Neither size nor weight of brain nor complexity of convolutions bears any constant relation to ignorance or intelligence, though doubtless an "unlimited microscopist" could find differences between the trained and the untrained brain. The brains of Beethoven, Gauss and Cuvier, although unusually large, have been matched in size and visible complexity by the brains of unknown and unlearned persons—persons who were richly endowed by nature but who had never learned to use their talents In all men the capacity for intellectual development is probably much greater than the actuality. The parable of the talents expresses a profound biological truth, men differ in hereditary endowments, one receives ten talents and another receives but one: but the used talent increases many fold, the unused remains unchanged and undeveloped. Happy is he who is compelled to use his talents; thrice happy he who has learned how to compel himself! We shall not live to see the day when human inheritance is greatly improved, though that time will doubtless come, but in the meantime we may console ourselves by the thought that we have many half-used talents, many latent capacities, and although we may not be able to add to our inheritance new territory we may greatly improve that which we have.

GENETICS AND ETHICS

Jennings has pointed out as one of the great tragedies of life the almost infinite slaughter of potential personalities in the form of germ cells which never develop. A more dreadful though less universal tragedy is the loss of real personalities who have all the native endowments of genius and leadership but who for lack of proper environmental stimuli have remained undeveloped and unknown; the "mute, inglorious Miltons" of the world; the Cæsars, Napoleons, Washingtons who might have been; the Newtons, Darwins, Pasteurs who were ready formed by nature but who never discovered themselves. One shudders to think how narrowly Newton escaped being an unknown farmer, or Faraday an obscure bookbinder, or Pasteur a provincial tanner. In the history of the world there must have been many men of equal native endowments who missed the slender chance which came to these. We form the habit of thinking of great men as having appeared only at long intervals, and yet we know that great crises always discover great men. What does this mean but that the men are ready formed and that it requires only this extra stimulus to call them forth? To most of us heredity has been kind—kinder than we know. The possibilities within us are great but they rarely come to full epiphany.

6. *Self Knowledge and Self Control.*—What is needed in education more than anything else is some means or system which will train the powers of self discovery and self control. Easy lives and so-called "good environment" will not arouse the dormant powers. It usually takes the stress and strain of hard necessity to make us acquainted with our hidden selves, to rouse the sleeping giant within us. How often is it said that the worthless sons of worthy parents are mysteries; with the best of heredity and environment they amount to nothing, whereas the sons of poor and ignorant farmers, blacksmiths, tanners and backwoodsmen, with few opportunities and with many hardships and disadvantages, become world figures. Probably the inheritance in these last named cases was no better than in the former, but the environment was better. "Good environ-

ment" usually means easy, pleasant, refined surroundings, "all the opportunities that money can buy," but little responsibility and none of that self discipline which reveals the hidden powers and which alone should be counted good environment. Many schools and colleges are making the same mistake as the fond parents; luxury, soft living, irresponsiblity are not only allowed, but are encouraged and endowed—and by such means it is hoped to bring out that in men which can only be born in travail.

The chief educational value of athletics is found in this, that it teaches self control. But in great athletic contests the self control of the spectators is usually inversely proportional to that of the players, and while excess of stimuli may lead to wholesome and beneficial reactions in the players it frequently leads to excess of stimulants and to other injurious reactions in the spectators. But college athletics has this much at least in its favor, it trains men who take part in the contests to do their best, to subordinate pleasure, appetite, the desire for a good time, to one controlling purpose; it trains them to attempt what may often seem to them impossible, to crash into the line though it may seem a stone wall, to get out of their bodies every ounce of strength and endurance which they possess. Such training makes men acquainted with their powers and teaches courage, confidence and responsibility. If only we could make young persons acquainted in some similar way with their hidden mental and moral powers what a race of men and women might we not have without waiting for that uncertain day when the inheritance of the race will be improved! Whatever the stimulus required, whether pride or shame, fear or favor, ambition or loyalty, responsibility or necessity, education should utilize each and all of these to teach men self knowledge and self control.

But it will be said that self control depends upon inheritance, that strong wills and weak wills are such because of heredity. It is true that one man may be born with a potentiality for self control which another man lacks, but in all men this potentiality becomes actuality only through development, one of the principal

GENETICS AND ETHICS

factors of which is use or functional activity. An amazing number of persons have but little self control. Is this always due to defective inheritance, or is it not frequently the result of bad habits, of arrested development? To charge defects at once to heredity removes them from any possible control, helps to make men irresponsible, excuses them for making the least of their endowments. To hold that everything has been predetermined, that nothing is self determined, that all our traits and acts are fixed beyond the possibility of change is an enervating philosophy and is not good science, for it does not accord with the evidence. It is amazing that men whose daily lives contradict this paralyzing philosophy still hold it, as it were in some water-tight compartment of the brain, while in all the other parts of their being their acts proclaim that they believe in their powers of self control: they set themselves hard tasks, they overcome great difficulties, they work until it hurts, until they can say with Johannes Müller, *Es klebt Blut an der Arbeit,* and yet in the philosophical compartment of their minds they can say that it was all predetermined in heredity and from the foundations of the world.

Whether all the phenomena of life and of mind can be explained on the basis of a purely mechanistic hypothesis or not, that hypothesis must square with the facts and not the facts with the hypothesis. It has always been true of those who "sat apart and reasoned high of fate, free will, foreknowledge absolute" that they have "found no end in wandering mazes lost." Whatever the way out of these mazes may be,—whether it be found in the varied responses of an organism to the same stimulus, to the introduction of memory, intelligence and reason as internal stimuli, or to some form of idealism which finds necessity not in nature but in the spectator, and freedom not in the spectator but in the agent,—it is true for those who do not "sit apart and reason high," but who deal merely with evident phenomena, that the way of escape is not to be found in denying the reality of inhibition, will and self control. Because we can find no place in our philosophy and logic for self determination shall we cease to be scientists and close our eyes to the evidence? The first duty of science is to appeal to fact and to settle later with logic and philosophy. Is it not a fact that the possibilities of our inheritance depend for their realization upon development, one of the most important factors of which is use, functional activity in response to stimuli? Is it not a fact that in many animals behavior is modifiable and that impulses may be inhibited and controlled? Is it not a fact that experience, intelligence, will are factors in human behavior and that by means of these men are often able to choose between alternatives and so to control their own activities as well as external phenomena? Is it not a fact that our capacities are very much greater than our habitual demands upon them? Is it not a fact that belief in our responsibility energizes our lives and gives vigor to our mental and moral fiber? Is it not a fact that shifting all responsibility from men to their heredity or to that part of their environment which is beyond their control helps to make them irresponsible?

This debilitating philosophy in which everything is predetermined, in which there is no possibility of change or control, in which there is hypertrophy of intellect and atrophy of will, is a symptom of senility whether in men or nations. We need to return to the joys of a childhood age in which men believed themselves free to do, to think, to strive, in which life was full of high endeavor and the world was crowded with great emprise. We need to think of the possibilities of development as well as of the limitations of heredity. Chance, heredity, environment have settled many things for us; we are hedged about by bounds which we cannot pass, but those bounds are not so narrow as we are sometimes taught and within them we have a considerable degree of freedom and responsibility.

> "That which we are we are,
> One equal temper of heroic hearts
> Made weak by time and fate, but strong in will
> To strive, to seek, to find, and not to yield."

IV. THE INDIVIDUAL AND THE RACE

There is a larger freedom and a greater responsibility than that which characterizes the individual. What the individual cannot do because of weakness, ignorance, self interest, short life, society can accomplish with the strength, wisdom and interest of all, and through long periods of time. There are many grades of organization from the bacterium to the vertebrate, from the germ cell to the man. Society is the last and highest grade of organization and its freedom and responsibility are to those of the individual very much as the freedom and responsibility of the developed man are to those of the germ cell from which he came. Out of the correlations, differentiations and integrations of persons has grown this higher type of organization which we call society.

1. *The Conflict between the Freedom of the Individual and the Good of Society.*—The freedom, power and responsibility of society are founded upon limitations of individual freedom for the good of the race. Among social animals, such as ants and bees, there is so much instinct and so little reason and freedom that there is practically no conflict between the individual and the race, but with the increase of intelligence and freedom among men there has developed an increasing conflict between the individual and society. So far as social limitations are artificial, selfish, for the good of a few rather than of all, this conflict of the ages, this struggle to be free has been the crowning glory of mankind. The struggle for freedom from tyranny in thought and speech, in religion, government and industry, no less than for the freedom that comes by the conquest of nature, is one of the greatest achievements of the human race.

But social restrictions on individual freedom are not all artificial and selfish. Some of them are absolutely essential not only to the welfare but even to the continued existence of the race, and when demands for individual freedom go to the extent of fighting against these racial obligations they become a serious menace to mankind.

2. *Perpetuation and Improvement of the Species the Highest Ethical Obligation.*—Among all organisms the race or species is of paramount importance. Race preservation, not self preservation, is the first law of nature. Among all organisms the perpetuation and welfare of the race are cared for by the strongest instincts. In very many species of animals reproduction means the death of the individual. The breeding instinct drives every male bee, every male and female salmon, to its certain death in order that the race may be perpetuated. Among the higher organisms the strongest of all the instincts are those connected with reproduction. But in the human species intellect and freedom come in to interfere with instinct. The reproductive instincts are not merely controlled by reason, as they should be, but to an alarming extent they are thwarted and perverted among intelligent people.

The struggle to be free is part of a great evolutionary movement, but the freedom must be a sane one which neither injures others nor eliminates posterity. The feminist movement in so far as it demands greater intellectual and political freedom for women will be a benefit to the race but in so far as it demands freedom from marriage and reproduction it is suicidal. The cry of Rachel, "Give me children or I die," has been turned by many modern women to, "I'd rather die than have children." If the demand for individual freedom blinds men and women to their racial obligations the inevitable decadence and extinction of their lines must follow. In every age and country where demands for personal freedom have been most insistent and extreme, where men and especially women have demanded freedom from the burdens of bearing and rearing children as well as from other natural social obligations, the end has been degeneration and extinction.

This has been the history of many talented races and families of mankind. The decay of the most gifted races of the ancient world, especially those of Greece and Rome, was not due primarily to bad heredity nor to bad material environment but rather to the growth of luxury and selfishness and unrestricted freedom; marriage became unfashionable, immorality was widespread, and then came sterility and extinction or mixture with inferior stock and degeneracy. And then the barbarian, the immigrant, the natural man, unspoiled by too much freedom and true to his instincts, came in to take the place of the more gifted race. Truly "there is a power not ourselves that makes for righteousness."

In these days when we talk of our race and our civilization as if they were necessarily supreme and immortal it is well to remember that there have been other races and other civilizations that regarded themselves in the same way.

"Assyria, Greece, Rome, Carthage, where are they?"

And what assurance have we that our race and our civilization will not run a similar course and come to a similar end? May we not surely predict that if we continue to put individual freedom and luxury and selfishness above social obligations our race and civilization will also see the writing on the wall, "Thou art weighed in the balances and art found wanting?" In these days when individuals are demanding more and more freedom it is well to remember that "the best use that man has made of his freedom has been to place limitations upon it." Again and again, age after age, men and families and nations have gone up to a climax of greatness and then have declined, while other unknown men have taken their places. Greatness has not for long perpetuated itself. An epitome of human history is contained in the words, "He hath put down the mighty from their seats and hath exalted them of low degree."

It may well be asked by those who are interested in breeding a better race of men whether such a thing is possible, whether the better race may not be lacking in vitality or fertility or morality and thus be doomed to an early end. Although this has been the fate of many gifted races of the past I do not think that it was a necessary fate. The history of domesticated animals and of cultivated plants, and especially the recent notable advances in genetics, indicate what eugenics might do for the human race. In time, under intelligent guidance, the worst qualities of the race might be weeded out and the best qualities preserved. This is the goal toward which intelligent effort should be directed. This should be the supreme duty of society and of all who love their fellow men.

But I think that notable human improvement can take place only upon two conditions: (1) The physical and intellectual improvement of the individual through environment and training must not interfere with his racial and ethical obligations. Individual freedom must be subordinated to racial welfare. (2) The promotion of human evolution must be undertaken by society as its greatest work. Not only has society greater freedom and greater power than the individual but it persists while men come and go.

Our hereditary lines are so interwoven with those of other races and will be so entangled with other lines in the future that any selfish or narrow policy of improving our family or class can have little permanent value. We shall rise only as our race rises. Indeed when we consider all the influences of our fellow men upon our development, when we consider our hereditary connections with multitudes of men and women of the past, when we think of the nexus of hereditary strands which are woven into our personalities and which will be continued through us to many future generations, we realize that after all the individual is not really a separate and independent being, but a minor unit in the great organism of humanity, and that his greatest duty is to transmit unimpaired and undefiled a noble heritage to generations yet unborn.

It is possible greatly to improve environment. Conditions of life are still hard and cruel for many. A vast amount of good human material is wasted in modern society. As civilization becomes more complex the quantity of human wreckage and wastage ever grows greater. Many useful lives and some great possibilities are blotted out by unfavorable environment. It is the duty of society as far as possible to conserve these lives and to develop these possibilities.

It is possible greatly to improve education, to make it a potent factor in development instead of a conventional veneer. In spite of innumerable educational reforms the essential reform has not yet been reached; mere refinements of bad methods are not real reforms. The essence of all education is self discovery and self control. When education helps an individual to discover his own powers and limitations and shows him how to get out of his heredity its largest and best possibilities it will fulfil its real function; when children are taught not merely to know things but particularly to know themselves, not merely how to do things but especially how to compel themselves to do things, they may be said to be really educated. For this sort of education there is demanded rigorous discipline of the powers of observation, of the reason, and especially of the will.

It is possible greatly to improve heredity: (a) By weeding out from the possibility of reproduction human stocks bearing serious defects. (b) By cultivating pride in good heredity and by discouraging voluntary infertility on the part of those who have a goodly heritage. (c) By increasing opportunities for early and favorable marriages. (d) By carefully conserving the best human mutations or inherited variations. In this way if in any way the better race will be produced.

The possible improvements of heredity are great, the possible improvements of environment and training are great, but whether men of the future will be better than those of the past or present is a question not only of genetics but also of ethics.

How better can I close this course of lectures than with the words of Francis Galton, one of the greatest students of human heredity and the founder of the science of Eugenics?

"The chief result of these inquiries has been to elicit the religious significance of the doctrine of evolution. It suggests an alteration in our mental attitude and imposes a new moral duty. The new mental attitude is one of a greater sense of moral freedom, responsibility and opportunity; the new duty which is supposed to be exercised concurrently with, and not in opposition to the old ones upon which the social fabric depends, is an endeavor to further evolution, especially that of the human race."

REFERENCES TO LITERATURE

The following list of books and publications includes only those works which are referred to most frequently in the preceding pages. Several of the books cited, particularly those by Plate, Morgan, Babcock and Clausen, contain extensive bibliographies. Those desiring to become more fully acquainted with books and articles dealing with the subjects of heredity and development are referred to the larger works which are listed here.

Books and Larger Works

Angell, Conklin, Ferris, Keller, Lull, Parker. The Evolution of Man. Yale University Press. 1922.

Babcock and Clausen. Genetics in Relation to Agriculture. New York, 1918.

Bateson, W. Materials for the Study of Variation, London, 1894.

Bateson, W. Problems in Genetics. Yale Univ. Press. 1913.

Bateson, W. Mendel's Principles of Heredity. 3rd Impression, Cambridge, 1913.

Baur, E. Einführung in die experimentelle Vererbungslehre. Berlin, 1911.

Cannon, W. B. Bodily Changes in Pain, Hunger, Fear and Rage. New York, 1915.

Castle, W. E. Heredity in Relation to Evolution and Animal Breeding. New York, 1912.

Castle, W. E. Genetics and Eugenics. Harvard University Press, 1920.

Castle, W. E.; Coulter, J. M.; Davenport, C. B.; East, E. M.; Tower, W. L. Heredity and Eugenics. Chicago, 1912.

Conklin, E. G. The Direction of Human Evolution. New York, 1922.

Correns, C. Die Neuen Vererbungsgesetze. Berlin, 1912.

Darbishire, A. R. Breeding and Mendelian Discovery, Cotton. London, 1911.

Darwin, C. Animals and Plants under Domestication. New York, 1887.

Davenport, C. B. Heredity in Relation to Eugenics. New York, 1911.

De Vries, H. Intracellular Pangenesis. Jena, 1889.

De Vries, H. Die Mutationstheorie. Leipzig, 1901.

De Vries, H. Plant Breeding. Chicago, 1907.
Doncaster, L. Heredity in the Light of Recent Research. Cambridge, 1911.
Driesch, H. The Science and Philosophy of the Organism. Gifford Lectures, London, 1908.
Ellis, H. The Task of Social Hygiene. London, 1912.
Ellis, H. The Problem of Race Regeneration. New York, 1911.
Forel, A. The Sexual Question. New York, 1908.
Galton, F. Inquiries into Human Faculty. New York, 1883.
Galton, F. Natural Inheritance. London, 1889.
Galton, F. Hereditary Genius. London, 1892.
Galton, F. Essays in Eugenics. London, 1909.
Goddard, H. H. The "Kallikak" Family. New York, 1912.
Goldschmidt, R. Einführung in die Vererbungswissenschaft. Leipzig, 1911.
Häcker, V. Allgemeine Vererbungslehre. Braunschweig, 1912.
Hertwig, O. Allgemeine Biologie. Jena, 1909.
Johannsen, W. Elemente der exakten Erblichkeitslehre, 2d Auf. Jena, 1913.
Kellicott, W. E. The Social Direction of Human Evolution. New York, 1911.
Lock, R. H. Variation, Heredity and Evolution. New York, 1911.
Loeb, J. Comparative Physiology of the Brain and Comparative Psychology. New York, 1900.
Loeb, J. The Dynamics of Living Matter. New York, 1906.
Loeb, J. The Mechanistic Conception of Life. Chicago, 1911.
Loeb, J. Artificial Parthenogenesis. Chicago, 1913.
Metchnikoff, E. The Nature of Man. Chicago, 1903.
Morgan, T. H. Heredity and Sex. New York, 1913.
Morgan, T. H. A Critique of the Theory of Evolution, 2nd ed. Princeton University Press. 1916. Revised and republished, 1925, as Evolution and Genetics.
Morgan, T.H. The Physical Basis of Heredity, 2nd ed. Philadelphia, 1919.
Morgan, Sturtevant, Muller and Bridges. The Mechanism of Mendelian Heredity. 2d. Ed. New York, 1923.
Mott, F. W. Heredity and Eugenics in Relation to Insanity. London, 1912.
Nägeli, C. Mechanische-Physiologische Theorie der Abstammungslehre. München, 1884.
Plate, L. Vererbungslehre. Leipzig, 1913.

Problems in Eugenics. Papers communicated to 1st International Eug. Cong. London, 1912.
Punnett, R. C. Mendelism. London, 1911.
Rignano, E. The Inheritance of Acquired Characters. Chicago, 1911.
Romanes, G. J. Darwin and After Darwin. Chicago, 1892.
Saleeby, C. W. Parenthood and Race Culture. New York, 1909.
Semon, R. Das Problem der Vererbungslehre erwörbener Eigenschaften. Leipzig, 1912.
Schäfer, E. A. The Endocrine Organs. London, 1916.
Spencer, H. Principles of Biology. New York, 1883.
Thompson, J. A. Heredity. Edinburgh, 1908.
Thorndike, E. L. Animal Intelligence. New York, 1911.
Walter, H. E. Genetics. New York, 1913.
Weismann, A. The Germ Plasm. New York, 1893.
Weismann, A. Essays on Heredity. Oxford, 1889.
Wilson, E. B. The Cell in Development and Inheritance. New York, 1900.
Woods, F. A. Heredity in Royalty. New York, 1906.

Monographs and Papers

Bailey, L. H. The Modern Systematist. Science, 46, Dec. 28, 1917.
Bardeen, C. R. Abnormal Development of Toad Ova fertilized by Spermatozoa exposed to the Roentgen Rays. Jour. Exp. Zool., 4, 1907.
Bateson, W. Presidential Address. British Ass'n Adv. Science, Australia, 1914.
Baur, E. Vererbungs und Bastardierungsversuche mit Antirrhinum. Zeit. f. induk. Abstam. 3, 1910.
Blakeslee, A. F. Chromosomal Duplication and Mendelian Phenomena in *Datura* Mutants. Science, 52, 1920.
Blakeslee, A. F. Mutations in the Jimson Weed. Jour. Heredity, 10, 1919.
Boveri, Th. Zellen Studien. Die Entwicklung dispermer Seeigel-Eier, etc. Jena, 1907.
Bridges, C. B. Triploid Intersexes in *Drosophila Melanogaster*. Science, Sept. 16, 1921.
Broman, I. Ueber geschlechtliche Sterilität. Wiesbaden, 1912.
Brooks, W. K. Are Heredity and Variation Facts? Address before 7th Internat. Zoological Congress, 1907.

Castle, W. E. Studies of Inheritance in Rabbits. Carnegie Inst. Wash. Publ. 11, 1909.

Castle, W. E. and Phillips, J. C. On Germinal Transplantation in Vertebrates. Carnegie Inst. Wash. Publ. No. 144, 1911.

Castle, W. E. and Phillips, J. C. Piebald Rats and Selection. Carnegie Inst. Wash. Publ. No. 195, 1914.

Castle, W. E. Piebald Rats and the Theory of Genes. Proc. Nat. Acad. Sci. 5, 1919.

Cattel, J. McK. The Causes of the Declining Birth Rate. Proc. First National Conference on Race Betterment. Battle Creek, 1914.

Conklin, E. G. Embryology of Crepidula. Journ. Morph. 13, 1897.

Conklin, E. G. The Cause of Inverse Symmetry. Anat. Anz., 1903.

Conklin, E. G. The Organization and Cell-Lineage of the Ascidian Egg. Jour. Acad. Nat. Sci. Phila., 1905.

Conklin, E. G. Experimental Studies on Nuclear and Cell Division. Ibid., 1912.

Conklin, E. G. The Mutation Theory from the Standpoint of Cytology. Science, 21, 1905.

Conklin, E. G. The Mechanism of Heredity. Science, 27, 1908.

Conklin, E. G. The Share of Egg and Sperm in Heredity. Proc. Nat. Acad. Sci., 2, 1917.

Correns, C. Zur Kenntnis der scheinbar neuen Merkmale der Bastarde. Ber. D. bot. Ges. 23, 1905.

Crew, F. A. E. Sex Reversal in the Fowl. Proc. Roy. Soc. B, 95, 1923.

Davenport, C. B. Inheritance in Poultry. Carnegie Inst. Wash. Publ. 53, 1906.

Davenport, C. B. Inheritance of Characteristics in Domestic Fowl. Carnegie Inst. Wash. Publ. 121, 1909.

Davenport, C. B. Heredity of Skin Color in Negro-White Crosses. Carnegie Inst. Wash. Publ. 188, 1914.

East, E. M. and Hayes, H. K. Heterozygosis in Evolution and Plant Breeding. Bureau Plant Industry, Bull. 243, 1912.

Elderton, E. M. and Pearson, K. A First Study of the Influence of Alcoholism on the Physique and Ability of the Offspring. Eugenics Lab. Mem. X, 1910.

Finlay, G. F. Studies on Sex Differentiation in Fowls. British Jour. Exp. Biol. 2, 1925.

REFERENCES

Fischer, E. Exper. Untersuchungen über der Vererbung erwörbener Eigenschaften. Allg. Z. f. Entomol. 6, 1901; 7, 1902.
Greenwood, A. W. Gonad Grafts in Embryonic Chicks, etc. British Jour. Exp. Biol. 2, 1925.
Goldschmidt, R. A further Contribution to the Theory of Sex. Jour. Exp. Zool. 22, 1917.
Gudernatsch, J. F. Feeding Experiments on Tadpoles. II. Amer. Jour. Anat. 15, 1914.
Guthrie, C. C. Further Results of Transplantation of Ovaries in Chickens. Jour. Exp. Zool. 5, 1908.
Guyer, M. F. Accessory Chromosomes in Man. Biol. Bull. 19, 1910.
Guyer, M. F. and Smith, E. A. Studies on Cytolysins. II, Transmission of Induced Eye Defects. Jour. Zool. 31, 1920.
Harrison, R. G. Experimentelle Untersuchungen, etc. Arch. f. Mik. Anat. 63, 1903.
Harshberger, J. W. Maize; A Botanical and Economic Study. Cont. from Bot. Lab. Univ. of Penna., 1893.
Hering, E. On Memory and the Specific Energies of the Nervous System. Chicago, 1897.
Hertwig, O. Neue Untersuchungen über die Wirkung der Radiumstrahlung auf die Entwicklung tierischer Eier. Sitz. d. Kgl. Preuss. Akad. d. Wissenschaften, 29, 1910.
Hertwig, O. Keimesschädigung durch chemische Eingriffe. Ibid., 30, 1913.
Hertwig, R. Ueber den derzeitigen Stand des Sexualitätsproblems. Biol. Centralblatt 32, 1912.
Hoppe, H. Die Tatsachen über den Alkohol, 4te Aufl. München, 1912.
Hoy, W. E., Jr. and Muller, H. R. Mitoses in the Epithelium of the Convoluted Tubules (of Man). In press.
Huxley, T. H. Evolution and Ethics. Romanes Lecture, 1893.
James, W. The Energies of Men. Science, 25, 1907.
Jennings, H. S. Behavior of the Lower Organisms. New York, 1906.
Jennings, H. S. Heredity, Variation and Evolution in Protozoa. Proc. Am. Philos. Soc. 47, 1908.
Jennings, H. S. Experimental Evidence of the Effectiveness of Selection. Amer. Nat. 44, 1910.
Jennings, H. S. Heredity and Personality. Science, 34, 1911.
Johnson, R. Marriage Selection. Jour. Heredity, 5, 1914.

Jordan, D. S. The Human Harvest. Boston, 1907.
Jordan, H. E. The Biological Status and Social Worth of the Mulatto. Pop. Sci. Monthly, 1913.
Kammerer, P. Direkt induzierte Farbanpassungen und deren Vererbung. Zeit. für Induk. Abstam. 4, 1911.
Kammerer, P., Inheritance of Acquired Characteristics. New York, 1924.
King, H. D. Studies on Sex Determination in Amphibians. Biol. Bull. 16 and 20, 1909, 1911. Jour. Exp. Zool. 12, 1912.
Lang, A. Vorversuche zu Untersuchungen über die Varietätenbildung von Helix hortensis und nemoralis. Festschr. f. Häckel. Jena, 1904.
Lillie, F. R. The Mechanism of Fertilization in Arbacia. Jour. Exp. Zool. 16, 1914.
Lillie, F. R. The Free-Martin; a Study of the Action of Sex Hormones in the Foetal Life of Cattle. Jour. Exp. Zool. 23, 1917.
Lotsy, J. P. La theorie du croisement. Arch. néerl. Sci. Exact. et Nat., Ser. 3B, T. 2, 1914.
MacDougal, D. T. Alterations in Heredity induced by Ovarial Treatment. Bot. Gaz. 51, 1911.
McClendon, J. F. Chemistry of the Production of One-Eyed Monsters. Am. Jour. Phys. 29, 1912.
McClung, C. E. The Accessory Chromosome, Sex Determinant? Biol. Bull. 3, 1902.
MacDowell, E. C. Size Inheritance in Rabbits. Carnegie Inst. Wash. Publ. 196, 1914.
MacDowell, E. C. Multiple Factors in Mendelian Inheritance. Jour. Exp. Zool. 16, 1914.
Macfarlane, J. M. Contributions to the History of Dionaea muscipula. Contributions Bot. Lab. University Pennsylvania, 1, 1892.
Mendel, G. Versuche über Pflanzenhybriden. Verh. naturf. Ver. Brünn, 4, 1866.
Montgomery, T. H. Human Spermatogenesis. Jour. Acad. Nat. Sci. Phila. 15, 1912.
Morgan, T. H. A Biological and Cytological Study of Sex Determination in Phylloxerans and Aphids. Jour. Exp. Zool. 7, 1909.
Morgan, T. H. The Theory of the Gene. Am. Nat. 51, 1917.
Muller, H. J. An Oenothera-like Case in Drosophila. Proc. Nat. Acad. Sci. 3, 1917.

REFERENCES

Mulsow, K. Der Chromosomencyclus bei Ancyracanthus cystidicola Rud. Arch. f. Zellf. 9, 1912.

Nettleship, E. Some Hereditary Diseases of the Eye. Trans. Ophthal. Soc. 29, 1909.

Nilsson-Ehle, H. Einige Ergebnisse von Kreuzungen bei Hafer und Weizen. Bot. Notiser für 1908.

Osborn, H. F. Cartwright Lectures on Contemporary Evolution in Man. Amer. Nat. 26, 1892.

Painter, T. S. The Spermatogenesis of Man. Jour. Exp. Zool. 37, 1923.

Pearl, R. The Mode of Inheritance of Fecundity in the Domestic Fowl. Jour. Exp. Zool. 13, 1912.

Pearl, R. The Experimental Modification of Germ Cells. Parts I, II, III. Jour. Exp. Zool. 22, 1917.

Pearson, K. Tuberculosis, Heredity and Environment. London, 1912.

Plate, L. Bemerkungen über die Farbenrassen der Hausmaus und die Schreibweise der Erbformeln. Zeit. f. induk. Abstam. 6, 1912.

Riddle, O. The Control of the Sex Ratio. Jour. Washington Acad. Sci. 7, 1917.

Rosanoff, A. J. The Inheritance of the Neuropathic Constitution. Jour. Am. Med. Assoc. 58, 1912.

Shull, G. H. The "Presence and Absence" Hypothesis. Am. Nat. 43, 1909.

Shull, G. H. Hybridization Methods in Corn Breeding. Am. Breeder's Mag. 1, 1910.

Shull, G. H. Duplicate Genes for Capsule-form in Bursa bursa-pastoris. Zeit. f. induk. Abstam. 6, 1914.

Shull, G. H. Inheritance of Sex in Lychnis. Bot. Gaz. 49, 1910.

Stevens, N. M. Studies in Spermatogenesis with Especial Reference to the "Accessory Chromosome." Carnegie Inst. Wash. Publ. 36, 1905.

Stockard, C. R. The Experimental Production of Various Eye Abnormalities, etc. Arch. f. vergleich. Ophthal. 1, 1911.

Stockard, C. R. An Experimental Study of Racial Degeneration in Mammals Treated with Alcohol. Arch. Internat. Med. 10, 1912.

Stockard, C. R. and Papanicolaou, G. N. Further Studies on the Modification of the Germ Cells in Mammals; the Effect of Alcohol on Treated Guinea Pigs and their Descendants. Jour. Exp. Zool. 26, 1918.

Sumner, F. B. An Experimental Study of Somatic Modifications and their Reappearance in the Offspring. Arch. Entw. Mech. 30, 1910.

Tennent, D. H. Studies in Cytology, I and II. Jour. Exp. Zool. 12, 1912.

Tennent, D. H. The Dominance of Maternal or of Paternal Characters in Echinoderm Hybrids. Arch. Entw. Mech. 29, 1910.

Tower, W. L. An Investigation of Evolution in Chrysomelid Beetles of the Genus Leptinotarsa. Carnegie Inst. Wash. Publ. 48, 1906.

Toyama, K. Maternal Inheritance and Mendelism. Jour. Genetics 2, 1913.

Weeks, D. F. The Inheritance of Epilepsy. Problems in Eugenics, 1.

Whitman, C. O. Animal Behavior. Biol. Lectures. Woods Hole, 1899.

Whitney, D. D. The Effects of Alcohol not inherited in Hydatina senta. Amer. Nat. 46, 1912.

Whitney, D. D. The Influence of Food in Controlling Sex in Hydatina senta. Jour. Exp. Zool. 17, 1914.

Whitney, D. D. Sex Controlled in Rotifers by Food. Abstracts of Papers Am. Soc. Zoologists, Dec. 1915.

Wieman, H. L. The Chromosomes of Human Spermatocytes. Amer. Jour. Anat. 21, 1917.

Wiesner, J. Die Elementar-structur und das Wachstum der lebenden Substanz. Wien, 1892.

Wilder, H. H. Duplicate Twins and Double Monsters. Amer. Jour. Anat. 3, 1904.

Wilson, E. B. Some Aspects of Cytology in Relation to the Study of Genetics. Am. Nat., 1912.

Wilson, E. B. Studies on Chromosomes. I-VIII. Jour. Exp. Zool. 2, 3, 6, 9, 13. Jour. Morph. 22.

Winiwarter, H. Etudes sur spermatogenèse humain. Archiv d. Biologie, 27, 1912.

Wolff, C. F. Theoria Generationis. 1759.

Woltereck, R. Beiträg zur Analyse der Vererbung erwörbener Eigenschaften; Transmutation und Präinduktion bei Daphnia. Verh. D. Zool. Ges., 1911.

GLOSSARY

ACCESSORY CHRO'-MO-SOME. An odd chromosome which is found in only half of the spermatozoa of certain animals; see "sex-chromosome."

A-CHRO'-MA-TIN. The non-staining substance of the nucleus as contrasted with the chromatin.

A-CHON'-DRO-PLA-SY. A condition in which the long bones cease to grow in length at an early age thus producing a dwarf with large body and head but short limbs.

ACQUIRED CHARACTER. A character, the differential cause of which is environmental.

ALLELOMORPH. Contrasting unit characters, or their corresponding genes.

ALTERNATIVE INHERITANCE. Galton's term for a doubtful kind of inheritance in which all characters are derived from one parent. In present use, Mendelian inheritance.

AM'-NI-ON. One of the embryonic membranes of higher vertebrates.

AM-PHI-OX'-US. One of the lowest and simplest animals having a notochord (backbone).

AN-EN-CEPH'-A-LY. The condition of a brainless monster.

ANIMAL POLE. That pole of an egg at which the polar bodies are formed.

AN'-LA-GE. The embryonic basis of any developed part.

A-OR'-TA. The great artery arising from the heart.

AR'-CHI-PLASM. The deeply staining plasm surrounding the centrosome.

AS'-CA-RIS. A genus of round worms which are intestinal parasites.

AS'-CA-RIS meg-a-lo-ceph'-a-la. A parasite in the intestine of the horse.

AS-CID'-I-AN. A "sea-squirt"; one of the lowest types having a notochord. or elementary backbone.

AS'-TER. The radiating figure surrounding the centrosome in a cell.

AS-SIM-I-LA'-TION. Conversion of food substances by an organism into its own living substance.

A-SYM'-ME-TRY. The condition where opposite sides are unlike.

AT'-A-VISM. The condition in which an individual resembles a grandparent, or a more distant ancestor, more than one of the parents.

BI'-O-PHORES. The ultimate units of life (Weismann).

BI'-O-TYPE. A group of individuals all of which have the same Genotype (Johannsen).

BI'-VA-LENT CHRO'-MO-SOMES. A pair of chromosomes, one maternal the other paternal, temporarily united.

BLAS'-TO-COEL. The cavity within a blastula.

GLOSSARY

BLAS′TO-DER′-MIC VES′-I-CLE. A hollow sphere, formed from the segmented egg of a mammal, which becomes attached to or embedded within the wall of the uterus.

BLAS′-TO-PORE. The mouth of a gastrula.

BLAS′-TU-LA. A mass of cells, usually in the shape of a hollow sphere, formed by repeated divisions (cleavages) of an egg.

BLENDING INHERITANCE. Galton's term for that kind of inheritance in which the characters of the parents seem to blend in the offspring.

BRACH-Y-DAC′-TYLISM. The condition of having abnormally short fingers or toes.

CELL. The fundamental unit of structure and function in all living things.

CEN′-TRO-SOME. The body at the center of radiations in a dividing cell.

CEPH′-A-LO-PODS. A class of mollusks which includes the squid, cuttle-fish and devil-fish.

CER′-E-BRAL GANG′-LI-ON. The brain of an invertebrate animal.

CHARACTER. Any feature or property of an organism.

CHOR′-DA. A cellular rod in vertebrate embryos which forms the basis of the backbone.

CHOR′-DATE. A member of the highest phylum of the animal kingdom, including all animals having a chorda or backbone.

CHO′-RI-ON. A tough membrane around an egg secreted by surrounding cells.

CHRO′-MA-TIN. The deeply staining substance of the nucleus.

CHRO′-MO-MERES. The linear series of chromatin bodies in a chromosome.

CHRO′-MO-SOMES. Deeply staining bodies found in the nucleus at the time of indirect division.

CHROMOSOMAL VESICLES. The swollen chromosomes of inter-division stages.

CIL′-I-A. Minute protoplasmic threads on the surface of a cell which produce movements in the surrounding medium by waving back and forth.

CLASS. The chief sub-division of a phylum.

CLEAV′-AGE. The division of the egg cell into many cells after fertilization.

CLEP-SI′-NE. A genus of leeches.

COE′-LOM. The body cavity.

CONTINUOUS VARIATION. A series of minute variations.

CORRELATIVE DIFFERENTIATION. Differentiation due chiefly to the interaction of different parts of an organism.

CRE-PID′-U-LA. A genus of marine gastropods.

"CRISS-CROSS" INHERITANCE. Morgan's term for that kind of inheritance in which maternal characters are transmitted to sons and paternal ones to daughters.

GLOSSARY

"CROSS-OVERS." The regrouping of linked characters, probably caused by interchange of genes between bivalent chromosomes.

CTEN'-O-PHORE. A jelly-sphere; a member of a phylum of marine animals standing above the jelly-fishes.

CY-CLO'-PI-A. A monstrosity in which both eyes have fused into a single one.

CY-TOL'-O-GY. The science which treats of cells.

CY'-TO-PLASM. The protoplasm of a cell outside of the nucleus.

DAL'-TON-ISM. That form of color-blindness in which one is unable to distinguish red and green; usually limited to males.

DAR'-WIN-ISM. The doctrine that evolution takes place through natural selection or the survival of the fittest.

DETERMINANTS. The units of heredity (Weismann).

DETERMINER. The differential cause or factor in a germ cell which determines the development of a character.

DEX'-TRAL SNAIL. The usual type of snail in which the shell coils from base to apex in a clockwise direction.

DIFFERENTIATION. The process of producing specific parts or substances from a general part or substance.

DI-HY'-BRID. The offspring of parents differing in two characters.

DI-O-NAE'-A. An insect-catching plant, the "Venus Fly-trap."

DIP'-LOID. The full number of chromosomes found in the fertilized egg and in all cells derived from this, except the mature germ cells.

DOMINANT CHARACTER. A character inherited from one parent which develops to the exclusion of a contrasting character of the other parent.

DROS-OPH'-I-LA. A genus of fruit-flies.

DU'-PLEX FACTORS OR CHARACTER. A condition where the determiners for a character are derived from both parents.

E-CHI'-NO-DERMS. A phylum of marine animals which includes star-fishes and sea-urchins.

E-COL'-O-GY. The science which deals with the relations of organisms to one another and to environment.

EC'-TO-DERM. The outer layer of cells of an embryo which gives rise to epidermis, sense organs and nervous system.

EM-BRY-OG'-E-NY. Early development of an egg leading to the formation of an embryo.

EN'-DO-DERM. The inner layer of cells of an embryo, which gives rise to the digestive cells of the alimentary system.

EN-DO-GEN'E-SIS (= development from within). The theory that the differential causes of development are within the germ cells, which are therefore complex.

GLOSSARY

Ep-i-gen'e-sis (= development from without). The doctrine that the germ is simple and homogeneous and that the differential causes of development are in the environment.

Equation-Division. An ordinary nuclear division in which each chromosome divides equally.

Eu-gen'-ics. The system of improving races by good breeding.

Eu-then'-ics. The system of improving individuals by good environment.

Ex-o-gas'-tru-la. A gastrula with the endoderm turned out instead of in.

Factor. A specific germinal cause of a developed character.

Fertilization. The union of male and female sex cells.

Fla-gel'-lum. A vibratile thread of protoplasm which serves as an organ of locomotion.

Fluctuations. Variations which are not inherited.

Fol'-li-cle Cells. Nutritive cells surrounding an ovarian egg.

Fraternal Twins. Twins produced from different eggs and showing different hereditary characters.

Functional activity. Use.

Gam'-ete. The mature male or female sex cell.

Gang'-li-on. A group of nerve cells.

Gas'-tro-coel. The digestive cavity of the gastrula.

Gas'-tru-la. A stage in development following the blastula, in which the embryo consists of an outer (ectoderm) and an inner (endoderm) layer of cells.

Genes. Factors, units, elements of germ cells which condition the characters of developed organisms (Johannsen).

Ge-net'ics. The science which deals with the origin of individuals and particularly with heredity.

Ge'-no-type. The germinal type with all its hereditary peculiarities. "The fundamental hereditary constitution of an organism" (Johannsen).

Germ-Plasm. The material basis of inheritance.

Germ-Track. The cell-lineage of the germ cells in a developing animal.

Germinal Units. Hypothetical parts of germ cells which are supposed to have certain specific functions in development.

Hae-mo-phil'-i-a. An abnormal condition in which the blood clots slowly.

Hap'-loid. The reduced number of chromosomes in the gametes.

Heredity. The appearance in offspring of characters whose differential causes are in the germ cells.

Heritage. The sum of those characters which are inherited by an individual.

Het-er-o'-sis. Increased vigor due to hybridity.

GLOSSARY

Het-er-o-zy-go'-sis. Hybridization; cross-breeding.

Het-er-o-zy'-gotes. Hybrids resulting from the union of gametes which are hereditarily dissimilar.

Ho-mo-zy'-gotes. Pure-breds resulting from the union of gametes which are hereditarily similar.

Hy'-brid. The offspring of parents which differ in one or more characters.

Identical twins. Twins which have come from a single egg and which show identical hereditary characters.

Id'-i-o-plasm. The germ-plasm or inheritance material.

Induction. A modification of the first filial generation caused by the action of environment on the germ cells of the parental generation. (Woltereck.)

Inherited Character. A character the differential cause of which is in the germ.

Instincts. Complex reflexes involving nerve centers.

Inverse Symmetry. Having the right half of one asymmetrical individual equivalent to the left of another; mirrored symmetry.

Irritability. Capacity of receiving and responding to stimuli.

Kar-y-o-ki-ne'-sis. See Mitosis.

La-marck'-ism. The doctrine that evolution takes place through the inheritance of acquired characters.

Lethal Factors. Factors which cause the early death of gametes or zygotes.

Linkage. Inheritance of characters in groups, probably due to the linkage of genes in a chromosome.

Localization. The gathering together of particular substances in definite parts of an egg or embryo.

Lol'-i-go. The squid, a genus of cephalopod mollusks.

Mar-su'-pi-als. A primitive group of mammals, including opossums and kangaroos, which carry the young in a pouch.

Mat-u-ra'-tion. The final stages in the formation of sex cells, characterized by two peculiar cell divisions.

Me-ris'-tic Variation. Variation in the number of parts.

Mes'-en-chyme. Loosely scattered cells of the mesoderm.

Mes'-o-derm. A layer or group of embryonic cells lying between ectoderm and endoderm.

Me-tab'-o-lism. Transformation of matter and energy within a living thing.

Mi'-cro-pyle. The minute opening in an egg membrane through which the spermatozoon enters.

Mi-to'-sis. Indirect nuclear division in which the nucleus is transformed into a spindle and chromosomes; the latter split and the halves move to the poles of the spindle where they form the daughter nuclei.

Modifying Factors. Factors whose principal influence is seen in modifying other factors or the characters to which they give rise.

Mon-o-hy'-brid. The offspring of parents differing in one character.

Mon'-o-tremes. The lowest group of mammals, including the duck-bill and the spiny anteater.

Mor-phol'-o-gy. The science which deals with structure and form.

Mus'-ca. A genus of flies including the house-fly.

Mu'-tant. A sudden variation or sport which breeds true.

Mu-ta'-tions. Inherited variations which are more or less striking; "sudden variations," "sports."

Nec-tu'-rus. A large salamander; the mud-puppy.

Nem'-a-tode. A round-worm or thread-worm.

Ne'-re-is. A marine annelid, or ringed worm.

Neural Groove. The groove on the dorsal surface of the embryo of a vertebrate which develops into the brain and spinal cord.

Neural Tube. A tube formed from the neural groove and giving rise to brain and spinal cord.

No'-to-chord. The cellular rod which forms the basis of the backbone.

Nu'-cle-us. The central organ of a cell, composed of chromatin and achromatin.

Nulliplex Factors or Character. A condition in which a character is absent because its determiner is found in neither parent.

On-tog'-e-ny. Development of an individual.

O'-o-cyte. The ovarian egg before maturation (formation of polar bodies).

O-o-gen'-e-sis. The development of an ovum from a primitive sex-cell.

O-o-go'-ni-a. The earliest generations of cells which produce ova; primordial egg cells.

O'-o-sperm. The fertilized egg after union of egg and sperm.

Order. The chief sub-division of a class.

Organization. Differentiation and integration, *i.e.*, different parts united into one whole.

Or-gan-og'-e-ny. The formation of various organs of the body.

Or-tho-gen'-e-sis. The doctrine that the course of evolution is definitely directed by intrinsic causes.

O-vi-par'-i-ty. Young brought forth as eggs, *i.e.*, in an early stage of development.

O'-vules. The female sex cells of flowering plants with the immediately surrounding parts.

GLOSSARY

O'-vum. The female sex cell.

Ox-y-chro'-ma-tin. That portion of the chromatin which does not form chromosomes.

Pan-gen'-e-sis. The hypothesis proposed by Darwin that every cell of the body gives off minute germs, "gemmules," which then collect in the sex cells.

Par-a-me'-ci-um. A ciliated protozoan.

Par-the-no-gen'-e-sis. Development of an egg without previous fertilization.

Particulate Inheritance. Galton's term for that kind of inheritance in which certain characters are derived from one parent and others from the other parent, *i.e.*, Mendelian Inheritance.

Pa-thol'o-gy. The science which deals with disease.

Phe'-no-type. The developed type in which some of the hereditary possibilities are realized while others remain undeveloped. "Developed, measurable realities" (Johannsen).

Phy-log'-e-ny. Evolution of a race or species.

Phyl-lox'-e-ra. A genus of plant lice.

Phy'-lum. One of the chief sub-divisions of the animal kingdom.

Phys-i-ol'-o-gy. The science which deals with function.

Plas'-to-somes. Threads or granules in the cytoplasm which are colored by certain dyes.

Polar Bodies. Two minute cells which are separated from the egg in its two maturation divisions.

Po-lar'-i-ty. The condition where two poles of a body differ; in eggs the two poles are the animal (formative) and the vegetative (nutritive).

Pol'-len. The male sex cells of flowering plants.

Pol-y-dac'-tyl-ism. The condition of having more than the normal number of digits on hands or feet.

Pol-y-hy'-brid. The offspring of parents differing in more than three characters.

Pre-for-ma'-tion. The doctrine that the fully formed organism exists in the germ, and that development is merely its unfolding.

Pre-in-duc'-tion. A modification of the second filial generation caused by the action of environment on the germ cells of the parental generation. (Woltereck.)

Pre-in-her'-it-ance. The transmission of characters developed in a previous generation.

Pre-po'-tency. The preponderance of one parent over the other in the transmission of hereditary characters.

Pri'-mates. The highest order of mammals including monkeys, apes, and man.

GLOSSARY

Primitive Sex Cells. The earliest recognizable progenitors of the sex cells in development.

Pro'-te-in. Complex organic substances containing nitrogen, *e.g.* white of egg.

Pro-te'-nor. A genus of the true bugs.

Pro'-to-plasm. The living material of an organism.

Pro-to-zo'-a. The simplest animals, usually consisting of a single cell.

Py-lo'-rus. The narrow opening between stomach and intestine.

Recessive Character. An inherited character which remains undeveloped when mated with a dominant character.

Reduction-Division. That maturation division in which the number of chromosomes is halved.

Reflexes. Relatively simple, automatic responses.

Response. Any activity of an organism called forth by a stimulus.

Reversions. The sudden reappearance of long-lost racial characters.

Segregation. The separation of contrasting parental characters in the offspring of hybrids, or of contrasting genes (allelomorphs) in the formation of gametes.

Self Differentiation. Differentiation due chiefly to intrinsic causes.

Sensitivity. Capacity of receiving and responding to stimuli.

Sex Chro'-mo-some. The "odd" or accessory chromosome which is supposed to determine sex.

Sex-limited. Any character which is found in one sex only.

Sex-linked. Any character, the determiner of which is associated with the determiner of sex.

Simplex Factors or Character. A condition where the determiner for a character is derived from one parent only.

Sin'-is-tral Snail. A type of snail in which the shell coils from base to apex in an anti-clockwise direction.

So'-ma. The body as contrasted with the germ cells.

So-mat'-ic. Pertaining to the body, as contrasted with "germinal" pertaining to the germ cells.

So'-ma-to-plasm. The body-plasm as contrasted with the germ-plasm.

So'-mite. A segment of the body of a segmented animal.

Sper-ma'-to-cytes. The mother and grandmother cells of spermatozoa.

Sper-ma-to-gen'-e-sis. The development of a spermatozoon from a primitive sex cell.

Sper-ma-to-go'-ni-a. Primordial sperm cells.

Sper-ma-to-zo'-on. The mature male sex cell.

Spindle. The nuclear division figure.

Spi'-reme. A coiled thread of chromatin which appears in the nucleus at the beginning of mitosis.

GLOSSARY

Spi-ril'-la. A spiral type of bacteria.

Sten'-tor. A ciliated protozoan.

Ster-e-o-i'-so-meres. Molecules having the same composition but different properties dependent upon varying spatial relations of their constituent atoms.

Stim'-u-lus. Anything acting on an organism which calls forth a response.

Sty-e'-la. A genus of Ascidians.

Sym'-me-try. The condition where opposite sides or poles are alike; bilateral, having equivalent right and left sides.

Syn-ap'-sis. The conjugation of maternal and paternal chromosomes preceding the maturation divisions.

Syn-dac'-tyl-ism. The condition of having webbed fingers or toes.

Te-neb'-ri-o. A genus of beetles, the larva of which is the common meal worm.

Ter-a-tol'-o-gy. The science which deals with monstrous or abnormal forms.

Tet'-rads. Bivalent chromosomes which appear 4-parted in the maturation divisions.

To-tip'-o-tence. The capacity of a cleavage cell to give rise to a whole animal.

Tox'-in. A poisonous substance particularly such as is produced by bacteria.

Tri-hy'-brid. The offspring of parents differing in three characters.

Troph'-o-blast. The outer layer of the blastodermic vesicle of a mammal.

Tro'-pisms. Automatic movements of organisms toward or away from a source of stimulus.

Unit Character. A character which is inherited as a whole and cannot be sub-divided.

Vegetative pole. The pole of an egg opposite the polar bodies.

Vil'-li. Processes which grow out from the embryonic membranes of a mammal and connect it to the walls of the uterus.

Vi-tel'-line membrane. A delicate membrane around an egg secreted by the egg itself.

Viv-i-par'-i-ty. Young brought forth "alive," *i.e.,* in an advanced stage of development.

Zy'-gote. The product of the union of male and female sex cells.

INDEX

Proper names and titles of sections are in small capitals; page references to illustrations in italic numerals.

Aborigines of Australia, 290
    New Zealand, 290, 301
    North America, 290
    Pacific Islands, 290
    South America, 290
    Tasmania, 290
    West Indies, 290
"Accessory" chomosome, 158, 180
Achondroplasy, 69
Achromatin, 8
    differential distribution, 148, in cell body, 207
ACKERT, Effects of Selection on Paramecium, 269
ACQUIRED CHARACTERS, INHERITANCE OF, 237-249
    Darwin on, 237
    Lamarck on, 237
    Weismann on, 238
    definition of, 238
    general objections to, 239
    specific objections to, 240-247
Adaptive responses, 46, 47
Adrenal gland, fed to tadpoles, 230
    influence on development, 234
African negro, 110, 294
    in Jamaica, 301
    in U. S., 301
Age of human race, 289
Albinism, 118
Alcoholism, 71, 119, 256 ✓
    cause of sterility, stillbirths, malformation, dwarfs, 218, 220, 246
    induction effect on rotifers, 246
    influence on germ cells, 218
Alkaptonuria, 119
Allelomorphs, 84. Multiple, 98
Alpine plants, non-inheritance of acquired characters, 243
"Alternative" inheritance, 73
Amalgamation of races, 300

American men of science, families of, 311
Amphiaster, 17
AMPHIOXUS, cleavage and differentiation, 144
    cleavage and gastrulation, 22
    isolated cleavage cells, 222
Ancestors, number of, 77, 78
Ancestry, pride of, 308
ANCYRACANTHUS, oogenesis, 155
    spermatogenesis, 154
Andalusian fowl, Blue, 106
Anencephaly, 229
Animal pole of egg, 11, 196
Annelid type of egg, 203
Ants and bees, 230, 315, 339
Artificial limitation of families, 311, 313
Artificial parthenogenesis, 134, 172, 221
ARTIFICIAL SELECTION, 266
    chief factor in production of domestic races, 266, 292
    creates nothing, 266, 272
    isolates pure lines, 266, 272
    lacking, 295
ASCARIS, fertilization of, 137
    "germ track," 145, 149
    sex differentiation in, 161
Ascidian egg, 142, 143, 146, 147, 203, 224, 225, 226
Aster, 17
ASTERIAS, fertilization of, 12
Astral radiations, 17
Atavism, 74, 82
ATHENS, 293
Athletes, prize, 300
Athletic, educational value of, 336
ATTICA, illustrious men of 293, 294
Automaton, 327
Autosomes, 170

INDEX

Backbone, development of, 23
Bacteria, reactions to light, 38
    to salt, 39
BAILEY, number of cultivated plants, 259
Baldness, inherited, 68
BALZAC, heredity a maze, 120
Barbarism, 256, 287, 296
BARDEEN, X-rays on spermatozoa, 218
BATESON, Blue Andalusian, 106
    brachydactyl hand, *115*
    "homozygote and heterozygote," 84
    on evolution, 283
BATESON and PUNNETT, on sweet pea hybrids, 102, *104*, 185
    linkage, 185
BAUR, factors for flowers of Antirrhinum, 103
    on leaf colors, 115
    induction effects of poor soil, 246
Beans, pure lines, 66, 267
Bees, and Ants, 168, 315, 339
    influence of food on development, 230
    workers, queens, drones, *231*
BEETHOVEN, 334
Behavior, dogs, cats, monkeys, 48
    fish and frog, 330
    lower organisms, 37-47
    modifiability of, 50
    Paramecium, 46, 47, *48*
    plasticity of, 51
    rigidity of, 51
    test of psychical processes, 36
    worms, star-fish, crustacea, vertebrates, 47-51
Biglow Papers, on Lamarckism, 237
"Biophores" of Weismann, 130
Birth Control, 314
Birthrate and deathrate, normally equal, 310
    both decreasing, 310
    decreasing most in best families, 311
    in Massachusetts, 311
Bivalent chromosomes, 153
BLAKESLEE, chromosomes of *Datura*, 282
Blastula, *22, 24, 25*
"BLENDING" INHERITANCE, 73, 108-112
    in length of ears in rabbits, 112
    in length of skull in rabbits, *113*
    in skin color of mulatto, 109, 111
"Blood lines," 268
Body and mind, 35
    parallel development, 55
BOVERI, chromosomes differ in value, 173
    cleavage of sea urchin egg, *13*
    dispermic eggs, 221
Brachydactylism, 69, *115*, 118
Brain, development, 23
    size and weight, 334
Breeder, methods of, 272
BRIDGES, intersexes in Drosophila, 169
    non-disjunction in Drosophila, *183*, 184
BROMAN, death of families, 313
BROOKS, 76
Buddhistic belief in transmigration, 33
BURBANK, new combinations of characters, 273
BUTLER, BISHOP, 321

Cards, comparison with chromosomes, 176
Canary birds, fed on red pepper, 230
CANNON, relations of adrenin to pain, hunger, fear and rage, 234
    physical substitute for warfare, 305
Capacity, greater than realization, 334, 335
Captivity, cause of infertility, 313
CASTLE, factors for coat colors of rabbits, 103
    recombinations of characters, *270, 271*
    size in rabbits, *113*
    value of selections, 268, 272
CASTLE and PHILLIPS, transplanted ovaries of guinea-pigs, *244, 245*
Cataract, hereditary, 67, 119
CATTELL, birthrate of college graduates, 309
    families of scientists, 311
Cause and effect, universality of, 320, 321
Causes, natural *vs.* final, 182
Celibacy, 295, 309
Cell characters inherited, 67

INDEX

Cell division, 16, 18, 19, 20, 125, 135, 139, 136, 145
  differential, 207
  non-differential, 145, 207
  significance of, 144, 148
Cell-Lineage, diagram of, 126
Cellular Basis of Heredity, 123
Centrifuged eggs, 228
Centrosomes, 8, 14, 17
  equal division of, 144, 207
Chances, definition of, 5
  infinity of in development, 306
Characters, developed not transmitted, 124
  Individual, 65
  Inheritance of acquired, 237-249
  inherited, definition of, 64, 238
  latent, 73
  New combinations, 72
  New, or Mutations, 74
  new in evolution, 276-286
  not independent, 64
  patent, 73
  Racial, 65
Child, on chromosomes, 182
Choice, conscious, 52
  of alternatives, 331
Chorea, 119
Chromatin, 8, 128, 136
  granules, 16, 138
  as germplasm, 128
Chromomeres, equal division of, 144, 207
Chromosomes, 16
  abnormal distribution, 182, 221, 281, 282
  accessory, 158
  bivalent, 153
  compared with digits, 151, 152, 156, 161
  compared with cards, 176
  compared with letters, 177
  conjugation of, 151
  daughter, 138
  diploid, 156
  distribution, 138
  division, 17, 144, 145
  of egg and sperm, 137
  haploid, 156
  identity, 140
  map of, 195
  maternal and paternal, 135, 137, 141, 175
  non-disjunction of, 183, 184
  number of, 16, 138
  number in man, 162, 163, 164
  permutations of, 173, 175
  "odd," 158
  reduction of, 153, 157
  seat of factors, 130, 179
  shuffle and deal of, 176
  tetrads, 153
  X and Y, 160, 161
Chromosomal vesicles, 17, 20, 136, 139
Chromosomal Inheritance Theory, 179-195
  applied to embryonic differentiation, 208
  Localization, 178, 195
Civilization, means good environment, 215
  vs. heredity, 255
  will it endure, 289, 341
Classes, heredity, 298
  exclusive, 299
Cleavage, of egg, 14, 21
  and differentiation, 138
  differential, 145, 207
  non-differential, 207
  significance of, 144, 148
Cleavage cells, differentiation of, 21, 140, 148
  isolated development of, 222, 224-226
Clepsine, behavior of, 51
Climatic effects not inherited, 243
Coeducation, 308
Cold, induction effect on Daphnia, 246
  induction effect on mice, 247
Coloboma, 69, 119
Color, of skin, hair, eyes, 109, 111, 118
Color-blindness, 120, 190, 191
Conjugation of homologous chromosomes, 151, 152
Consciousness, 53, 54
  continuity of, 54
  loss of, 54
  subconscious, 53
Contributors, 77-79
Control of, alternatives, 329
  development and evolution, 4

INDEX

HUMAN EVOLUTION, 292
  meaning of, 328
  nature, 4
  phenomena, 259, 328
  self, 328, 335
CORRELATIONS OF GERM AND SOMA, 178
"Correlative differentiation," 235
CORRENS, rediscovery of "Mendel's Law," 82
  on Mirabilis, 87, 98, 106
  leaf colors, 115
COTTON, insanity not directly inherited, 71
Creationism, 33
Creative Synthesis, 31, 59, 205
CREPIDULA, maturation and fertilization, 135
  individuality of germ nuclei, 141
  exogastrula of, 229
Cretin, 233
CREW, sex reversal, 169
Criminality, 255, 320
"Cross Overs," 193, 194
CTENOPHORE, egg, 203
CULTIVATED PLANTS, 259
  number of species, 259
Culture, grades of, 287
CUVIER, 334
Cyclopia, 230
Cytoplasm, 8
  differentiates, nuclei do not, 145
  distribution of, 140
  is somatoplasm, 128
CYTOPLASMIC CORRELATIONS, 196
  differentiations, 145, 148
  inheritance, 196
  localization, 142
  movements, 140

Daltonism, sex-linked, 189, 190
DAPHNIA, effect of cold on, 246
DARWIN, hypothesis of pangenesis, 124
  on domestic pigeons, 82
  inheritance of acquired characters, 237
  linked characters, 185
  prepotency, 82
  reversion, 82
  "sports," 74
  zeal, 334
  organism a microcosm, 210
  theory of Natural Selection, 266
DAVENPORT, degrees of relationship, 78
  extra toe in fowls, 107
  inheritance of skin color, 109-111
  Mendelian inheritance in man, 116, 119
  transplanted ovaries, 242
  "weakness with strength," 307
  white $\times$ black leghorns, 107
DAVENPORT AND WEEKS, epilepsy inherited, 71
DAVIS, on Oenothera hybrids, 277
Deaf-mutism, 69, 119
Death, of families, 312
Deathrate, declining, 310
DECANDOLLE, number of food plants, 259
Declaration of Independence, 214
Decline of families and nations, 341
Defectives, growing burden of, 296
  alarming increase of, 303
Defects, educational influence of, 252
Democracy and human equality, 214
DESCARTES, 214
"Determinants" of Weismann, 130
Determiners, 99, 130
  combinations of, 102
  differential causes, 101
DETERMINISM,
  and RESPONSIBILITY, 327
  definition of, 327
  not FATALISM, 328
  not predeterminism, 328
  of ENVIRONMENT, 324
  of HEREDITY, 321
  of personality, 327
  scientific, 328
Development, a series of responses, 231
  alternatives in, 329
  definition of, 133
  is transformation not new formation, 177
  mosaic, 223
  not reversible, 328
  of function, 30
  of personality, 55, 319, 328
  potentialities of, 325
  physiology of 216
  various aspects of, 55
  viviparous, 26

INDEX

DEVELOPMENT OF BODY, 6-32
  OF MIND, 32-56
  tabular summary, 56
DEVELOPMENTAL RESPONSES, 218
  movements, 231
  AFTER FERTILIZATION, 221
  BEFORE FERTILIZATION, 218
  DURING FERTILIZATION, 221
DE VRIES, action of selection, 267, 272
  fluctuations, 75
  induction effects of poor soil, 246
  intra-cellular pangenesis, 206
  "mutation theory," 75, 276
  mutations, 74, 275, 279
  Oenothera mutants, 276, *278*, 279-282
  on nuclear control of differentiation, 206
  pangenes, 130
  rediscovery of "Mendel's Law," 82
Diabetes, 119
DIDEROT, 214
Digits compared to chromosomes, 151, 153, 156, 161
Differential division of Cytoplasm, 145, 148
  of cells, 145, 206
Differentiation, 31
  "correlative," 235
  definition of, 133
  due to interaction of cell parts, 178, 205
  measure of, 208
  nuclear control of, 205
  "self," 236
Dihybrid, 92, *93*
Dimples, inheritance of, 66
DIONAEA, reactions of, *44*
Diploid number of chromosomes, 156
Disease, inheritance of, 69
  slight resistance to, 70
Dislocation of organs in centrifuged eggs, *227*, *228*
Dispermic eggs, 183, 221
Divines, poor health, 252
Division period, 148
Dogs, different races, 261
  psychological characters inherited, 70
DOMESTIC ANIMALS, 259
  degree of change, 259

how produced, 262, 273
number of species, 259
progenitors, 259
regressive mutants, 282
DOMINANCE, MODIFICATIONS OF, 106
  Blue Andalusian, 106
  echinoderm hybrids, 107
  extra toe in fowls, 107
  incomplete, 106
  in red × white Mirabilis, *87*
  nature of, 107
  not fundamental, 108
  plain × banded snails, 107
  red × white cattle, 106
  reversible, 107
  sex-limited characters, 170
  sex-linked characters, 185
  white × black leghorns, 107
Dominant characters, 84
  "extracted," 86
  ratio to recessive, 84, 88
  races, 290, 291
Double monsters, *223*, *227*, *229*
DRIESCH, 236
DROSOPHILA, chromosomes of, *183*, *191*, *192*
  "cross overs," 193, *194*
  modifying factors, 105
  lethal factors, 105, 191
  mutants, *284*, *285*, *286*
  rapid breeding of, 116
  sex-linked characters, *186*, *187*
DRYDEN, 71
Duplex Factors, 98
Duty, 319, 322
  of science, 338
Dwarfs, true, 118
  caused by alcohol, *220*
Dynamic equilibrium, 8

Ear, development, 23
EAST, heterosis, 273, *274*, 275
ECHINODERM type of egg, *203*
ECHINUS, first cleavage, *14*
Ectoderm, 23
Education and heredity, 308
  capacity for, 250
  definition, 251, 343
  good and bad, 251, 252
  habit formation, 326
  limiting activities, 252

INDEX

more potent in man, 249
needs of, 336
possible improvements, 343
Egg and sperm, 10
hereditary inequality of, 199
Egg nucleus, 21
Egg organization, types of, *203*
ELSBERG, plastidules, 130
Emboitement, 57
EMBRYOGENY, 23
Embryology, experimental, 216
Embryonic differentiation, processes in, 204, 308
Embryos, double, *222, 223, 229*
half and three quarter, *224-226*
Endoderm, 23
ENDOGENESIS AND EPIGENESIS, 58
"Energies of Men," 334
ENGLEMANN, *38*
English sparrow in U. S., 310
Engrammes, 246
Environment, acting at sensitive period, 266
definition, 216
direct action on germ cells, 246-248
and education, 252
good and bad, 251-254, 335
INFLUENCE IN PRODUCING NEW RACES, 265, 266
influence on ontogeny, 214, 216
influence on phylogeny, 214, 280
larger influence on man, 249
non-specific, 172
possible improvements, 343
social institutions, 214
Epidermolysis, 118
EPIGENESIS, 58, 177, 205, 324
Epilepsy, 71, 119
Equality of Man, 214, 322
Equatorial plate, 16, 17
Equation division, 157
ETHICAL OBLIGATION, 340
Ethics, 343
Eugenicist, methods of, 297
Eugenical predictions uncertain, 305
rules as to defects, recessive, 307
serious, 307
slight, 287, 307
EUGENICS, 295
CONTRIBUTORY, 308
DECLINING BIRTHRATE, 310
definition of, 296
ideals, 297, 298
negative, 302
only hope, 322
positive, 305
EUTHENICS, 249
EVANS, chromosomes of man, 163
Evolution, control of, 292
experimental, 215, 283
progressive, 289
promotion of, 292, 343
requires new characters, 276, 280, 283
retrogressive, 189
EVOLUTION OF MAN, 287, *288*
contemporary, 287
control of, 292
future, 289, 292
intelligence in, 291
natural selection in, 292
prehistoric, 287
Exogastrula, *229*
Experience, factor in behavior, 330
learning by, 49, 330
Experimental evolution, 287
medicine, 4
EXPERIMENTAL, STUDY OF INHERITANCE, 82
"Extracted" Dominants or Recessives, 86
Eyes, development, 23
color, 66, 118
lacking, 232
fused together, *230*

Facial features, inheritance of, 66
Hapsburg type, 118
FACTORS OF DEVELOPMENT, 56-60
Factors, added in progressive mutations, 283
chemical comparisons, 101, 131, 283
definition of, 101, 130
differential, 101
distribution in maturation and fertilization, *180*
dominant and recessive, not modified by union, 243
drop out in regressive mutations, 282
extrinsic and intrinsic, 60
for color developers, 102

INDEX

of rabbits and mice, 103
of sweet peas, 102, *104*
for pigment, 102
lethal, 105, 191
location in cell, 131, 178, *181*
Mendelian, 101. *180*
modifying, 105
multiple, 102
nature of, 101, 105, 130
no formation *de novo*, 283
not undeveloped characters, 101
origin of new, 282, 283
relations to characters, 178
sex determining, 164
FAHLENBECK, noble families of Sweden, 312
Family names, extinction of, 312
FARADAY, 335
Fat stains, effects on next generation, 115, 249
"Fate of part, function of position," 236
of organization, 237
Fecundity, inherited, 68
Feeble-mindedness, 71, 119
Feminist movement, 340
Fertility, of lower types, 296
FERTILIZATION, 11, *12*, 134, *137*
a stimulus to development, 134
heterogeneous, 221
union of germplasms, 136
"Fewer and better children," 311
FINLAY, gonad grafts, 169
FLUCTUATIONS, 75, 280
FOL, fertilization of starfish, *12*
Food, influence on development in tadpoles, canaries, bees, 230, *231*
FOOT and STROBELL, on chromosomes, 182
on sex-limited characters, 189
FOREL, effect of alcohol on germ cells, 221
FORMATION OF SUBSTANCES IN CELLS, 205
FORMULAE, INHERITANCE, 95
Fowls, races of, *262*, *263*
transplanted ovaries, 242
FREEDOM, and determinism, 327
BIRTH AND GROWTH OF, 330
definition of, 331
development of, 330
from reproduction, 340

greatest in man, 330
not absolute, 327
not uncaused activity, 331
of action, 53
of individual, 319, 339
of society, 339
Friedrich's Disease, 119
Frog, behavior of, 330
double embryos, 222, *223*
early development, *24*, *25*
Function and structure inseparable, 31
correlated, 32, 68
development of, 30
FUNCTIONAL ACTIVITY, 231
in human development, 250

GALTON, age of marriage, 302, 309
"Ancestral Inheritance," 77, *78*
artistic faculty, 64
characters, 64
definition of eugenics, 296
diseases, 64, 77
eugenical policy, 300
eye-color, 64, 76
"Filial Regression," 79, 80
genius inherited, 71, 77
heredity *vs.* civilization, 255
intermarriage of scholars, 300
kinds of inheritance, 73
method of, 63
nature and nurture, 213
on Ancient Greeks, 293, 294
on identical twins, 253
pioneer in heredity, 64
poor health of divines, 252
religious significance of evolution, 343
on "sports," 74
statistical study of inheritance, 76-82
stature, 64, 79
weight of seeds, 64
Gamete, 10
Gardener, methods of, 300
Gastrula, *22*, *25*
GATES, Oenothera chromosomes, 281
GAUSS, 334
"Gemmules," of Darwin, 124, 130
Generalized types, 299
Generations, parental and filial, 84, 86
symbols of, 86
Genes, 130

INDEX

Genius, and physical defects, 252
    hereditary, 71
    unstable nervous organization, 71
Genotype, 92, 95, *96*, 97, 127, 272
Geographical isolation, 300
Geotropism, of seedling, *42*
GERMAN EMPEROR, number of ancestors, 78
GERM CELLS, 6, 133
    alive, 9
    and body cells, 125, 127
    complexity of, 210
    possibilities determined in, 321
    potential personalities in, 321, 334
    reactions of, 38, *42*, 330
    specific, 172, 177
Germ nuclei, 21
    Individuality of 140, *141*
Germplasm, in nucleus, 128
    Theory of Weismann, 127, 238
GERMPLASM AND SOMATOPLASM, 126, 127
"Germ track," diagram of *126*, *149*
GERMINAL BASES OF MIND, 36
GERMINAL CONTINUITY, 124
Glandular secretions, effects of, 232
Glaucoma, 69, 119
GODDARD, feeble-mindedness inherited, 71
GOLDSCHMIDT, sex determination, 169
    sex hormones, 169
Gonia, 148
Grafts, not modified by stock, *241*, *242*
Great men, in crises, 335
GREECE, decay of, 340, 341
GREEKS, ancient, 293, 294
GREENWOOD, gonad grafts, 169
Growth period, 150
GUDERNATSCH, effects of food on tadpoles, 230
Guinea-pigs, recombinations of characters, *270*, *271*
    transplanted ovaries, 242, *244*, *245*
GUTHRIE, transplanted ovaries, 242
GUYER, on chromosomes of man, 162
GUYER and SMITH, inheritance of induced eye defects, 247, 248
Gynandromorphs, 169
Gypsy moth, sex determination, 169

HABITS, definition, 251
    good and bad, 251, 330

HAECKEL, plastidules, 130
HAEMOPHILIA, 68, 120, 189
Hair, color, *117*, 118
    form, 118
    white forelock, *117*
Half castes, of Australia, 301
    of New Zealand, 301
Haploid number of chromosomes, 156
Hardships, educational value of, 251 335
HARRIS, induction effects of poor soil, 246
HARRISON, on transplanted limbs, 236
    graft of tadpoles, *241*, *242*
HARVEY, epigram, 6
    epigenesis, 58
HATSCHECK, cleavage and gastrulation of AMPHIOXUS, *22*
Hereditary lines, interwoven, 342
HEREDITARY RESEMBLANCES, 65
    DIFFERENCES, 72
Heredity, and memory, 246
    and variation, 65, 75
    confused ideas of, 123
    definition of, 133
    mechanism of, 131, 133, 171
    more potent than environment, 314
    possible improvements, 343
    theories of, 133
    share of egg and sperm in, 199
    usually unchanged by environment, 249, 321
HEREDITY AND DEVELOPMENT, 132
HEREDITY, ENVIRONMENT, TRAINING, 253
HERING, Organic memory, 45
Heritage, definition of, 133
HERMAPHRODITES, 168
HERTWIG, O., discovery of fertilization, 134
    human ovum, 9
    idioblasts, 130
    influence on germ cells of X-rays, radium, chemicals, drug habit, 218, 221
HERTWIG, R., modification of sex ratio, 166
Heterosis, 276
Heterozygosis, 273
Heterozygotes, 84, *93*, *96*
HIPPOCRATES, 124

INDEX

Homo sapiens, 287, 290
   neanderthalensis, *288*
Homozygotes, 84, *93*, *96*
"Homunculus," 57
Hoppe, Effect of alcohol on germ cells, 221
Hormones, effects of, 232
Human embryo, development, *27, 28, 29*
Human evolution, Control of, 292
   slow, 293
Human faculties, definition, 250
Human Heredity, no improvement in, 292
   no pure lines, 116, 306
Human oosperm, early development, *27, 28, 29*
   ovum, *9, 10*
   spermatozoa, *11*
Humidity, influence on mutation, 218
Huntington's Chorea, 119
Huxley, Evolution and Ethics, 255
Hybrid races, quality of, 301
Hybridization, 273
   human, 301
Hybrids, increased vigor, 273
Hypertrophied heart, not inherited, 241
Hypophysis, effects on development, 234
Hypotrichosis, 118
Hysteria, 71, 119

Ideals, individual and social, 298
Identity, sense of, 54
"Idioblasts" of Hertwig, 130
Idioplasm, of Nägeli, 128
Immigration, 293, 301
   laws, 303
Immunity, transmission of, 115
Impulses, conflicting, 50, 330
Inbreeding, 299
Individual and Race, 339
   minor unit, 342
Individual Characters, 65
   Morphological, 66
   Physiological, 67
   Psychological, 70
   Teratological, 69
Individuals and their characters, 63
Individuals Unique, 75
"Induction," effect of colored soil, 247

poor soil, cold, alcohol, 246, 247
   not inherited, 247
Inequality of all men, 322
Infancy, prolonged in man, 250, 325
Infertility, causes of, 312
Inheritance of Acquired Characters, 237
   statement of problem, 239
   lack of evidence, 240
   no influence of stock on graft, 242
   dominants and recessives remain pure, 243
   climate effects not inherited, 243
Inheritance, "alternative," 73
   "blending," 73
   of baldness, 68
   cell characters, 67
   dimples, 66
   facial features, 66
   fecundity, 68
   genius, 71
   instincts, 70
   intellectual capacity, 71
   left-handedness, 68
   longevity, 68
   moral tendency, 71
   obesity, 68
   "particulate," 73
   pathological characters, 69
   physiological characters, 67
   psychological characters, 70
   sex-limited and sex-linked, 74, 120, 170, 185
   stature, 66, 79, 80
   temperament, 71
   teratological characters, 69
   through cytoplasm, 196
   tuberculosis, 70
   will, 71
Inheritance Factors, 100, 130
   are differential causes, 101
   formulae, 95
   units, 130
   location of, 131, 178
Inheritance material, 130
   seat of, 131, 178
Inheritance Units, 130
   location of, 131
Inhibition, 50, 330
Insanity, 71, 119
Instincts, 39

altruistic, 332
  inherited, 70
  origin of, 42, 43,
  reproductive, 340
INTELLECT, 45-49
Intellectual capacity, inherited, 71
  genius, 71, 118
  mediocrity, 118
Intelligence, factor in control, 329, 331
  in evolution of man, 291
Intelligence, from Trial and Error, 48
Interaction of parts, 231
Internal secretions, effects of, 232, 235
INTERSEXES, 168
Intra-cellular pangenesis, 206
INVERSE SYMMETRY, 197, 200, 201, 202
Irritability, 9
ISOLATION OF SUBSTANCES IN CELLS, 206
  in protozoa, 208

JAMES, WILLIAM, 304, 334
JENNINGS, action of selection, on Paramecium, 267
  on Difflugia, 269
  behavior of Paramecium, 47
  behavior of Stentor, 50
  inheritance of size in Paramecium, 67
  on Galton's law, 81
  on potential personalities, 335
  rapid breeding of Paramecium, 116
  training of star-fish, 51
JEROME, ST., 33
Jews, mixture with Gentiles, 301
JOHANNSEN, action of selection, 267
  genotype and phenotype, 127
  inherited weights of seeds, 66
  "pure lines," 266-269
JOHNSON, marriages of college women, 309

KAMMERER, inheritance of acquired characters, 247
KEIBEL, development of human embryo, 27, 28, 29
KEITH, internal secretions and racial characters, 235
Keratosis, 118
KING, modification of sex ratio, 166

KORSCHELT and HEIDER, Symmetry of egg of Musca, 197
LANE, development of vision, 31
LAMARCK, on inheritance of acquired characters, 237
LAMARCKIAN HYPOTHESIS, 246
Lamarckism, 32
LANG, snail hybrids, 107
LAUGHLIN, ancestors and contributors, 80
Laws on Eugenics, 303
Learning by experience, 49, 330
Left-handedness, inheritance of, 68
Lens, cataract, 67, 119
  development of, 235
  displaced, 119
  weight of, 66
LEPTINOTARSA, selection in, 267, 281
  mutations in, 218
Lethal factors, 105
  sex-linked, 191
Life, artificial production of, 214, 215
  conditions limited, 324
  definition of, 6
  maze of, 329
LILLIE, on fertilizin, 173
  on "free martin," 168
  sex determination, 168
Limbs, transplanted, 236
LINCOLN, 306
LINKAGE OF CHARACTERS, 185
LITTLE, Inheritance of cancer, 70
LOCALIZATION PATTERN, 198
  in eggs of ctenophore, flat-worm, echinoderm, annelid-mollusk, chordate, 203
Localization of substances in cells, 142
LOCKE, JOHN, 214
LOEB, J. Artificial parthenogenesis, 134, 172
  tropisms, 43
LOEB, L. Inheritance of cancer, 70
Logic, as test of truth, 323
LOLIGO, symmetry of the egg, 197
Longevity inherited, 68
LOTSY, source of variation, 273, 279
LOWELL, BIGLOW PAPERS, 237
LUTZ, chromosomes of Oenothera, 282
Luxury, cause of infertility, 313

MACFARLANE, on Dionaea, 44

INDEX 373

McCLENDON, production of one-eyed monsters, 172
McCLUNG, on sex determination, 159, 180
 on mammalian chromosomes, 162
McCRACKEN, maternal inheritance, 114
McDOUGAL, influence of chemicals on ovules, 218
MacDOWELL, size in rabbits, 112
 Selection in Drosophila, 269
McGREGOR, Restoration of Pithecanthropus skull, 288
Male babies, greater mortality, 167
MALTHUS, theory of, 302
Man, controls destiny, 289
 dominant races of, 290, 291
 evolution of, 287
 extermination of, 290, 291
 extinct types of, 288
 freer than animals, 330
 mongrel race, 306
 place in nature, 3
 prehistoric, 288
 races of, 290
 slow breeding of, 116
 species of, 287, 290
 value of races of, 290
MAORIS of New Zealand, 290, 301
Marriage, age of, 302, 309
 selection, 307, 308
Marsupials, 26
MASSART, reactions of SPIRILLA, 39
Materialism, 35
"Maternal impressions," 30
Maternal Inheritance, 113, 204
Matter and mind, 35
MATURATION PERIOD, 153
 divisions, 153, *154*, *155*
MECHANISM OF DEVELOPMENT, 204
"Mayflower" descendants, 313
Maze, of heredity, 76, 120
 life, 329
MECHANISM OF HEREDITY, 133, 171
Mechanistic hypothesis, 330
Mediocrity, tendency to, 79
MEMORY, 43
 associative, 45
 organic, 45
MENDEL, abbot of Brünn, 82
 dominant and recessive characters, 84

 dominant recessive ratios, 84
 experiments on peas, 64, 83, *85*
 inheritance formulae, 95
 inheritance units, 101
 method of work, 63, 83
 neglect of discoveries, 83
 purity of germ cells, 88
 study of characters, 64
MENDELIAN ASSOCIATION AND DISSOCIATION, 272
Mendelian factors and chromosomes, 131, 179, 180, 181
MENDELIAN INHERITANCE, diagrams of, *86*, *87*, *90*
 IN MAN, 116-120
 Table of, 118-120
MENDELIAN PRINCIPLES, 99
 DOMINANCE, 99, 106
 MODIFICATIONS AND EXTENSIONS, 100
 SEGREGATION, 99
 UNIT CHARACTERS, 99, 100
Mendelian ratios, simple, 84, 85
 other ratios, 90, 91
 unusual ratios, 108
 Monohybrid, 84, 91, *92*
 dihybrid, *92*, *93*
 trihybrid, 95, *96*, *97*
 dominant-recessive, 84
 departures from, 108
MENDELISM, 82-120
MENDELSSOHN, reactions of Paramecium, *40*
Meniere's disease, 119
Mentality, influence of education, 214
Mesoderm, 23
Metabolism, 8
METCHNIKOFF, disharmonies in man, 255
Metempsychosis, 33
Microscopic particles, smallest visible, 129
MIDDLETON, Effects of Selection on Stylonychia, 269
MIESHER, On stereoisomeres of albumin, 173
Mind and body, 35
MIND, DEVELOPMENT OF, 32-56
MIND, GERMINAL BASIS OF, 36
MIRABILIS, white-red cross, *87*, *98*, 106
Mitosis, 16, *18*, *19*, *20*, 138

INDEX

significance of 131, 145, 148
"Mneme" theory, 246
Modifiability of behavior, 50
Molecular constitution, stereo-isomers, 173
Molecules, largest known, 129
Monasticism, 295, 308
Monohybrid, 91, 92
Monotremes, 26
Monstrous development, 217, 325
    cause of, 217
MONTGOMERY, on Chromosomes of man, 162
Moral qualities inherited, 71
MORGAN, gynandromorphs, 169
    chromosomes, of Drosophila, 192
    "cross-overs," 193, 194
    map of chromosomes, 195
    mutations of Drosophila, 185, 282
        284, 285, 286.
    sex chromosome, 169, 170, 184
        186, 192
    sex determination in Phylloxera, 167
    sex-linked inheritance, 185, 186, 187, 188, 189
    rapid breeding of Drosophila, 116
MORGAN AND BRIDGES, modifying factors, 105
    lethal factors, 105
    Drosophila mutants, 284-286
Morphological characters, 66
    tests, 32
Mosaic development, 223
Moth and flame, 330
Mother and Child, 26
MOTT, insanity inherited, 71
Mouse, maturation and fertilization, 139
Movements, within eggs and cleavage cells, 38, 135, 136, 142, 231
    effects of stopping, 232
    random, 43
    of spermatozoa, 38
Mulattoes, skin color, 109-111, 111
    in Jamaica 301
    in U. S., 301
MÜLLER, JOHANNES, 337
MULLER, mutations of Oenothera, 279
MULSOW, 154, 155
Multiple factors, in oats and wheat, 109
    skin color, 109-111
    size, 111, 112
Multiple sclerosis, 119
MUSCA, symmetry of egg, 197
Muscular atrophy, 120
Mutation Theory, 74, 276
MUTATIONS, 73, 276
    AND FLUCTUATIONS, 75, 276, 277
    in Drosophila, 282-286
    in Leptinotarsa, 281
    in Oenothera, 276, 279
    progressive and regressive, 282
    origin of, 280
Mutilations, not inherited, 240
Myopia, 69

NÄGELI, idioplasm, 128
    non-inheritance of alpine habit, 243
Natural selection, 266, 272, 291, 292,
    nullified, 295
Nature, definition of, 321
    man part of, 321
    MECHANISTIC CONCEPTION OF, 320
    vs. nurture, 213
    stability of, 289
    VOLUNTARISTIC CONCEPTION OF, 319
NECTURUS, behavior of, 51
Neo-Darwinism, 35
Neo-Lamarckism, 35, 246
NETTLESHIP, hereditary cataract, 66, 67
Neural plate, groove, tube, 225, 226, 232
Neuritis optica, sex-linked, 120
Neuropathy, 119
New England families dying out, 311
NEWCOMB, practical eugenics, 298
NEWTON, 335
Night blindness, sex-linked, 120
NILSON-EHLE, multiple factors, 109
NON-MENDELIAN INHERITANCE, 108
Notochord, 142, 228
Nuclear division, indirect, see Mitosis
Nuclear inheritance theory, 179
Nucleus, does not differentiate, 145
    and cytoplasm concerned in heredity, 204
Nulliplex character, 98
Nutrition and Development, 232

INDEX

Obesity inherited, 68
OBSERVATIONS ON INHERITANCE, 63
"Odd" chromosome, 158
OENOTHERA, mutants, 276-279
  chromosomes of, 282
Oneness of Life, 3, 36
Ontogeny and Phylogeny, 5
Oocytes, 150
  of rabbit, 150
Oogenesis of Ancyracanthus, 155
Oogonia, 148
Oosperm, 14
  double cell, 14
  individuality of, 14
  infection of, 115
Organ-forming substances, 59
  in Styela, Amphioxus, frog, 142, 144
Organism of humanity, 342
Organization, 6, 8
ORGANOGENY, 23
ORIGIN OF SEX CELLS, 148
  DIVISION PERIOD, 148
    Primitive sex cells, 148
    Oogonia, 148
    Spermatogonia, 148
  GROWTH PERIOD, 150
    Oocytes, 150
    Spermatocytes, 150
  MATURATION PERIOD, 153
ORTHOGENESIS, 215
OSBORN, Cartwright Lectures, 287
Otosclerosis, 69, 119
Ovaries, transplanted, 242, 243
OVIPARITY, 26
Oviparous development, 14, 26
Ovules, 10
Oxychromatin, differential distribution, 207
  in cell body, 205

PAINTER, number of chromosomes in man, 163, 164
"Pangenes" of deVries, 130, 206
Pangenesis, hypothesis of, 124, 237
Panmerism, 129
PARAMECIUM, avoiding reaction, 48
  behavior of, 46
  reactions to heat and cold, 40
  races differing in size, 67
  races of, 66

  rapid breeding of, 116
  selection in, 267
  trial and error, 47
Parthenogenesis, 134
"Particulate" inheritance, 73
Partition walls between cells, 206
PASTEUR, 335
PATHOLOGICAL CHARACTERS INHERITED, 69
PEARL, action of selection, 267
PEARL AND PARSHLEY, modification of sex ratio, 166
PEARSON, law of reversion, 77
  inheritance of tuberculosis, 70
  statistics, fault of, 81
Peas, Mendel's experiments on, 83
  monohybrids, 91, 92
  dihybrids, 92, 93
  trihybrids, 95, 96
Permutations in distribution of chromosomes, 173, 175
Personality, determined by heredity, 321, 324, 327
  development of, 5, 56, 326
  infinity of chances in, 306
  not predetermined, 328
  potential, 334
  prediction impossible, 306
PHENOMENA OF DEVELOPMENT, 5
Phenotype, 92, 95, 96, 97, 273
  vs. genotype, 127
Phototropism of Seedling, 41, 42
PHYLLOXERA, degeneration, of male-producing spermatozoa, 167
Physiological characters, inheritance of, 67
  division of labor, 31
  processes, 8-10
  states, 50, 330
  tests, 32, 173
  units, 130
Pigeons, behavior of, 51, 52
  numerous races, 260, 261
PITHECANTHROPUS ERECTUS, 288
Pituitary gland influence on development, 234
"Plasomes" of Wiesner, 130
Plasticity of behavior, 51, 52
"Plastidules" of Elsberg and Haeckel, 130
Plastosomes, equal division of, 207

INDEX

PLATE, ancestors of German Emperor, 78
    factors for coat colors of mice, 103
    Mendelian inheritance in man, 116-120
PLATO, on transmigration, 33
Polar bodies, 11, 157
Polarity, 196, *197*
    of Styela egg, 142-147
Polydactylism, 69, *114*
Polyhybrid, 92
Population, normally stationary, 310
    of Europe, 310
Poultry, selection for egg production 267, 268
PREFORMATION, 57, 177, 283, 323
"Preinduction," in Daphnia, 246
"Preinheritance," 114, 204
Prepotency, 82
PRESENCE-ABSENCE HYPOTHESIS, 97
Primitive sex cells, *126*, 148
Principles of good breeding, violation of, 295
Propagation of worst, 295
PROTENOR, sex differentiation in, *159*
Protoplasmic and Cellular Organization, 6
Psychical Anlagen, 36
Psychical development, table of, 56
PSYCHOLOGICAL CHARACTERS INHERITED, 70
PUNNETT, 83, 103, *104*
"Pure lines," 266-269
Puritans and Cavaliers disappearing, 311
Purity of germ cells, 88, 99
PYTHAGORAS, on transmigration, 33

Race amalgamation, 300
    extermination, 290
    improvement, 4, 292, 340, 343
    preservation, 340
Radium, disintegration of atom, 283
    influence on spermatozoa, 218
RANA, grafted tadpoles of, *241*, 242
Reactions, of germ cells, 38, 42
    machine-like, 330
REASON, 45-50
Reception cone, *12*, 136
Recessive characters, 84
"extracted," 86
ratio to dominant, 84, 88
Reduction of chromosomes, 153-157
REFLEXES, 39, 40, 42
Regeneration, in eggs and adults, **227**
Reproduction, 8, 30
Responses, useful, 46
    varied, 50
Responsibility, 319, 331
    definition of, 331
    of society, 332, 339
    varied, 332
Retinal degeneration, 119
RETZIUS, human spermatozoa, 9, *11*
Reversible changes not inherited, 247, 266
Reversion, 74, 82, 266
RICHARDS, mitosis in Fundulus, 20
Rickets, not inherited, 239, 240
RIDDLE, Sex determination, 169
Rigidity of behavior, 51
RIGNANO, "Centro-epigenesis" theory, 246
ROMANES, 260-265
    cattle, *265*
    fowls, *262, 263*
    pigeons, *260*, *261*
    swine, *264*
ROME, decay of, 340, 341
ROSANOFF, insanity inherited, 71
Rotifiers, induction effect of alcohol, 246
ROUSSEAU, 214

Salamanders, effects of colored soil on, 247
Savagery, 256, 287
Scholars, prize, 300
SCHULTZE, double frog embryos, *223*
Science, duty of, 337
Segregation, 99
    apparent lack of, 108-112
    unusual ratios, 108
    blending of color, 109
    blending of size, 112
    fundamental to Mendelism, 108
    of Mendelian factors, 99, 108
    of substance in cells, 142, 206
SELECTIVE BREEDING, only method of improving race, 292
    Spartan method, 292

INDEX

Selection, value in Evolution, 266, 272
Selective Mating, 306
SELF KNOWLEDGE AND SELF CONTROL, 335, 343
"Self differentiation," 235
Self discovery, 335
SMITH, ADAM, 214
SEMON, "Mneme" theory, 246
Sensitive periods, 218
SENSITIVITY, 37
  differential, 37
  general, 38
  of embryo, 38
  of germ cells, 37
Sex, a Mendelian character, 91, 164, 165
  influence of food and temperature, 167
Sex cells, fundamentally alike, 10
SEX DETERMINATION, 157
  in human embryo, 166
  in man, 162, 163
  in Ascaris, *161*
  in Protenor, *159*
  in Tenebrio, *160*
  CHROMOSOMAL DETERMINATION, 158
    XO type, 158
    XY type, 160
  ENVIRONMENT INFLUENCE, 165
  alteration of sex ratios, 165
  HERMAPHRODITES AND INTERSEXES, 168
  MCCLUNG on, 159
  WILSON on, 160, *161*
  STEVENS on, 160
Sex glands, effects on development, 232
Sex hormones, 168
Sex-limited inheritance, 74, 170
SEX-LINKED INHERITANCE, 74, 120, 185-191
Sex ratio, modification of, 166-168
Sexual reproduction, value of, 174
Share of egg and sperm in heredity, 199-204
Share of chromosomes and cytoplasm, 204
SHULL, A. F., sex determination in Rotifers, 167
SHULL, G. H., leaf colors, 115

heterosis, 276
Oenothera mutants, 279
Sex determination in *Lychnis*, 169
Significance of cleavage, 144, 148, 206
  of mitosis, 131, 145, 151
Simplex character, 98
Skin color, 109
  mulatto, 109, 111
  influence of light on, 214
Slow breeding of man, 116, 315
SOBOTTA, fertilization of mouse, *139*
Social inheritance *vs.* germinal, 254
Social institutions, deal only with environment, 254
Society, highest grade of organization, 339
  power of, 296
  responsibility of, 339
  supreme duty of, 340-342
Soil, poor, induction effect on plants, 246
SOMATIC DISCONTINUITY, 125
Somatoplasm, in cell body, 129, 148
SPARTA, destruction of unfit, 292
Special senses, origin of, 38
Specialized types, 300
SPECIFICITY OF GERM CELLS, 171, 177
  of protoplasm, 172
SPENCER, physiological units, 130
Sperm centrosome, 136
Sperm nucleus, 11, 21
Spermatocytes, 150
Spermatogenesis of Ancyracanthus, *154*
Spermatogonia, 148
Spermatozoon, 9, *11*, *12*
  formation of, 157
Spina bifida, 229
Spinal cord, development, 23
Spindle, mitotic, 17, *18*, 138
SPIRILLA, reactions to chemicals, 39
"Sports," 74
Star-fish, isolated cleavage cells, 222
Statistical methods *vs.* Physiological, 80, 81
STATISTICAL STUDY OF INHERITANCE, 76-81
Stature, inheritance of, 66, 79, 80
  tendency to mediocrity, 80
  influence of food on, 214
STENTOR, modifiable behavior, 50

INDEX

Sterile insects, 315
Sterility, 292, 311, 313
Sterilization, 303
STEVENS, on sex determination, 160
STIMULI, GENERAL VS. SPECIFIC, 217
   CHEMICAL AND PHYSICAL, 217, 331
   conflicting, 50
   definition, 50, 216
   DEVELOPMENTAL, 216
   non-specific, 217
   external and internal, 50, 330
   range of, 331
   rational, social, ethical, 331
   SUMMATION OF, 43
STOCKARD, alcohol on guinea-pigs, 218, 219, 220
   experimental cyclopia, 172, 229, 230
STRASBURGER, phototropism of seedling, 41
Structures and functions, reciprocal relations, 31, 35, 68
STYELA, anterior half-embryos, 225
   dislocated organs, 228
   egg substances, 142
   gastrula and larva, 147
   half and three-quarter embryos, 224
   maturation, fertilization and cleavage, 143, 146, 147
   posterior half-embryos, 226
SUMNER, effect of cold on mice, 247
Superman, 298, 299, 315
SWEDEN, extinct noble families, 312
Swine, wild and domestic, 265
Symmetry, 196, 199
Synapsis, 151, 152
Syndactylism, 69, 118

Tadpoles, fed on thyroid, thymus, adrenal, 230
TALENTS, OUR UNUSED, 333
   parable of, 334
Temperament, inheritance of, 71, 321
Temperature, influence on mutation, 218
   influence on cell division, 217
TENEBRIO, sex differentiation in, 160
TENNENT, modified dominance in echinoderm hybrids, 107
Teratological characters inherited, 69
TERTULLIAN, 33

Tetrads, 153
THOMPSON, cross of yellow × green peas, 85
   diagram of Galton's 1st Law, 78
Thomsen's disease, 119
THORNDIKE, behavior of dogs, cats, monkeys, 48
"Thoroughbreds," 299
Thymus gland, fed to tadpoles, 230
Thyroid, effects on development, 233
   gland, fed to tadpoles, 230
Tissue cells, 7
Totipotence, of cleavage cells, 222, 226, 236
Tower, action of selection, 267
   mutations in Leptinotarsa, 218, 281
TOXOPNEUSTES, fertilization of egg, 12
Toyama, maternal inheritance, 114
Traducianism, 33
Training of animals, 51
TRANSMISSION HYPOTHESIS, 124
"Trial and Error," 46
Trihybrid, 95, 96
Triplets, hereditary, 68
Trophic correlations, 232
TROPISMS, 40, 42
TSCHERMAK, rediscovery of "Mendel's Law," 82
Tuberculosis, inheritance of, 70
Turbellarian type of egg, 203
Twins, fraternal, 228
   hereditary, 68
   identical, 76, 229, 253, 327

Ultra-microscopic units, 129
Uniqueness of every individual, 75, 177
UNIT CHARACTERS, 99, 100
UNITS OF LIVING MATTER, 129
   ultra microscopic, 129
   units of growth and division, 129
   units of heredity, 100-106, 129-131
Unity of organism, 55
Universal laws, 323
Use and disuse, effects of, 232
   effects not inherited, 237
Useful responses, 46
Uterus, attachment of oosperm to, 27

Variability, caused by environment, 280

INDEX

Variations, 72
   continuous, 74
   discontinuous, 74
   fluctuations 74, 280
   meristic, 74
   mutations, 74, 280-286
   "sports," 74
Varied responses, 51
Vegetative pole of egg, 11, 196
VIVIPARITY, 26
Viviparous development, 26

WALTER, diagram of Galtonian inheritance, 73
   filial regression, 80
Wars, effects of, 304
   shake off social heredity, 256
Wassermann test, 173
WATASE, symmetry of egg of Loligo, 197
WEEKS, 71
Weidal test, 173
WEISMANN, determinants and biophores, 130
   germplasm theory, 126, 130
   hereditary and environmental variations, 75
   on differential division of chromosomes, 145, 205
   inheritance of acquired characters, 238
   nuclear control of differentiation, 206
   reduction of chromosomes 157
WENRICH, chromosomes of Phrynotettix, 152
What is Life?, 6
WHITMAN, behavior of CLEPSINE, 51
   NECTURUS, 51
   pigeons, 51, 52
   freedom and choice, 52, 53
   sex of pigeons, 169

WHITNEY, induction effects of alcohol, 246
   sex determination in rotifers, 166
WIEMAN, chromosomes of man, 163
WIESNER, plasomes, 130
WILDER, Duplicate twins and double monsters, 229
WILL, 50-53
   absolutely free, 320, 331
   defined, 331
   good and evil, 320
   inherited, 71, 334
   nature, expression of, 319
   RESPONSIBILITY AND, 331
   supreme faculty, 333
   training of, 333
WILSON, cell division, 18, 19
   distribution of factors, 180
   of chromosomes, 181
   dwarf and double Amphioxus embryo, 222
   fertilization of sea-urchin, 12
   on sex determination, 159, 161
WINIWARTER, on chromosomes of man, 162
   oocytes of rabbit, 150
WOLFF, "Theoria Generationis," 58
WOLTERECK, induction effects of, 246
   preinduction, 246
Women's Colleges, influence on marriage, 309
WOODS, "Heredity in Royalty," 72

X-rays, influence on spermatozoa, 218
X and Y chromosomes, 170, 184

Yolk influence on size of egg, 10

ZELENY AND MATTOON, Selection in Drosophila, 269
Zygote, 10

RET'D NOV 21 1985

RET'D MAY 11 1987